Science
MAGNIFIER
™

BuildingBlocks®
of Science

CAROLINA
www.carolinacurriculum.com

Marty
Science
MAGNIFIER™
Mai
Chase
Tomás
Kari
Rain
Kelvin
Will
Rocket
BuildingBlocks® of Science
CAROLINA
www.carolinacurriculum.com

www.carolinacurriculum.com

Published by Carolina Biological Supply Company

Send all inquiries to:
Carolina Biological Supply Company
Curriculum Division
2700 York Road
Burlington, NC 27215-3387

ISBN-13: 978-1-4350-0591-4

First printing.
Printed in the United States of America.

Acknowledgments

Reviewers

Dr. Michael Klentschy
San Diego, CA

Ron DeFronzo
Warren, RI

Tom Peters
Clemson, SC

Dr. Sandy Ledwell, NBCT
Montgomery, AL

Juanita Juárez, English Language Learners Consultant
Austin, TX

Development Team

Cindy Morgan, Director, Product and Development,
Carolina Curriculum, a Division of Carolina Biological
Supply Company

Marsha W. Jones, Developer,
Carolina Curriculum, a Division of Carolina Biological
Supply Company

Dr. Donovan Leonard
Appalachian State University
Boone, NC

Contents

Science Basics

Life Science

Earth Science

Physical Science

Resources for Studying Science

Marty the Meerkat is the mascot for the Science Magnifiers. He has left his home in the Kalahari Desert in southern Africa to join the kids in their quest for scientific knowledge. Just like his meerkat buddies in the Kalahari, Marty might pop up anywhere!

Will loves skateboarding and all sorts of physical activity. Even though some people might think he's a slacker, he's really very smart, especially about physics. You got to know about motion and force when you're running the half pipe, dude! He's curious and kind of mischievous.

Rain is a crunchy-granola type of girl. She is a vegetarian and environmentalist, and her favorite science is Earth science. She and her family love the beach, so she is concerned about global warming and rising sea levels. She also has her dog, **Rocket**, to help her out.

Tomás is a very athletic young man. He plays baseball and soccer and is an avid hiker and rock climber. He is very responsible—in addition to his regular studies, he's very concerned with safety. (His dad is a fire fighter.) He loves Earth science, and wants to work for NASA as a planetary geologist studying Earth and the other planets.

Appendices

Chase is a computer nerd. He thinks of himself as a 21st century cyber dude! He's really interested in life science, especially bioengineering. If he could get his robot, Kelvin, to work as well as a person, he'd never have to take out the trash again!

Mai is a nice young lady who is into music and hip-hop dance. She almost always wears her music player, but she's also kind of a "science roadie" for the group—if they need a piece of equipment, Mai can always provide one. Life science is her favorite science and she really digs insects and other animal life.

Kari is as cute as a button, and she loves the martial arts. Breaking all those boards in karate class gave her a great appreciation for physics. She is very inquisitive and carries a science notebook just about everywhere. When she grows up she wants to write about science for the newspaper.

I've Got A Question

MY LITTLE BROTHER LUKE, IS DRIVING ME CRAZY WITH A MILLION QUESTIONS. "WHAT IS THAT?" "HOW DOES IT WORK?" "WHY IS THIS COLD?"
YOU MEAN SCIENCE QUESTIONS?

YOU OUGHT TO BE PROUD.
YEAH, YOUR LITTLE BROTHER IS ASKING GREAT SCIENCE QUESTIONS AT SUCH A YOUNG AGE.
WHAT DO YOU MEAN?
WELL, LUKE WANTS TO KNOW HOW THINGS WORK, AND SCIENCE IS A WAY TO UNDERSTAND THE WORLD–

-AND HOW IT WORKS.
-OR DOESN'T.

DO YOU WANT TO KNOW WHY A MOSQUITO BITE ITCHES?
WELL, IT'S BECAUSE WHEN A MOSQUITO BITES YOU, SOME OF ITS SALIVA-

MOSQUITO SPIT!?!!

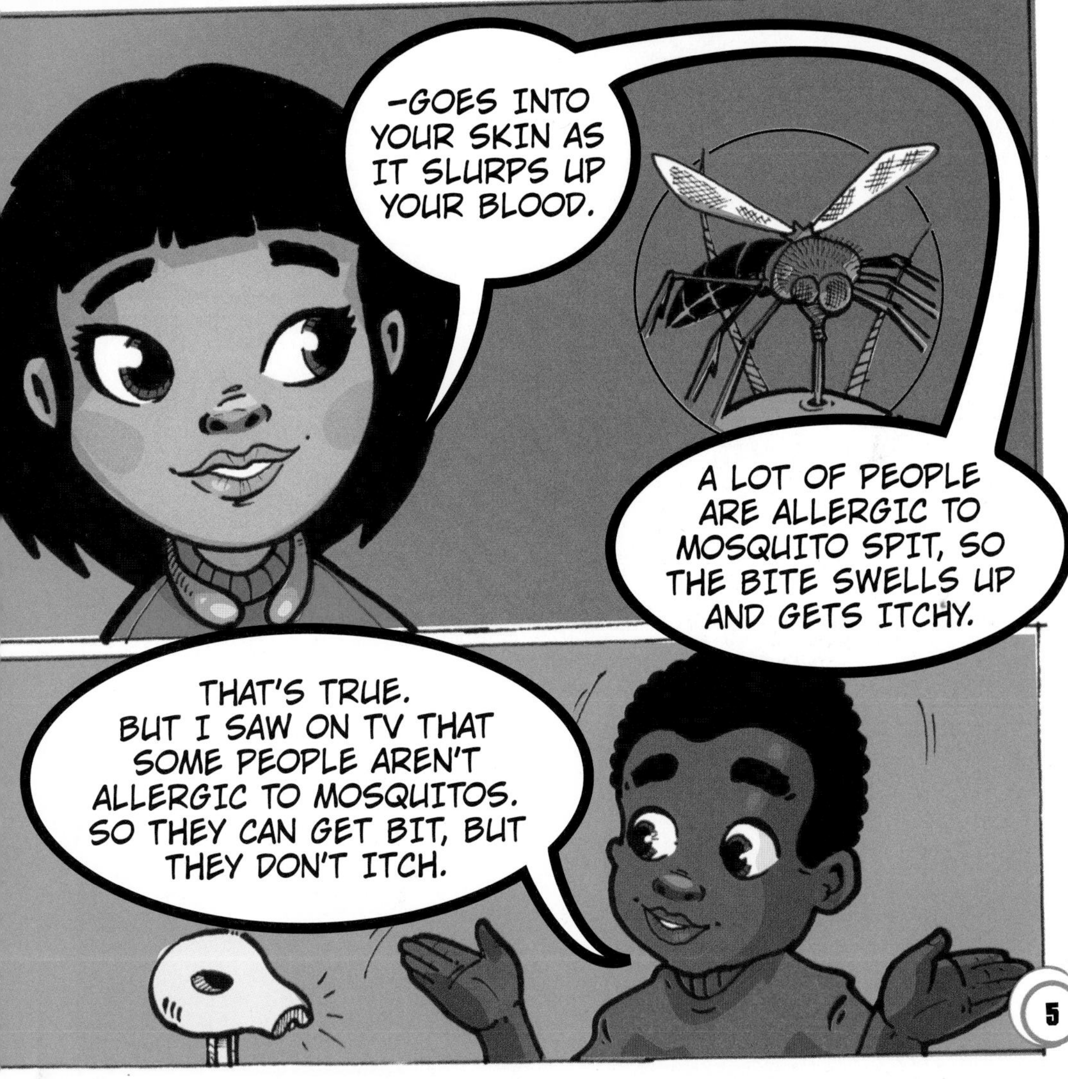
-GOES INTO YOUR SKIN AS IT SLURPS UP YOUR BLOOD.
A LOT OF PEOPLE ARE ALLERGIC TO MOSQUITO SPIT, SO THE BITE SWELLS UP AND GETS ITCHY.
THAT'S TRUE. BUT I SAW ON TV THAT SOME PEOPLE AREN'T ALLERGIC TO MOSQUITOS. SO THEY CAN GET BIT, BUT THEY DON'T ITCH.

HEY, LUKE! DO YOU WANT TO KNOW HOW DEEP THE OCEAN IS?
THE VERY DEEPEST POINT ON THE SURFACE OF EARTH IS DEEP IN THE PACIFIC OCEAN. THE MARIANA TRENCH IS 10,924 METERS (35,840 FEET) BELOW SEA LEVEL.

SCHOOL BUS
MR. COLE, HOW LONG IS THIS BUS?
FORTY FEET LONG.

0.000
WOW! THAT MEANS YOU COULD STACK NEARLY NINE HUNDRED SCHOOL BUSES END TO END ON THE BOTTOM OF THE TRENCH BEFORE THEY POKED OUT THE TOP OF THE WATER!

HEY, LITTLE BROTHER. I BET YOU DIDN'T KNOW THAT ONE OF YOUR FAVORITE THINGS IN THE WORLD HAS A GREAT SCIENTIFIC PRINCIPLE BEHIND IT...

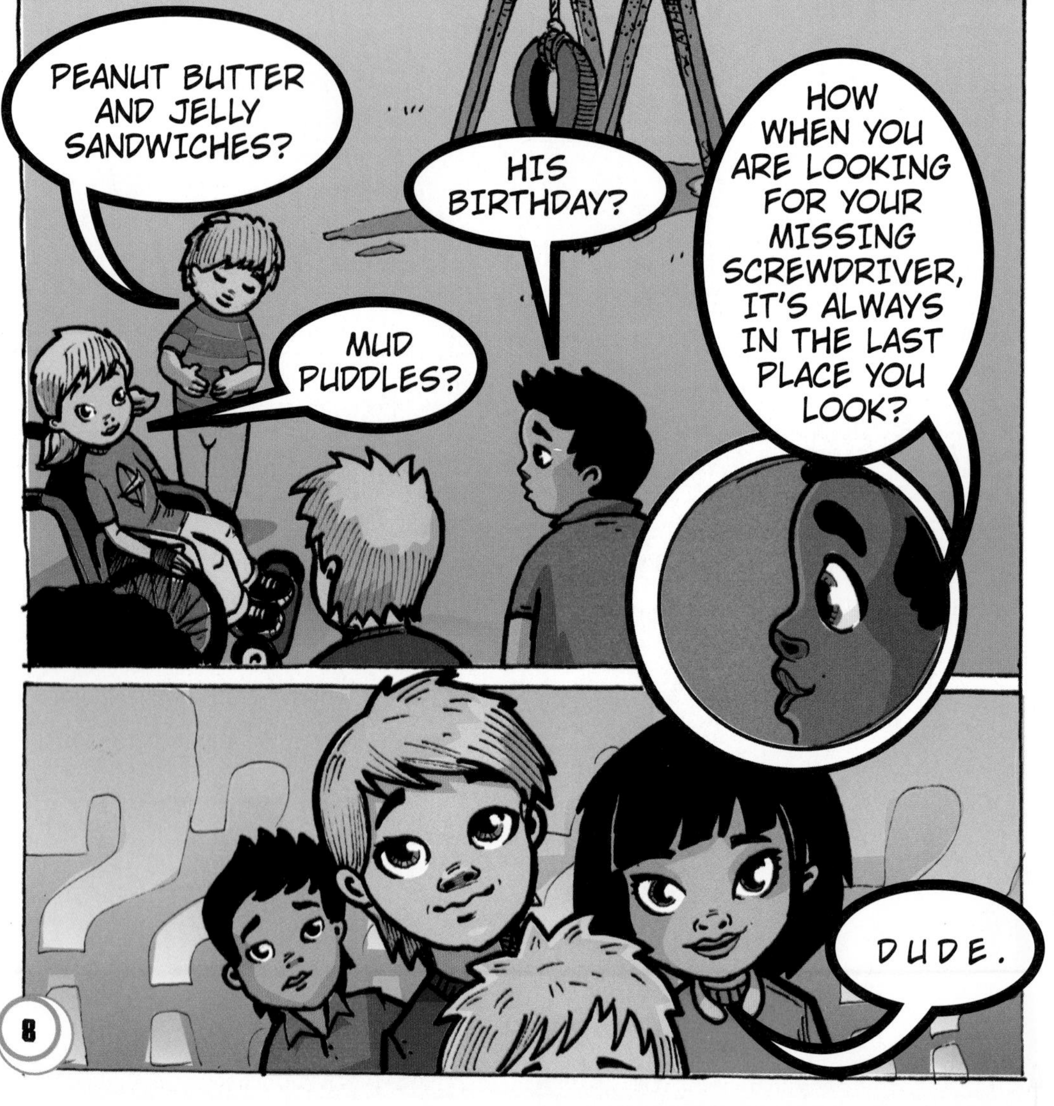
PEANUT BUTTER AND JELLY SANDWICHES?
MUD PUDDLES?
HIS BIRTHDAY?
HOW WHEN YOU ARE LOOKING FOR YOUR MISSING SCREWDRIVER, IT'S ALWAYS IN THE LAST PLACE YOU LOOK?
DUDE.

NO.
THE SWING?
YES! THE HUMBLE TIRE SWING IS, IN FACT, A PENDULUM. THIS FRENCH GUY–
FOUCAULT!
–YEAH, FOUCAULT USED A PENDULUM TO SHOW THAT THE EARTH SPINS AROUND ON ITS AXIS.

SO, IS IT OKAY FOR ME TO KEEP ASKING YOU QUESTIONS, WILL?
SURE THING, BUD. WE CAN FIND THE ANSWERS TOGETHER.

WELCOME TO THE
Science MAGNIFIER™
THIS IS ONE GREAT SCIENCE BOOK! YOU CAN FIND OUT ALL SORTS OF FUN FACTS IN HERE, BUT BETTER STILL- YOU CAN LEARN HOW TO ASK AND ANSWER REALLY COOL SCIENCE QUESTIONS.
AS YOU READ THE BOOK, YOU WILL SEE SOME SPECIAL FEATURES THAT ARE BUILT IN TO HELP YOU LOOK AT SCIENCE IN INTERESTING WAYS.

Explore More Let your inner scientist out! You can do it with a little help from the instructions in the Explore More feature. Set up your own investigations, make predictions, answer questions, and write down your very own results.

Make the Connection How does this work? Will it work that way every time? Get out your science notebook, write down those science questions, and Make the Connection. This feature can help you take what you've learned in new directions.

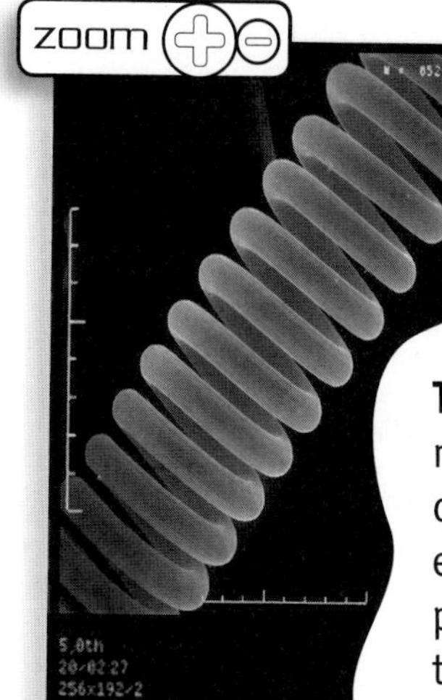

Thinking BIG™

www.carolinacurriculum.com/ThinkingBig

Thinking BIG There is a special kind of microscope that can take pictures of some of the smallest things. It's called a scanning electron microscope, or SEM, for short. The pictures it takes are called micrographs. When you see this feature, look closely at the picture and read the clues to see if you can guess what the object is.

And there is a bonus: Go online to www.carolinacurriculum.com/ThinkingBig and see micrographs of even more cool things. You can even use the zoom feature to change the magnification. Go from teeny-tiny to super-small with just a click!

Quick Question?

Quick Question These are questions sprinkled throughout the book. Use them to see if you understand what you read. If you want to check you answers, look in the Answer Key that begins on page 408.

Safe Science Whenever you see this symbol, read carefully so you can stay safe. Science investigations sometimes use things that can behave in surprising ways, so always work with safety in mind.

Is That a Fact?

Is That a Fact? You know how you sometimes hear a story that's been told over and over and the ideas get tangled up or are just plain wrong? This feature will help set the record straight!

Did You Know? What good is a book if it doesn't have super-cool tidbits of information? Well, this book has bunches of them! Look for this feature when you're in the mood for surprising science facts.

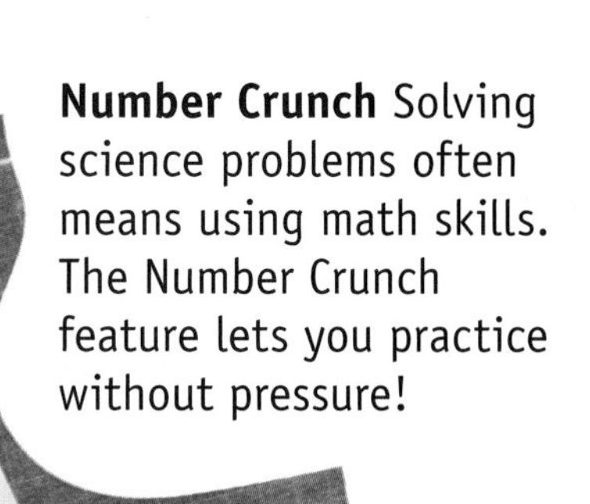

Number Crunch Solving science problems often means using math skills. The Number Crunch feature lets you practice without pressure!

UP FRONT

THE FIRST SECTION OF THE BOOK COVERS THE SCIENCE BASICS. LEARN ABOUT HOW TO CONDUCT SCIENTIFIC INVESTIGATIONS, KEEP A SCIENCE NOTEBOOK, AND WORK WITH LAB EQUIPMENT. THERE ARE ALSO GOOD SECTIONS ON BASIC IDEAS, OR PRINCIPLES, THAT COME UP WHEN YOU STUDY SCIENCE. CHECK IT OUT!

WHAT'S IN THE BACK?

EVERYBODY NEEDS A LITTLE BACKUP. IN THE BACK OF THIS BOOK, YOU'VE GOT HELPFUL SECTIONS LIKE A LIST OF NOTABLE SCIENTISTS AND INVENTORS, A GLOSSARY THAT DEFINES IMPORTANT SCIENCE WORDS, AND AN INDEX, SO YOU CAN FIND EXACTLY WHAT YOU'RE LOOKING FOR.

Doing the Work of Scientists

What Do Scientists Do?

There are many types of scientists.

Some scientists study living things. This scientist is looking at a sample of cells.

Some scientists study Earth, space, and weather. This scientist is tracking the path of a hurricane.

Some scientists study matter and energy. This scientist is inspecting a wind tunnel.

Scientists try to learn new things. Then, they share these new ideas with others. To do this, all scientists use scientific methods. **Scientific methods** are steps scientists follow to answer questions.

1. Ask a **scientific question.**
2. Develop a **hypothesis.**
3. **Plan** out your investigation carefully.
4. **Collect** your data.
5. **Interpret** your data.
6. **Explain** your results.
7. **Share** your results.
8. Think of **new questions.**

Scientists want to know more about the world around us for many reasons. Some want to solve a problem. Some scientists want to find an explanation. Others want to understand something better.

Dr. Louis Pasteur discovered a way to slow down the growth of germs. His discovery made milk safer to drink.

Every inquiry begins with a question. Not all questions are scientific questions. A scientific question requires three things: 1) Facts need to be noticed and recorded. 2) The answer must be something that you can measure. 3) The answer cannot be yes or no.

Nonscientific question	Scientific question
Are all flowers pretty?	Which flowers need more water—red ones or yellow ones?
Should we explore outer space?	How much brighter is a full moon than a quarter moon?
What is the best type of music?	I wonder if listening to music will help a runner run faster?

Ben Franklin showed that lightning is an electrical current.

The word science comes from the Latin word sci, which means "know". Scientists try to know more about the world.

Work Like a Scientist

What steps do you do take to find the answer to a **scientific question**? You can use **scientific methods.**

First, **develop your hypothesis**. A **hypothesis** is an educated guess about how things will work. A hypothesis should include a cause and an effect and an explantation: "If I do this, then this will happen because . . . ."

HYPOTHESIS

If I add salt to water it will take longer to freeze because salt is used in winter to melt ice on roads.

Next, carefully **plan your investigation**. A scientist wants experiments to be safe and successful. As a scientist, you must answer many questions before beginning your investigation. Write down what you will do first. Then, write down the rest of the steps you will take in your investigation.

Suppose your hypothesis is that water will take longer to freeze if salt is added. What will you observe and measure? Write down your plan. Record what you will do first, second, and so on.

You can plan to freeze water without salt and measure how long it takes to freeze. You can plan to add salt to the same amount of water under the same conditions and find out how long salt water takes to freeze.

Once you have a plan, you need to figure out how you will organize and **collect data**. Scientists make sure that their data are accurate. Accurate data are free from errors. Carefully recording observations in a science notebook helps a scientist make accurate observations.

Trial 1

	°C Freezer	°C Water	Water	Salt	Time
Salt	−17°	20°	500mL	6 g	266 min
No salt	−17°	20°	500mL	none	65 min

Trial 2

	°C Freezer	°C Water	Water	Salt	Time
Salt	−17°	22°	500mL	6 g	287 min
No salt	−17°	22°	500mL	none	83 min

Trial 3

	°C Freezer	°C Water	Water	Salt	Time
Salt	?	?	500 mL	6 g	?
No salt	?	?	500 mL	none	?

Scientists also want to make sure their results are consistent. Consistent results are similar each time an experiment is repeated. Repeat an experiment at least three times to ensure that the results are consistent.

Next, interpret your data. What have you learned from your results? Were your results consistent? How do the results support or not support your hypothesis?

Now, **explain your results.** In this case the results support your hypothesis: Water takes longer to freeze when salt has been added.

After your experiment you should **share your results**. You might make a graph or table. You might write a report. You could demonstrate your experiment.

Finally, think of **new questions** that resulted from your investigation.

Keeping a Science Notebook

What Is a Science Notebook?

One of the most important science tools you have is your science notebook. This is a place where you will be able to write down all sorts of important things you are curious about and how you found answers to your questions.

Your science notebook can be a spiral notebook or a pad of paper or even loose-leaf paper that you keep in a folder or binder. Your teacher may ask you to use a certain type of notebook. Just be sure that you have plenty of pages.

Write science questions in your science notebook. Are you curious about something that you see in the world around you?

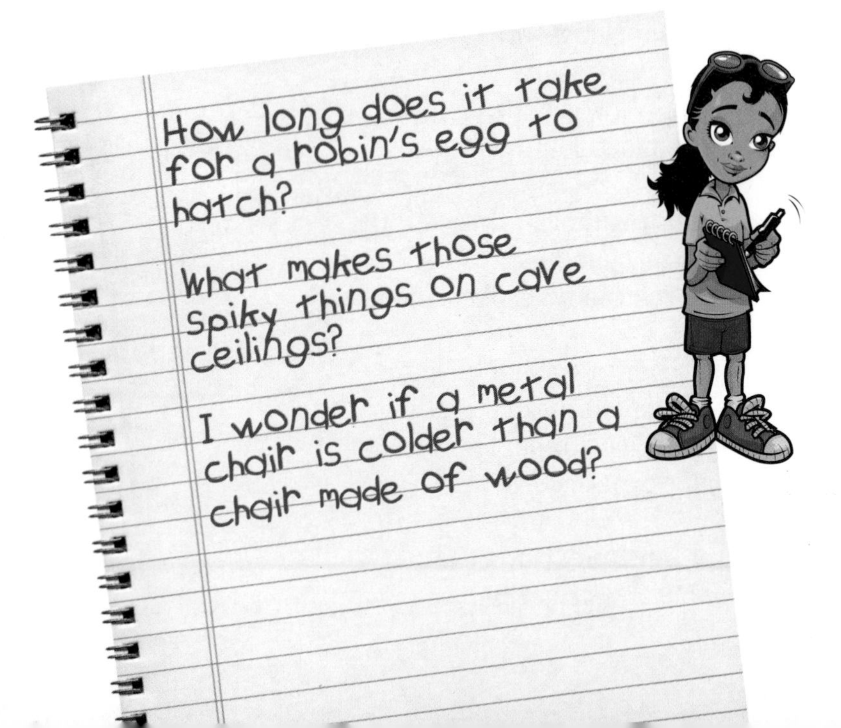

Remember, you don't have to use only words in your notebook—you can draw pictures, too. When you draw pictures, label them so that others will know what you have drawn. If you have a picture from an old magazine, you can glue it into your notebook to help you focus your thoughts.

Sharing your ideas and questions with your friends, family, or teachers is also a good way to use your notebook. Someone you know might have the same questions and you may decide to work together. Scientists do this all the time! It's called **collaboration** (co LA bor AY shun).

One way of getting answers to your science questions is by making **observations** (ob sur VAY shuns). This means that you look at something in a special way.

That special way of looking at things is called an experiment or an investigation. Your plan for carrying out your investigation and testing your hypothesis should be written down in your science notebook. Everything that you observe during your experiment should go into your science notebook, too.

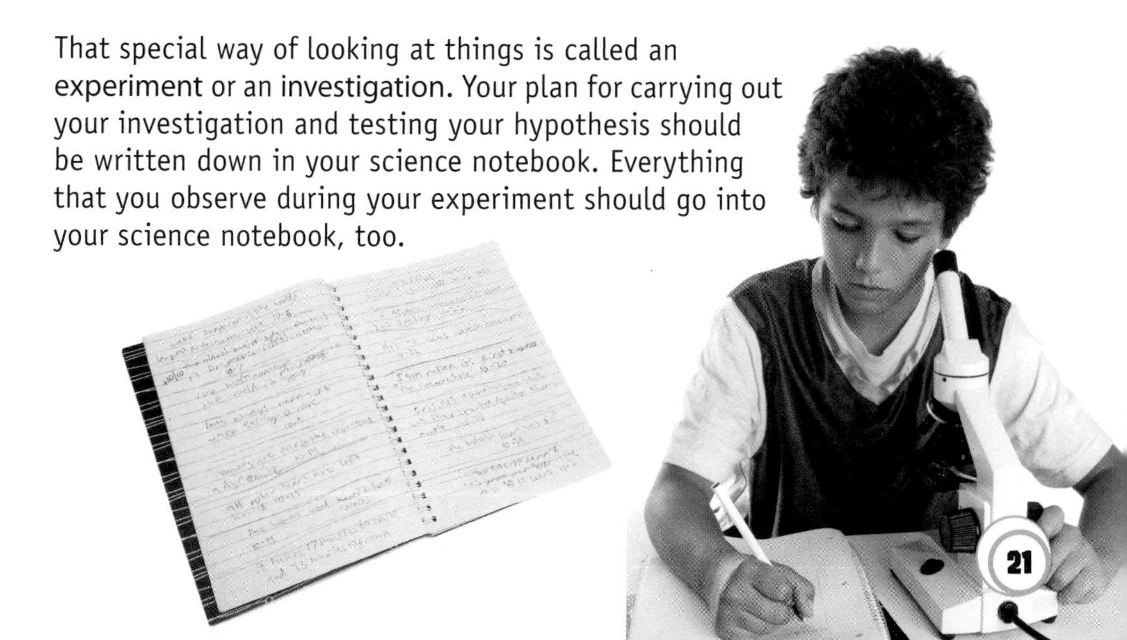

Organizing Data

What Are Data?

Scientists collect information during investigations. These pieces of information are called **data**. Some data are what you see and notice. These data are observations. Other data are measurements. Tools are used to make measurements.

Common Measuring Tools in Science

Data	Measuring Tool
weight	scale
mass	balance
width, height, length	tape measure, ruler, meterstick
temperature	thermometer
volume	beaker, graduated cylinder
elapsed time	stopwatch

As you collect scientific data, write it down. Everyone should be able to make the same observation. For example, everyone would agree that an Asian elephant has an arched back. But if you say "the elephant is heavy," it could mean many things. You must make a measurement. You must weigh the elephant to collect data on which everyone would agree.

Record the date and time of your observation. Put your name on the record sheet. Make notes about the purpose of your observation. Then, record your observation.

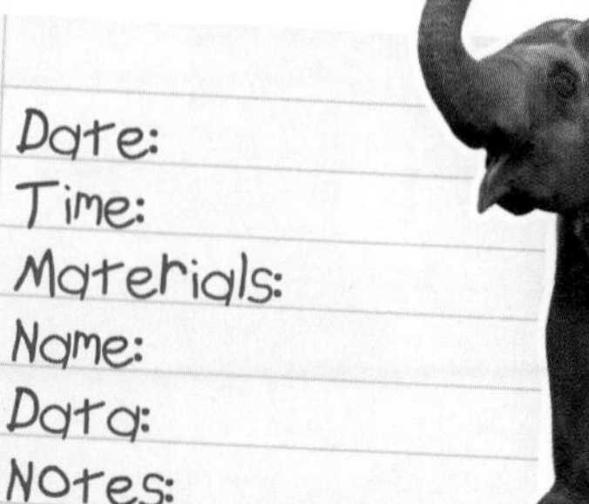

Observation: An Asian elephant has an arched back. Measurement: This Asian elephant weighs 10,000 pounds.

Using a Bar Graph

A bar graph lets you compare numbers quickly. The length or height of the bars shows the numbers you are comparing. For example, a bar graph will help you compare the weights of the heaviest land animals in Africa.

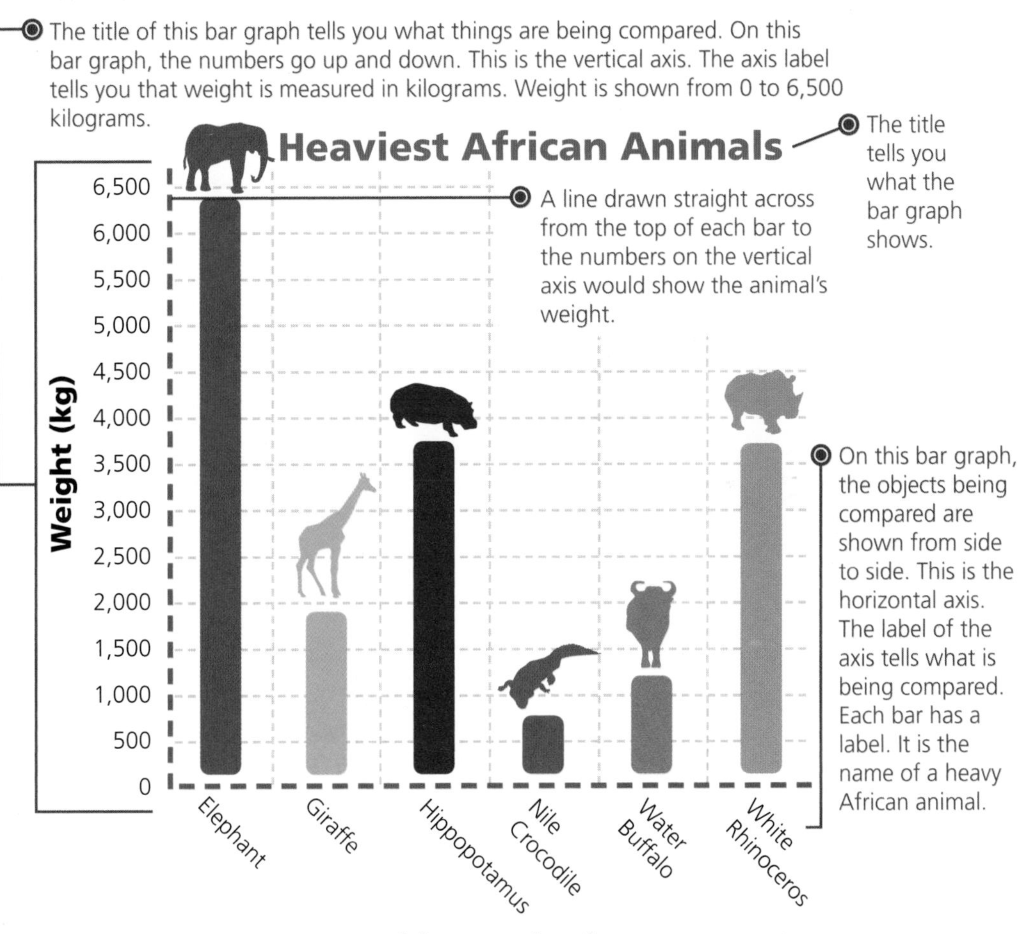

When you look at the graph you can see that the giraffe weighs between 1,500 kilograms and 2,000 kilograms. The giraffe weighs more than the Nile crocodile and less than the white rhinoceros.

Using Pie Graphs

Pie graphs show parts of a whole. The graph is a circle. The circle is always equal to 100%. A pie graph makes it easy to see which part is the biggest. Pie graphs help us compare amounts as parts of the total. Which part has the largest slice of the pie? Which has the smallest?

This pie graph shows what elements Earth is made of. Our planet is made up of many elements. All of the elements together add up to 100%. The pie graph shows about how much of the planet is made up of each element.

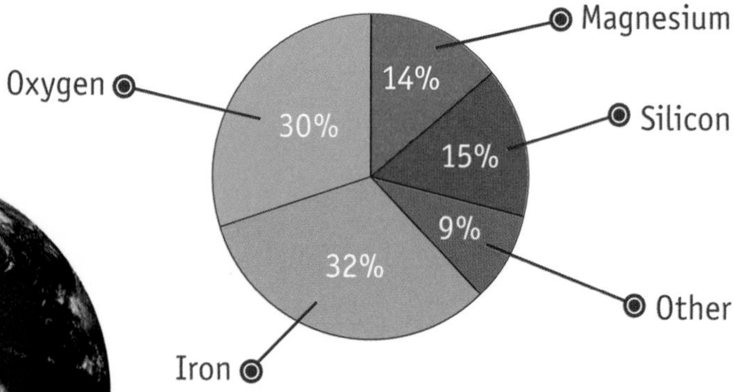

Magnesium 14%
Oxygen 30%
Silicon 15%
Other 9%
Iron 32%

To us, Earth seems to be mostly water in oceans, rivers, streams, and lakes. But we live just on the surface. Iron makes up the largest part of Earth. Most of the iron is in Earth's center. The core is 88% iron. The pie graph shows the parts of the entire planet. Just four elements make up more than 90% of Earth!

Why the Pie?

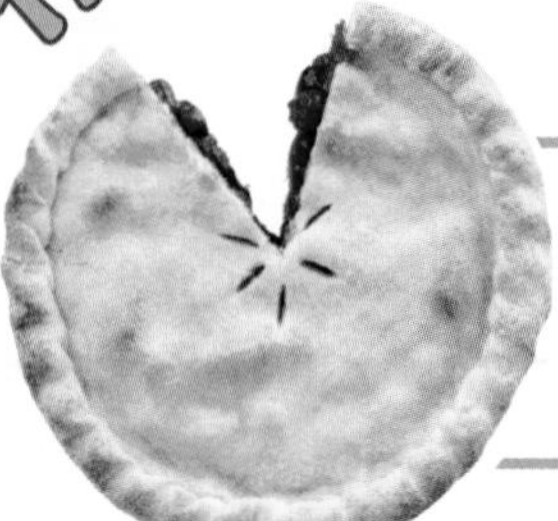

Pie graphs get their name from their shape. They look like a pie. Each part of the whole looks like a slice from a pie.

Using Line Graphs

Line graphs show how data change over time. The line shows if something went up, went down, or stayed the same.

How does the average high temperature change from month to month in Columbia, South Carolina?

Month	Avg. high temp. (°F)
January	55°
February	58°
March	68°
April	76°
May	84°
June	88°
July	91°
August	90°
September	85°
October	76°
November	67°
December	58°

The title tells you what the line graph shows.

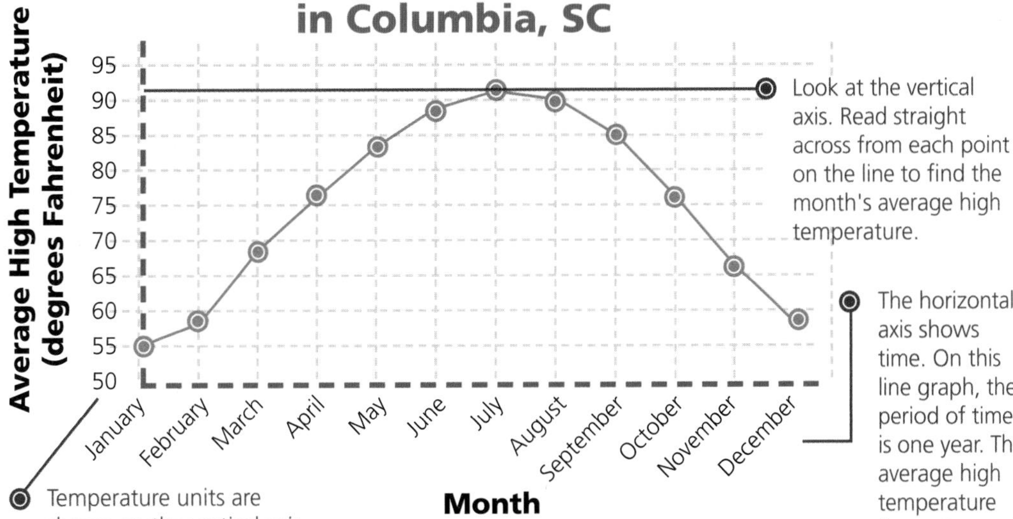

Look at the vertical axis. Read straight across from each point on the line to find the month's average high temperature.

The horizontal axis shows time. On this line graph, the period of time is one year. The average high temperature for each month is shown.

Temperature units are shown on the vertical axis of this line graph. Average high temperatures in Columbia are between 55°F and 91°F.

The shape of the line shows the pattern of temperatures over a year in Columbia, South Carolina.

Working Safely

Safety Rules

Scientific investigations are fun. They are interesting. But sometimes they can be dangerous. You must be careful, and always follow safety rules.

10 Rules for Safe Science

1. Always follow instructions. Do not skip steps.
2. Wait for permission to begin. Your teacher will tell you when to start.
3. Ask an adult for help if you need it. Ask for help if you do not understand any instructions.
4. Don't fool around.
5. Use equipment properly. Use it only the way it is meant to be used.
6. Make sure you and the work area are neat and clean. Tie back long hair to keep it away from your investigation.
7. Always wear the right safety equipment. Wear eye protection when needed. Do not wear loose clothing. Wear closed-toe shoes when you are working with liquids or heavy objects.
8. Know what to do in an emergency. Where can you quickly find water? Where is an adult to help?
9. Never taste anything unless you are told to do so by your teacher.
10. Dispose of some materials and recycle others. Follow your teacher's directions.

Safety in the Class and Laboratory

Many of your science experiments will be done at school. Follow these tips for keeping your work area safe.

TIPS FOR A SAFE LAB/CLASSROOM

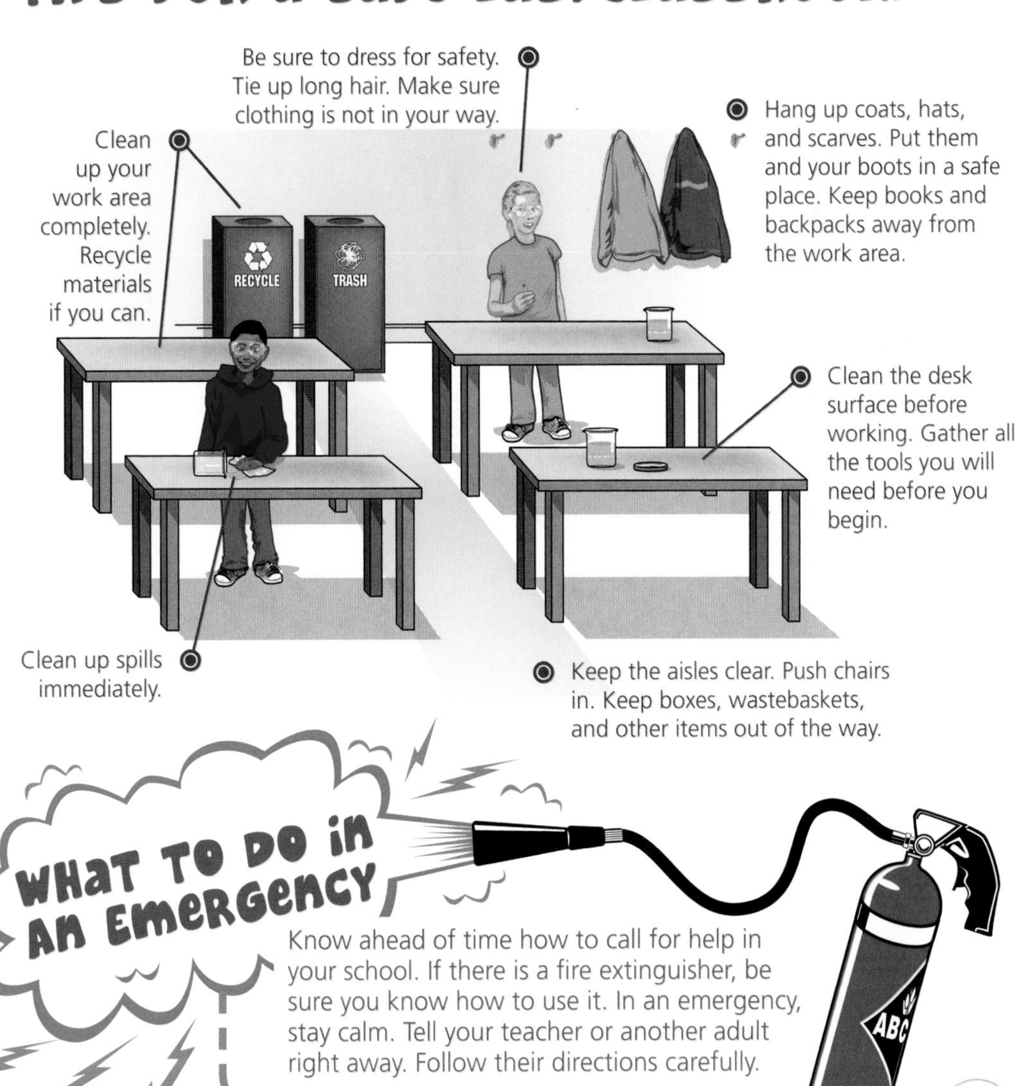

WHAT TO DO IN AN EMERGENCY

Know ahead of time how to call for help in your school. If there is a fire extinguisher, be sure you know how to use it. In an emergency, stay calm. Tell your teacher or another adult right away. Follow their directions carefully.

Safety When Experimenting at Home

Do you ever do scientific investigations at home? Many of the tips for safety are the same as you follow at school. But there are a few special safety tips to follow when you work at home.

Tips for Safe Investigations at Home

- Dress safely. Keep long hair tied back. Be sure that your clothes are not loose and hanging in your way.
- Keep your work area clean and clear of obstructions.
- Leave enough open desk space so that you can take notes.
- Get your tools and equipment together before you begin.
- Be sure others in your house know you are doing an investigation. Tell an adult what you plan to do before you start.
- Carefully select the materials you will use.

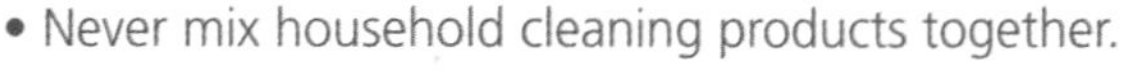

- Never mix household cleaning products together.
- Never experiment with electrical outlets, plugs, appliances, lamps without light bulbs, or other dangerous items.
- Clean up spills immediately.
- Completely clean up your work area. Recycle materials if you can.

Safety When Experimenting Outdoors

Working outside has a new set of safety challenges. There are special safety tips you should follow.

Working Safely Outside

- Always protect your skin. Wear appropriate clothing—long pants and long sleeves. Wear shoes. Wear a hat if you are going to be in the sun. Always put on sunscreen and insect repellent.
- Wear closed-toe shoes when investigating and experimenting outdoors.
- Do you have allergies or asthma? If so, be sure an adult knows about it before you work outside.
- Don't ever touch a plant you don't recognize. Wash your hands after you handle any plants or animals.

- Stay with a partner at all times. Do not wander away from the group. Always stay in sight of your teacher or other responsible adults.
- Before you do an experiment or collect plants or animals, review your investigation idea with your teacher or another adult.
- Pay special attention to ticks. Work with your partner to check your clothing and skin. Look very closely at your shoes, socks, and the legs of your pants. Be sure you know what a tick looks like.

Ticks are very small. If you see one on your skin, tell your teacher or another adult right away. Let the adult remove the tick.

Science Tools

What Is a Science Tool?

As you explore a scientific question, you must collect data. You may record your data in your science notebook. The answer to the question must be something that you can measure. Other scientists should be able to make the same measurements and get the same results.

You can use many different tools in a scientific investigation. You may use cameras, prisms, hand lenses, and microscopes. You might use clocks, stopwatches, and computers. You can measure mass or weight with pan balances, triple-beam balances, and spring scales. Specimens are caught in collecting nets. Many kinds of tools help you to make accurate measurements.

The **English system** is often used to make day-to-day measurements. Most often, you will use the **metric system** to make scientific measurements. The metric system is easy to use. Just divide or multiply by 10, 100, or 1,000 to change the unit of measurement.

To Measure What?	Metric System	English System
length, width, height, depth	meter (m) or centimeter (cm)	yard (yd), foot (ft) or inch (in)
weight	newton (N)	pound (lb)
mass	gram (g) or kilogram (kg)	no unit
temperature	Celsius (°C)	Fahrenheit (°F)
volume	liter (L)	quart (qt)
distance	kilometer (km)	mile (mi)

Tools and Measurements

It is important to make accurate measurements. You must use the right tools to make accurate measurements. A beaker won't show the temperature of the liquid it holds. A prism cannot make a spider easier to see.

What You Can See Magnify small objects with a hand lens or microscope. Use a camera to record details of a location or a scientific discovery.

A hand lens reveals details that are hard to see with your eyes alone. Adjust focus by moving the lens closer or farther away.

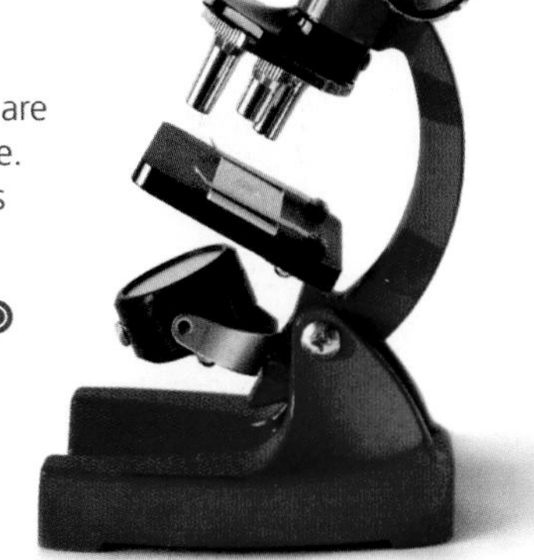

Some objects are too small to see with your eyes. These objects can often be viewed through a microscope.

Temperature A thermometer measures temperature.

In science, temperature units are degrees Celsius (°C). On a Celsius thermometer, 0° is the freezing point of water. Water boils at 100°C. In everyday life in the United States, the temperature scale used is degrees Fahrenheit (°F). On this scale, water freezes at 32°F.

Volume Volume is the amount of space something takes up. Liquid volume is measured using beakers and graduated cylinders.

Beakers and graduated cylinders come in many sizes. Marks on the side of the containers show volume. Often, the marks show differences of 10 milliliters (mL).

Measuring spoons and cups are most useful for cooks. They are not as helpful in science.

continued

Length, Width, Height, Depth Which measuring tool would best measure a 50-meter race? How thick is your thumbnail? How tall is your brother? How deep is a swimming pool? Measuring tools can help you answer these questions.

Many measuring devices, like this tape, show both inches and centimeters. A measuring tape is used to measure length, width, or depth that is greater than a few meters. Some measuring tapes are very long.

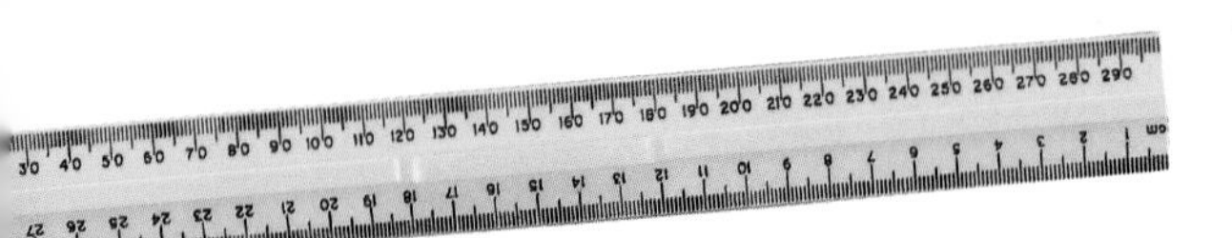

Use metric rulers and metersticks to measure short lengths and distances. Many rulers are marked in inches as well as in centimeters.

Time You must know distance and time in order to measure speed. Once you have measured distance, how will you measure the time the race takes? You will need to measure time in many other scientific investigations. Time is measured in seconds, minutes, hours, and days.

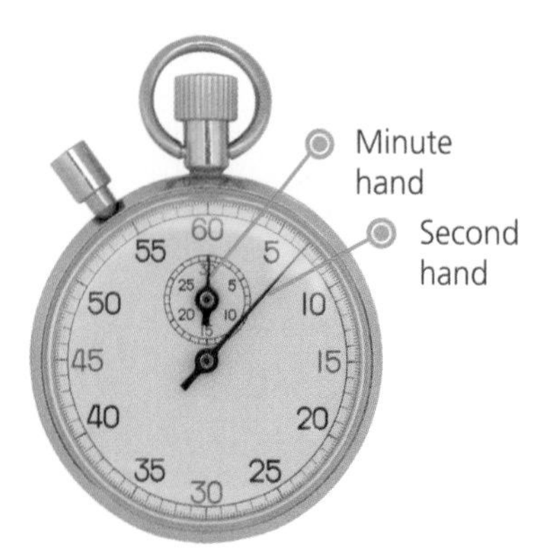

A stopwatch helps you measure minutes and seconds.

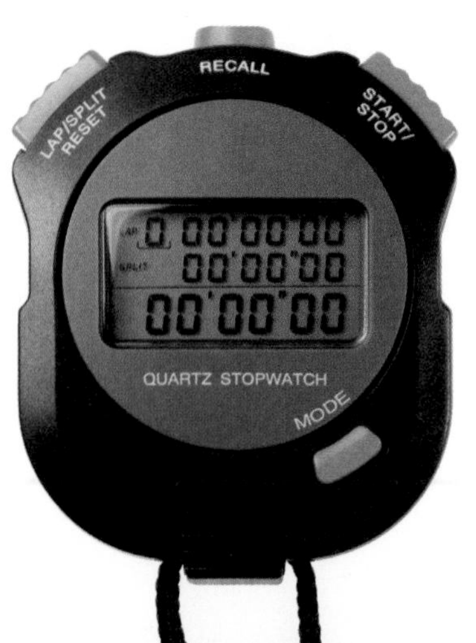

A digital stopwatch gives a precise number for fractions of a second.

Mass and Weight Many people wrongly think mass and weight are the same. Mass is the amount of matter in an object. Weight is the pull of gravity on the object.

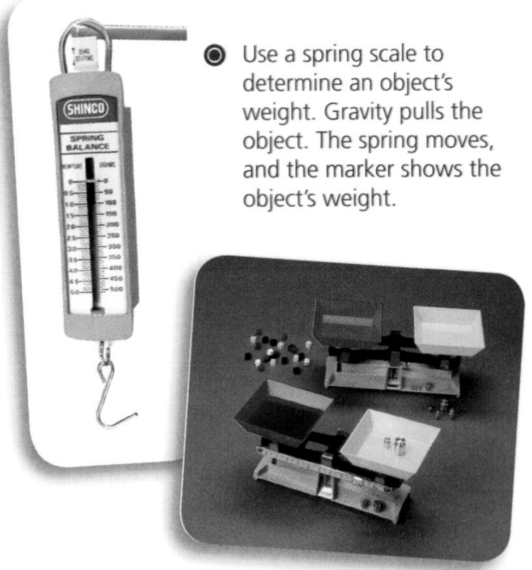

Use a spring scale to determine an object's weight. Gravity pulls the object. The spring moves, and the marker shows the object's weight.

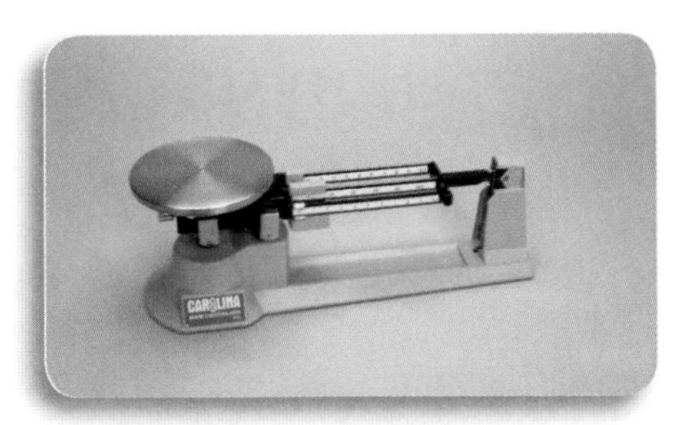

A triple-beam balance also measures mass. The riders are moved along the beam away from the pan. When the pointer is at zero (0), the position of the riders shows total mass.

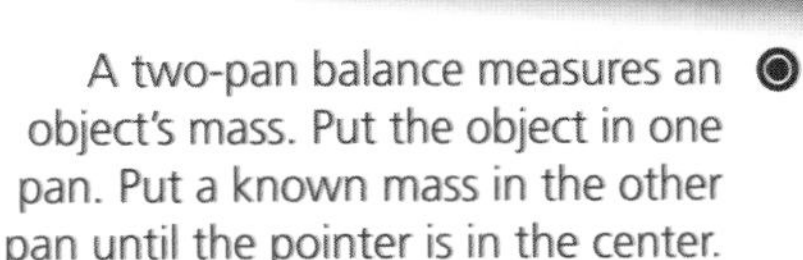

A two-pan balance measures an object's mass. Put the object in one pan. Put a known mass in the other pan until the pointer is in the center.

SCIENCE ALERT:

MAKE ACCURATE MEASUREMENTS!

Accurate data are very important. Accurate data are correct data. Always measure twice. Never guess about the data you include. Never change data to match the results you want. Repeat your test or observation to make sure your data are correct.

Technology

What Is Technology?

Scientists study the problems and needs that people have. Then, they try to solve these problems. The use of scientific discoveries to solve problems is called **technology**. Using technology can help us in our everyday lives.

You probably wear a backpack to school. It is light and easy to carry, thanks to technology.

Backpacks used to be made of leather or heavy cloth. They could weigh more than 50 pounds. They were very hard to carry. The invention of ultralight gear solved this problem.

Technology and science work hand in hand. Sometimes, the partnership helps everyone.

In the mid-nineteenth century, the microscope was a new technology. Louis Pasteur used a microscope to see germs. This helped him prove that germs cause disease.

The word technology comes from two Greek words. Techne means "art or skill." The suffix -ology comes from the word logos, meaning "study."

Life Science and Technology

Life science is the study of the living world. Life scientists study the plants, animals, and other living things around us. Technology has helped us learn more about the natural world.

Problem	Early Approach	Technological Advancement	New Situation
Poor soil	Handheld digging sticks	The plow	More food and less work

One very important place where technology has helped humanity is in farming. Technology has given us new types of wheat and rice. It has helped feed the world.

Seaweed is an important food source in many countries. People used to collect wild seaweed. Today, ocean farmers can grow their own seaweed, thanks to new technology.

Bacteria Make Waste Useful

Millions of tons of food are thrown away every year. This waste is hard on the environment. Life scientists asked questions and found a way to help with this problem. Bacteria can be used to turn food waste into fuel called *biogas*. Biogas can be used to run a car engine and can be made much more quickly than petroleum!

Earth Science and Technology

Earth science is the study of Earth, space, and weather. Advances in Earth science technology help us understand our planet better.

Suppose you want to go on a long hike tomorrow. Will it be sunny and warm? Will it rain? Technology helps us predict what tomorrow's weather will be.

Architects and engineers use technology to help them design buildings. Technology helps keep buildings standing during hurricanes and earthquakes.

Problem/Need	Early Approach	Technological Advancement	New Situation
Weather forecasting	Observing wind patterns	Weather satellites	Accurate data, reliable predictions

Understanding Hurricanes

This view of a hurricane was taken from a satellite. Satellite images help scientists know how fast storms are traveling. They help give people on land enough time to prepare for hurricanes.

Physical Science and Technology

Physical science is the study of matter and energy. Physical scientists explore the physical world. They also use technology to solve problems.

Problem/Need	Early Approach	Technological Advancement	New Situation
Transportation	Internal combustion engine (hurts environment, nonrenewable fuel)	Electric and gas powered hybrid cars	Better for the environment, uses less fuel, uses renewable fuel

Many of the tools and toys we use every day came from technology that physical scientists developed. Often, one technology replaced another. At each step, technology provided better tools or more convenience. For example, the telegraph let people communicate across long distances. Then, the telephone helped every home connect to others. Today, cell phones let us carry our phones wherever we go.

In 1522, Ferdinand Magellan became the first person to sail around the world. It took three years. Today, we can communicate over that same distance in seconds on the Internet.

Systems

What Is a System?

An organized group of different objects that work together to make a whole is called a **system**. A system can be very large, like the universe, or very small, like a cell. All of the parts of a system depend on each other to work properly. If one part of the system is taken away, the system may not be able to do eveything it needs to do.

A flower is a system. It is made up of parts that are essential to the plant's health and its ability to reproduce. If one part is removed, the flower may be unable to produce seeds. It might even die.

Flowers
These are the reproductive parts of a plant. Flowers make seeds that might grow into a new plant.

Leaves
These are the parts of the plant where food is made. Photosynthesis happens in the leaves. This is where the plant uses light energy to turn carbon dioxide and water into sugar for food.

Stems
Stems support the upper parts of plants, and they also have tubes inside them. Water and nutrients from the soil move up the tubes from the roots. Food from the leaves moves down the tubes to the roots. Stems also store food for the plant.

Roots
Roots hold plants in the ground and take up water and minerals from the soil. The roots may also store food.

If the leaves of this sunflower were removed, it could not make food. Without roots or a stem, the plant could not take up water. The sunflower could not make seeds without its flowers.

Large Natural Systems

Imagine you are hiking along a river in Yellowstone National Park. You see a forest of pines in the distance. You hear birds in the aspen trees near the riverbank. You see elk on the edge of the woods. You see a beaver pond in the river. All of these form a large natural system called an **ecosystem**.

An ecosystem is like any other system. It is made up of parts. All of the parts work together when the ecosystem is working the way it should. In the diagram below, the arrows show how the parts of an ecosystem are connected to one another.

Aspen Ecosystem

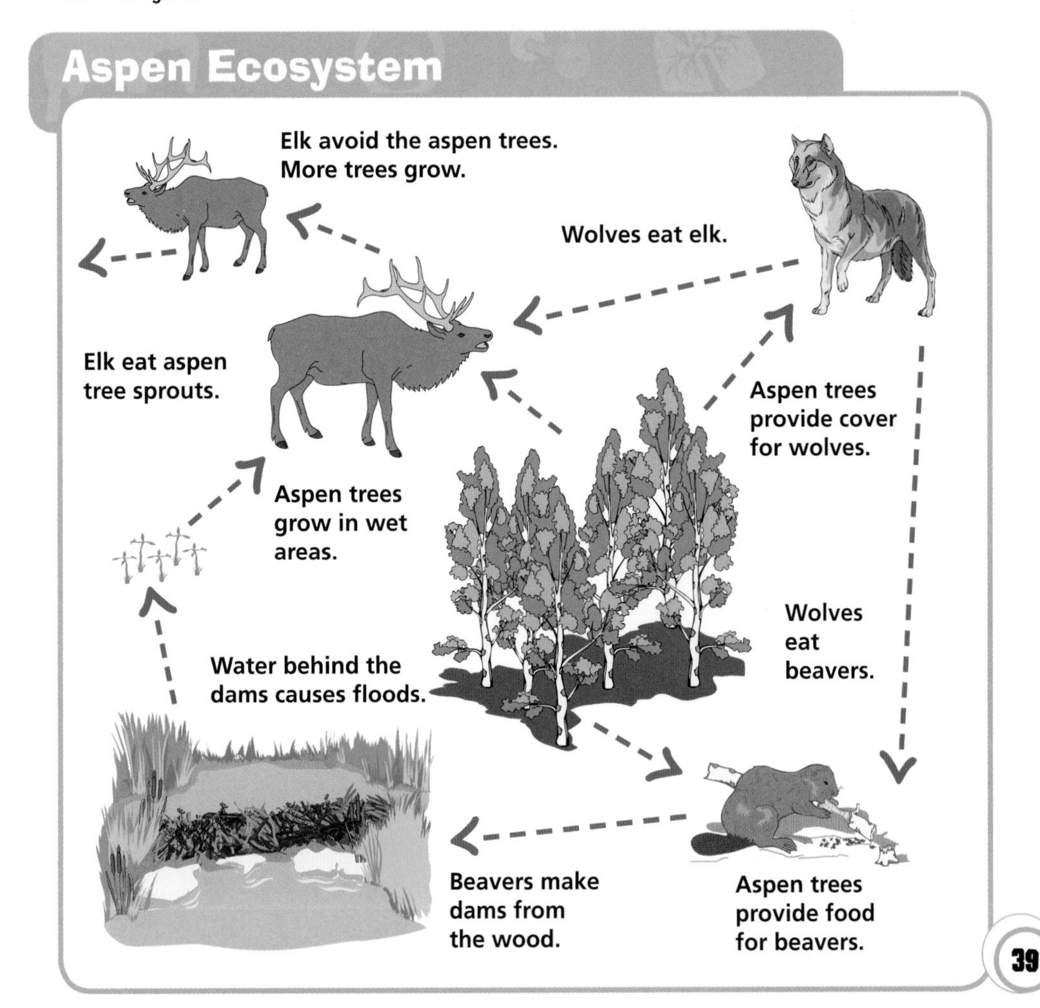

The Solar System

Every living thing on Earth is part of a system. Even nonliving parts of Earth, like minerals and gases, are parts of systems. The planet Earth itself is part of a system—the solar system!

The gravitational force between the Sun and the planets holds the planets in orbit.

A solar system is made up of a star and all of the planets and other objects that orbit the star. We call the star in our solar system the Sun. The solar system includes the Sun and eight planets that orbit the sun. The system also includes dwarf planets and other objects that revolve around the Sun, and the moons and objects that revolve around the planets.

The Sun provides light and heat to all the planets. It makes it possible for life to exist on Earth.

Radiation from the Sun affects weather on Earth and the other planets in the solar system.

The gravitational force between Earth and the Moon holds the Moon in orbit. The gravity of the Moon also causes ocean tides on Earth.

The word solar comes from the ancient Latin word sol, meaning sun. A sun is a star.

Nonnatural Systems

A machine, such as a bicycle, may also be a system. The individual parts of a bicycle cannot do what all of the parts together can do. You cannot ride a bicycle frame without the other parts.

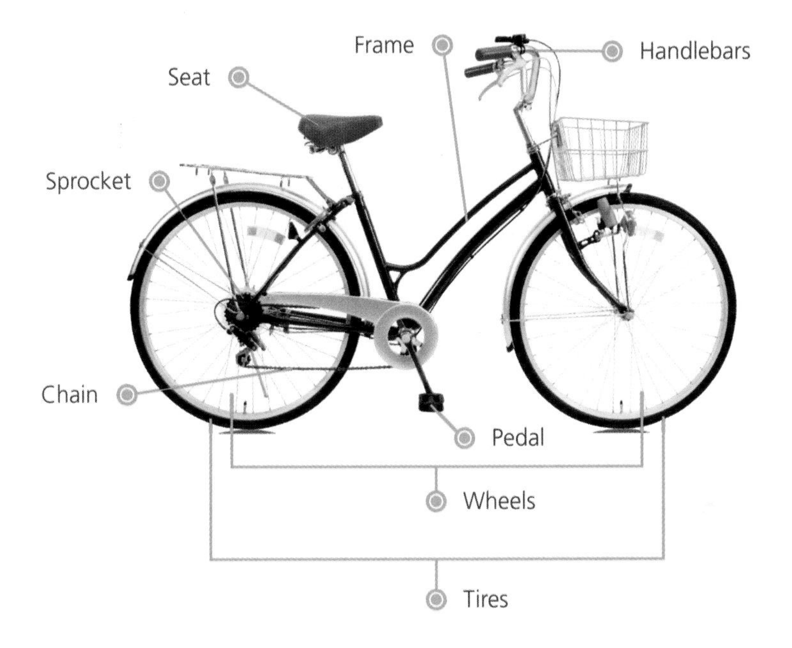

Likewise, the parts of the system cannot function without the whole system. You cannot ride a bicycle without the frame. The chain does nothing without the sprockets and the pedals. The wheels will roll, but you cannot ride them.

One part of the bicycle can be removed, and it will still function as a bicycle. What part is that? The seat is not essential to the system—but it would be very uncomfortable to ride standing on the pedals all the time, wouldn't it? Most parts of the bicycle are essential to the system.

Patterns

What Are Patterns?

Something that is repeated over and over again is called a **pattern**. Sometimes a design is repeated. Polka dots, stripes, and plaids are patterns. Sometimes an activity is repeated. Your school follows a pattern. Classes begin and end at the same time every day. You have science class at the same time every week.

For example, seasons are a pattern that happen every year. As Earth orbits the Sun, seasons change. Earth moves around the Sun in a pattern that we can predict. It takes 365 and 1/4 days for Earth to orbit completely around the Sun. This pattern completes a year, and we can create a calendar because of this pattern.

When it is winter in South America, it is summer in North America.

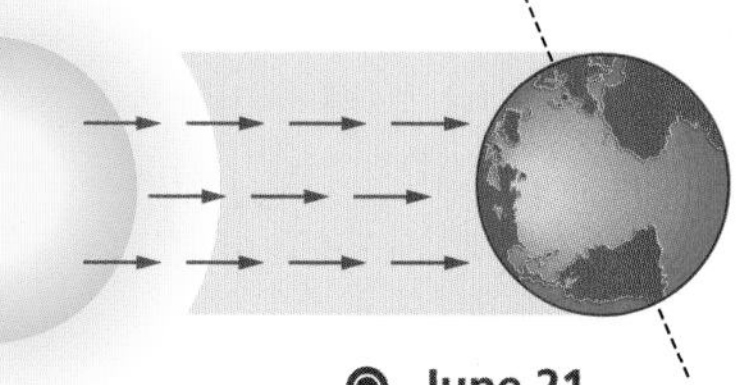

June 21
Summer begins in the Northern Hemisphere and winter begins in the Southern Hemisphere.

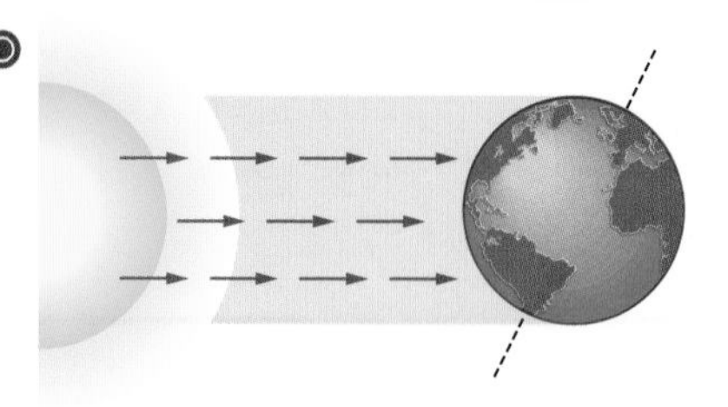

December 21
Winter begins in the Northern Hemisphere and summer begins in the Southern Hemisphere.

Patterns and Natural Cycles

The water cycle is a natural pattern. Water falls to the ground as rain, sleet or snow. It evaporates into the air from the surface and from plants and animals. Then, it condenses and falls back to the ground.

Other important cycles form patterns all around us. One important pattern is the nitrogen cycle. Nitrogen is necessary for life on Earth. No living thing can grow or repair its cells without nitrogen. The amount of nitrogen does not change. It is used over and over again, just like water.

Lightning causes nitrogen in the air to mix with other substances to form nitrogen compounds. Rain carries nitrogen to the ground.

Animals eat plants or other animals. Nitrogen goes back into the soil when plants and animals decompose.

Bacteria change nitrogen gas from the air into nitrogen compounds.

Bacteria change nitrogen compounds back into nitrogen gas. Nitrogen enters the atmosphere.

Patterns Over Time

Some long-term patterns help us see into the past. For example, tree rings show a tree's growth. Each year, a new layer grows on the outer part of the tree. If we know when a tree was cut, we can count the rings toward the center to see how old the tree is.

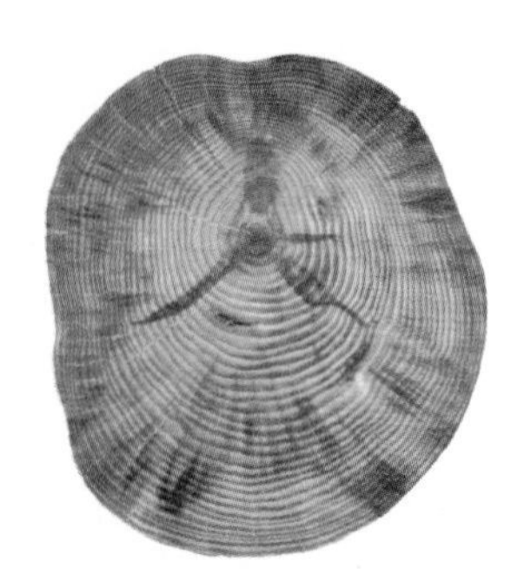

Tree rings reveal climate changes. Each ring shows some of the conditions the tree lived in during the year the ring was added. Thin rings can mean years of little water. Thicker rings can mean a wet year.

Another natural pattern shows much longer periods of time. Layers of sedimentary rock go back millions of years. For example, sandstone can form when sand piles up in layers at the bottom of an ocean. Over time, the layers become very deep. Pressure makes the layers harden into rock. As Earth's surface changes, the rocks become part of our landscape.

Now we can see the layers. The layers show time. Layers formed on top of other layers. The layers above are newer than the layers underneath.

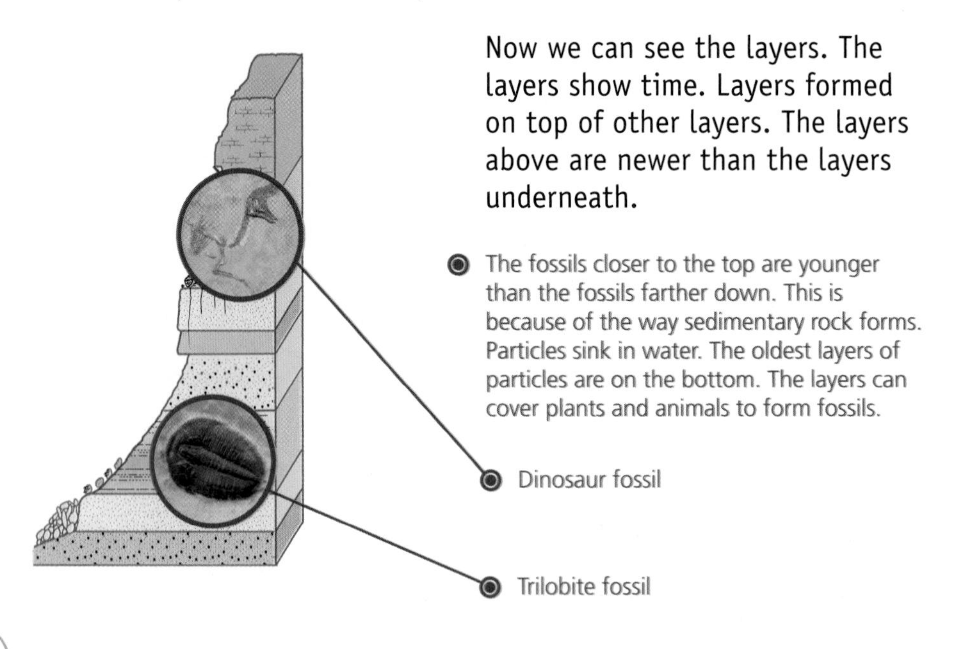

The fossils closer to the top are younger than the fossils farther down. This is because of the way sedimentary rock forms. Particles sink in water. The oldest layers of particles are on the bottom. The layers can cover plants and animals to form fossils.

Patterns Everywhere

Crystals are made when molecules connect in a regular pattern. The same pattern is repeated over and over in the material. Gem crystals form when liquid minerals and gases cool slowly, deep underground. Ice cubes and snowflakes are crystals of frozen water.

Small crystals of sodium chloride (sea salt) form when salt water evaporates.

Amethyst crystals formed from silica dissolved in water.

Answers on pages 408–419!

Thinking BIG™

Record your ideas in your science notebook.

- **Write one question about this picture.**
- **Describe a pattern you see. Is there more than one pattern?**
- **This is an organism. What might the rest of the organism look like?**

Solve IT

1) I have four of these.
2) My body has three parts.
3) First, I'm an egg on a plant.
4) I eat and grow and change.
5) Then, I'm all aflutter.

Models

How Do Models Help Us Understand Nature?

Sometimes scientists make models of objects or systems. A **model** takes the place of something that you cannot see. It helps you explain what an object or a system looks like, or how it works. There are many things we cannot understand without using models.

You can use a model to understand features of Earth. We can use a model to understand earthquakes. Models help us understand what faults deep beneath the ground would look like.

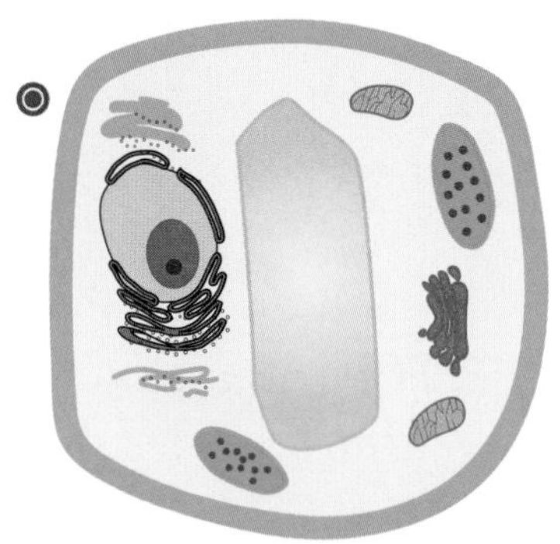

This diagram is a model of a plant cell. It is much larger than an actual cell.

This globe is a model of Earth. It is much smaller than the actual planet.

Most of the scientific models you will make will be physical models. A **physical model** can be a drawing or diagram. It can be a demonstration. It can be something you build. A physical model can be larger or smaller than the thing it represents.

To predict how rivers cause erosion, you could build a model of a river. For example, fill a box with sand and set it on a slope. Next, pour water into the box at one end. Water will drip from the top of the sand slope to the lower edge of the box. As it carves a path to the bottom, the water will erode the sand as a river might erode a hillside.

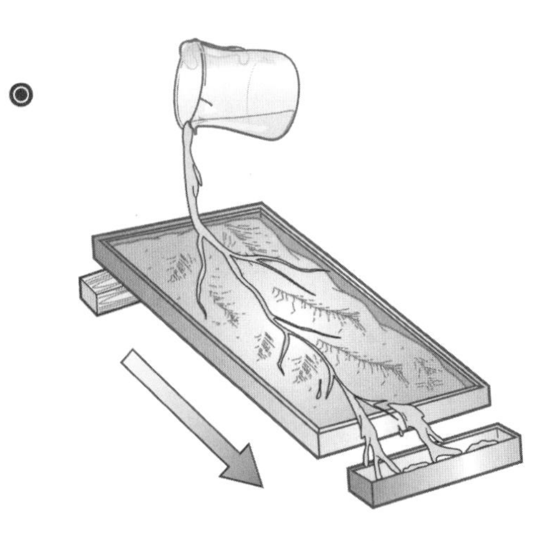

Make a Model of a Tornado

You can make a simple model to show what happens during severe weather.

1. You will need a jar with a lid. Fill the jar three-quarters full with water.
2. Add a spoonful of clear liquid dishwashing soap.
3. Add a little bit of colored confetti or glitter.
4. Close the lid tightly. Shake the jar for a few seconds. The liquid soap should form foamy clouds at the top of the jar.
5. Swirl the jar and watch what happens (you may need to swirl more than once).
6. What does the confetti represent in the model?

Conceptual and Mathematical Models

You don't need to see inside of something to make a model of it. Sometimes you can make a guess about what something looks like or how it works. When you think about how something might work you are making a **conceptual model**.

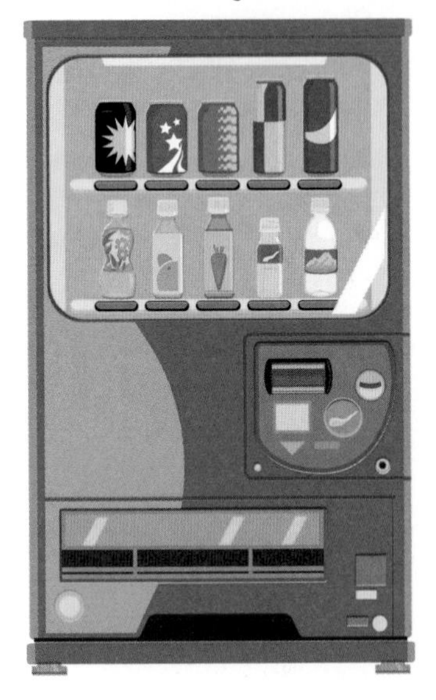

Try to imagine what the inside of a vending machine looks like. The vending machine has cans of juice inside. It also has a computer. The computer knows the amount of money that goes in the slot. When you punch buttons, the computer knows what you want to buy. A metal coil turns. The can falls to the bottom.

Scientists may make a **mathematical model** to make predictions or demonstrate how something works. They may make their own calculations. They may collect a lot of data and make a computer model.

Every year, hurricanes threaten coastal areas. Scientists try to predict how strong the storms will be, and whether they will reach land.

Scientists use computer models to help them make predictions. They collect data about winds, temperature, and rainfall to help them make their models. All of this information can change quickly. When the data changes, the computer model changes too.

Data for computer models are collected by sensors on the ground and by satellites orbiting Earth.

What Models Reveal

Models help us understand complicated systems. For example, an ecosystem is very complex. It includes all of the living things in an area. It includes all of the nonliving things too. And it includes how all of those things act together. This is too complex to see all at once. Only a model can combine all of the parts of an ecosystem.

This is so complicated, it is even hard to get an idea of an ecosystem. To help us understand the idea, we make models. One model is a food web.

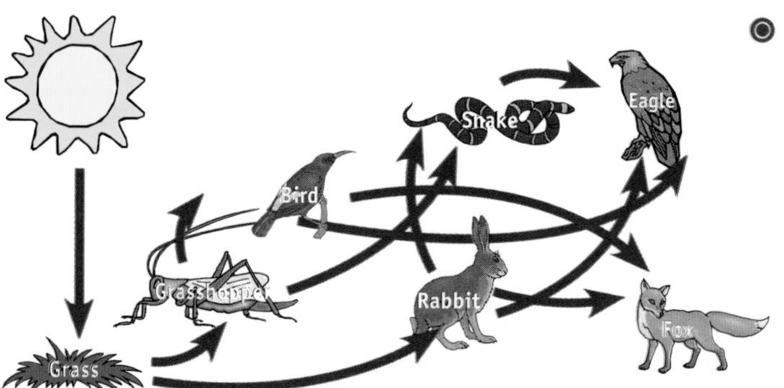

A food web is a model of one part of an ecosystem. You can see how complex it is. A food web includes animals that eat plants. It includes animals that eat plants and other animals. It shows animals that eat only other animals. And it shows how all of the living things interact.

In a model of a food web, you can ask "what if?" questions. What if the grass died? How would it harm the eagle?

Make the Connection

Check your answers on pages 408–419!

Build a Food Web

Find old magazines with photographs of animals and plants. Cut out five or six photos to make a food web model for your area. Glue the photos to a sheet of paper. Draw lines to connect plants and animals to one another. Show the feeding relationships. The arrows should point in the direction that energy moves—a seed feeds a mouse, which feeds a snake.

Atoms, Molecules and Elements

What Are Atoms, Molecules, and Elements?

How is a plant similar to the Moon? How is your hand similar to a pencil? If you said that all these things are made of matter, you are right! The building blocks of matter are atoms. But if all matter is made of atoms, why do a plant and the Moon look so different from each other?

An **atom** is the smallest part of a substance that has the same properties as the substance.

An **element** is a pure substance. Elements cannot be broken down into other substances.

The atoms of elements are like the letters in a word. Like letters, atoms can be joined together to make something bigger.

When two or more atoms join together they form a **molecule.** A molecule acts as one piece of matter. A molecule is like a word. The arrangement of its atoms determines its chemical and physical properties. So the chemical and physical properties of the molecule are like the definition of a word.

Think of an atom of a substance as a letter.

ATOMIC

Different atoms joined together form substances. Different letters joined together form words.

ANIMAL

An atom is always the same, no matter where it is found. A letter is the same letter, no matter where it is found in a word.

Parts of an Atom

All of the matter in the universe is made up of atoms—even your body. A human body is made up of about 7 billion billion BILLION atoms. About 67% of the atoms are hydrogen. Twenty-five percent of the atoms are oxygen. Hydrogen and oxygen atoms make up water, the most common substance in your body. Ten percent of the atoms in your body are carbon.

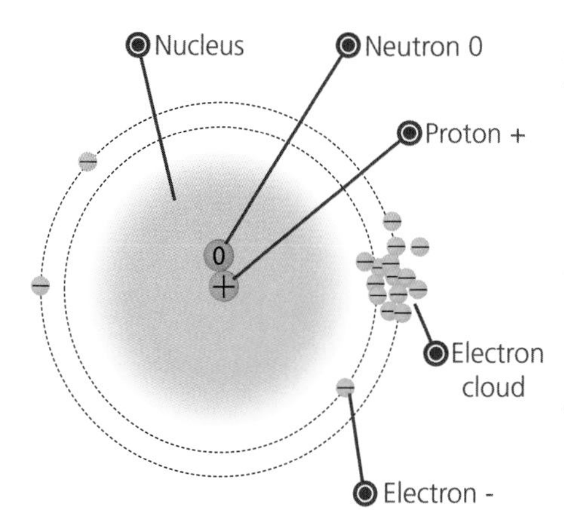

Atoms are made of **protons** (PROH tahnz), **neutrons** (NOO trahnz), and **electrons** (ih LEK trahnz).

Protons and neutrons are in an atom's nucleus. The **nucleus** is the center of an atom.

An atom usually has the same number of protons and neutrons. Neutrons add **mass** to an atom.

The electrons swirl around the nucleus. The area of the atom that contains the electrons is called the **electron cloud**. The number of electrons in an atom is usually the same as the number of protons.

If you could break hydrogen, oxygen, and carbon atoms apart, you would see that all of the protons, neutrons, and electrons are the same. How can you tell which atom is which? You need to count the number of protons.

The total number of protons in an atom is called the **atomic number.** The atomic number tells you what kind of atom it is. For example, hydrogen has an atomic number of one. All atoms with one proton are hydrogen atoms. Carbon has an atomic number of six. All atoms with six protons are carbon atoms.

Electric Charge of Atoms

Why do electrons orbit the nucleus? According to the **law of electric charges**, positive charges push away from each other. Negative charges also push away from each other. A positive charge and a negative charge pull toward each other.

Protons and electrons have electric charges. The charge on an electron is –1. The charge on a proton is +1. Neutrons have a charge of zero. This means the charges on the protons give the nucleus a positive charge. Opposite charges attract, so the positively charged nucleus attracts the negatively charged electrons. The attraction between the nucleus and the electrons holds the atom together.

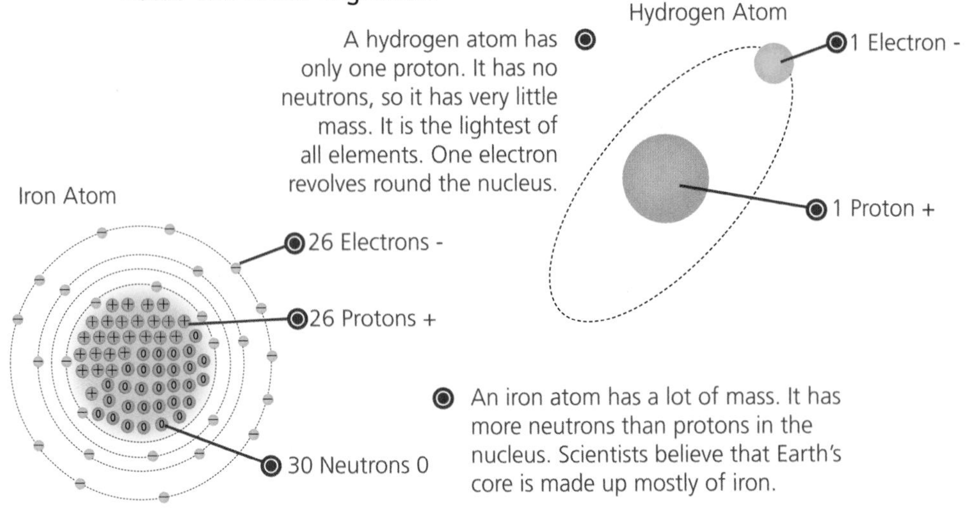

A hydrogen atom has only one proton. It has no neutrons, so it has very little mass. It is the lightest of all elements. One electron revolves round the nucleus.

An iron atom has a lot of mass. It has more neutrons than protons in the nucleus. Scientists believe that Earth's core is made up mostly of iron.

But wait! If all of the protons in an atom have the same charge (+), why do protons hold together in the nucleus? Protons do repel each other because they have the same charge. A special nuclear force holds the protons and neutrons together.

Molecules and Compounds

Atoms can join together to make up molecules. A **molecule** can be made of a single element. A molecule can also be made of atoms of two or more different elements. A molecule made of two or more elements is called a **compound**.

A molecule is the smallest part of a compound. All the elements in a compound are present in the same amounts all through the substance.

Unlike mixtures, which can be separated, atoms in a compound are very hard to separate. The properties of compounds are very different from the properties of their elements.

Sugar (sucrose) is a molecule and a compound. It is made of carbon, hydrogen, and oxygen atoms. Graphite is a molecule but not a compound. It is 100% carbon. The carbon in sugar looks and acts very differently from the graphite in a pencil.

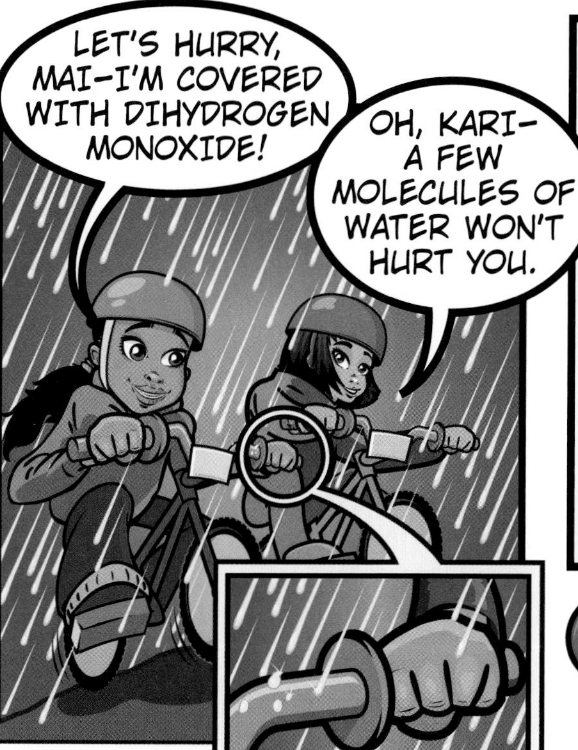

A WATER MOLECULE HAS TWO HYDROGEN ATOMS (THE DIHYDROGEN) AND ONE OXYGEN ATOM (THE MONOXIDE). THE MOLECULES IN A WATER DROPLET ALL STICK TOGETHER. THEY PULL THEMSELVES INTO THE SMALLEST POSSIBLE SURFACE AREA–A DROP. ALL OF THE MOLECULES ON THE OUTSIDE ARE HELD TOGETHER BY SURFACE TENSION.

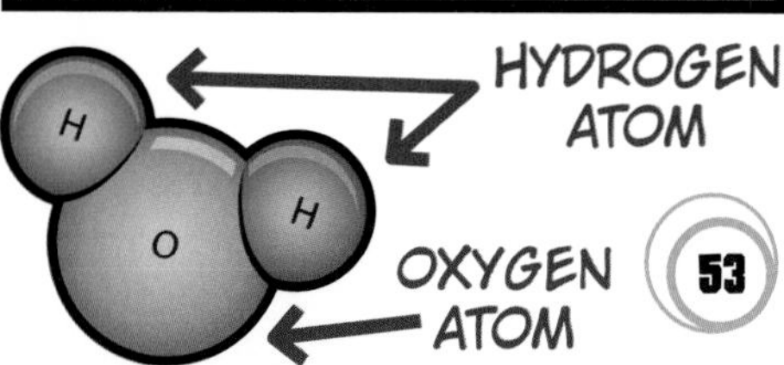

Elements

All of the matter in the universe is made up of **elements**. Atoms of elements are the same wherever they are found. For example, all of the iron in the universe comes from exploding stars. Iron (Fe) exists on meteorites. It exists on Earth, on Mars, and on other planets. Wherever it is found, the atoms in iron are exactly the same.

This chunk of iron is part of a meteorite that fell to Earth in 1947.

Iron ore is mined to make steel.

Each element has a name. It is also identified by a one- or two-letter **chemical symbol**. Some chemical symbols are letters from the name of the element. H is the chemical symbol for hydrogen. He is the symbol for helium. Other chemical symbols are letters from the Latin word for the element. For example, the Latin word for gold is aureum. The chemical symbol for gold is Au.

Scientists use the **periodic table of the elements** to keep track of all the elements. Each cell in the table is a different element. The periodic table shows the atomic number, the chemical symbol, and the name of each element. The table is arranged according to the atomic number of the element. The table is also arranged by the properties of the elements.

Organisms and most substances are made of only a few of these elements. Scientists, however, are still looking for new elements. When one is found, it goes into the periodic table.

The Periodic Table

- Each space in the periodic table contains the atomic number of the element, the chemical symbol for the element, and its name.

Key

Atomic number: 79 — Symbol: Au — Name: Gold

Hydrogen · Metals · Metalloids · Noble gases · Nonmetals

- Au is the chemical symbol for gold.

1 H Hydrogen																	2 He Helium
3 Li Lithium	4 Be Beryllium											5 B Boron	6 C Carbon	7 N Nitrogen	8 O Oxygen	9 F Fluorine	10 Ne Neon
11 Na Sodium	12 Mg Magnesium											13 Al Aluminum	14 Si Silicon	15 P Phosphorus	16 S Sulfur	17 Cl Chlorine	18 Ar Argon
19 K Potassium	20 Ca Calcium	21 Sc Scandium	22 Ti Titanium	23 V Vanadium	24 Cr Chromium	25 Mn Manganese	26 Fe Iron	27 Co Cobalt	28 Ni Nickel	29 Cu Copper	30 Zn Zinc	31 Ga Gallium	32 Ge Germanium	33 As Arsenic	34 Se Selenium	35 Br Bromine	36 Kr Krypton
37 Rb Rubidium	38 Sr Strontium	39 Y Yttrium	40 Zr Zirconium	41 Nb Niobium	42 Mo Molybdenum	43 Tc Technetium	44 Ru Ruthenium	45 Rh Rhodium	46 Pd Palladium	47 Ag Silver	48 Cd Cadmium	49 In Indium	50 Sn Tin	51 Sb Antimony	52 Te Tellurium	53 I Iodine	54 Xe Xenon
55 Cs Cesium	56 Ba Barium		72 Hf Hafnium	73 Ta Tantalum	74 W Tungsten	75 Re Rhenium	76 Os Osmium	77 Ir Iridium	78 Pt Platinum	79 Au Gold	80 Hg Mercury	81 Tl Thallium	82 Pb Lead	83 Bi Bismuth	84 Po Polonium	85 At Astatine	86 Rn Radon
87 Fr Francium	88 Ra Radium		104 Rf Rutherfordium	105 Db Dubnium	106 Sg Seaborgium	107 Bh Bohrium	108 Hs Hassium	109 Mt Meitnerium	110 Ds Darmstadtium	111 Rg Roentgenium	112 Cp Copernicium	113 Uut Ununtrium	114 Uuq Ununquadium	115 Uup Ununpentium	116 Uuh Ununhexium		

Lanthanides	57 La Lanthanum	58 Ce Cerium	59 Pr Praseodymium	60 Nd Neodymium	61 Pm Promethium	62 Sm Samarium	63 Eu Europium	64 Gd Gadolinium	65 Tb Terbium	66 Dy Dysprosium	67 Ho Holmium	68 Er Erbium	69 Tm Thulium	70 Yb Ytterbium	71 Lu Lutetium
Actinides	89 Ac Actinium	90 Th Thorium	91 Pa Protactinium	92 U Uranium	93 Np Neptunium	94 Pu Plutonium	95 Am Americium	96 Cm Curium	97 Bk Berkelium	98 Cf Californium	99 Es Einsteinium	100 Fm Fermium	101 Md Mendelevium	102 No Nobelium	103 Lr Lawrencium

WE SHARE THE PLANET WITH MAPLE TREES AND SQUIDS AND SAGE BRUSH AND MEERKATS AND PENGUINS AND DRAGON FLIES... BILLIONS OF LIVING THINGS!

Life Science

What is life science? **Life science** is the study of all living things. Living things include elephants, people, and potatoes. Living things also include Spanish moss, camels, and sponges. (A sponge isn't just something you clean with—it's also an animal!) A living thing is something that grows, moves, and reproduces. Living things can be as large as a whale or as small as a bacteria. Living things are all around us.

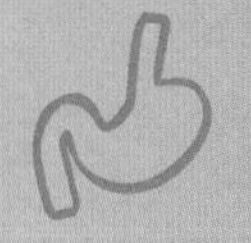

Cells, Tissues, and Organs

What Is a Cell?

Imagine looking at yourself through a microscope. What would you see? Would you see your eyes, ears, and nose? Would you see your hair and skin? Would you see your face? Would you even recognize yourself? You would not see any of these things. You would see your tiniest body parts—your cells.

Everything that is alive is made up of one or more cells. A **cell** is the basic unit of a living thing. A cell contains everything it needs to grow and reproduce. A cell can make new cells. A cell makes copies of itself. This is called **cell division**.

SAR11 is a kind of bacteria that lives in the oceans. Each single-celled bacterium is 0.0002 mm long. A teaspoon could hold billions of them!

History Makers

Robert Hooke (1635–1703)

Robert Hooke was a British inventor and scientist. One of his greatest contributions to science was the discovery of cells. One day he looked at a slice of cork with his microscope. He thought it looked like a honeycomb with holes. He called the holes "cells." The cells in cork are dead. Hooke had really seen their cell walls. Hooke's discovery led to the understanding that all living organisms are made of cells.

Cells can be any shape or size. An ostrich egg is a single cell. The egg is about 15 cm (almost 6 in) long and weighs about 1.4 kg (about 3 lb).

Unicellular Organisms and Multicellular Organisms

Living things can be made up of only one cell, or trillions of cells, or of any number of cells.

A **unicellular** (yoo nih SELL yoo ler) **organism** has only one cell. The single cell does everything needed to stay alive. Unicellular organisms make copies of themselves through cell division.

Unicellular organisms are usually very small. We can only see them with a microscope.

Yeasts are unicellular fungi. Yeast cells produce carbon dioxide as waste. When yeast is mixed into bread dough, bubbles of carbon dioxide make the dough rise.

Bacteria are unicellular. When too many *Escherichia coli* cells divide inside a human, it makes the person sick.

Multicellular (mul tih CELL yoo ler) **organisms** have more than one cell. Some algae and fungi are multicellular. All plants and animals are multicellular. Multicellular organisms are usually larger than single-celled organisms. They can replace worn-out cells through cell division so they live longer than organisms with only one cell.

Multicelluar organisms can be any size. The tree, the kitten, and the parasite *(Trichinella spiralis)* are all multicellular.

Multicellular organisms have cells that do only one job. Cells that do a particular job are called **specialized cells**. Humans have many kinds of specialized cells. Specialized human cells include bone cells, muscle cells and blood cells. A tree has specialized cells like root cells and leaf cells.

Parts of a Cell

Multicellular organisms have cells that do only one job. Specialized cells that do different jobs do not look the same. Even though the jobs cells do can differ, the structures inside each cell are similar.

All cells are surrounded by a **cell membrane**. It controls what enters and leaves the cell.

Cytoplasm (SY toh plaz um) is the fluid inside the cell membrane. The cell organelles are in the cytoplasm.

The **organelles** (or guh NELZ) do the work of the cell. Plant and animal cells have many of the same kinds of organelles:

Nucleus The **nucleus** (NOO klee uhs) is the command center of the cell. The nucleus sends information to other organelles. All of the information a cell needs to make new cells is in the nucleus. Not every cell has a nucleus. A red blood cell loses its nucleus as it develops.

Mitochondria The **mitochondria** (mite oh KAHN dree uh) break down food to make energy. Cells that need lots of energy, like muscle cells, have a lot of mitochondria.

Vacuoles Fluid-filled sacs called **vacuoles** (VAK yoo ohlz) are surrounded by membranes. Nutrients are stored inside the sacs while they are being digested. Cell wastes are held inside vacuoles until the cell can get rid of them.

Other organelles in the cell produce proteins and move them to cells where they are needed.

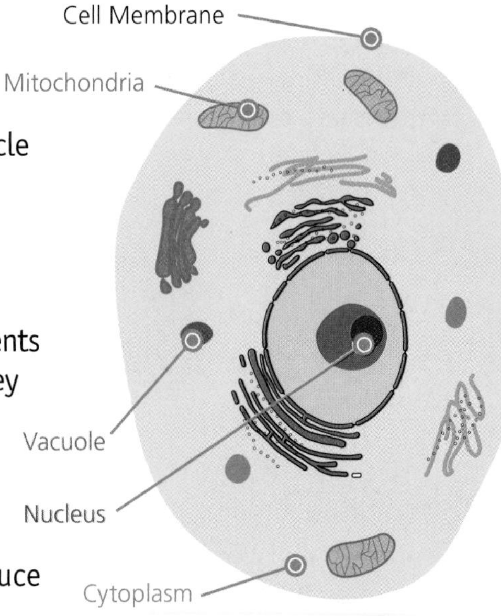

animal cell

Plant cells have some organelles that animal cells do not have:

Cell Wall A plant cell has a **cell wall** outside the cell membrane. The cell wall gives the cell extra support.

Chloroplast A special organelle called a **chloroplast** (KLOR oh plast) makes food for the plant. Chloroplasts contain **chlorophyll** (KLOR oh fil). Chlorophyll captures energy from sunlight. The energy combines with carbon dioxide and water to make a kind of sugar called **glucose** (GLOO cohs).

Large Cenral Vacuole Plant cells have a **large central vacuole** that stores water, salts, and carbohydrates. It also helps support the plant. It does this by filling with water.

If the large central vacuoles in plant cells dry out, the plant wilts.

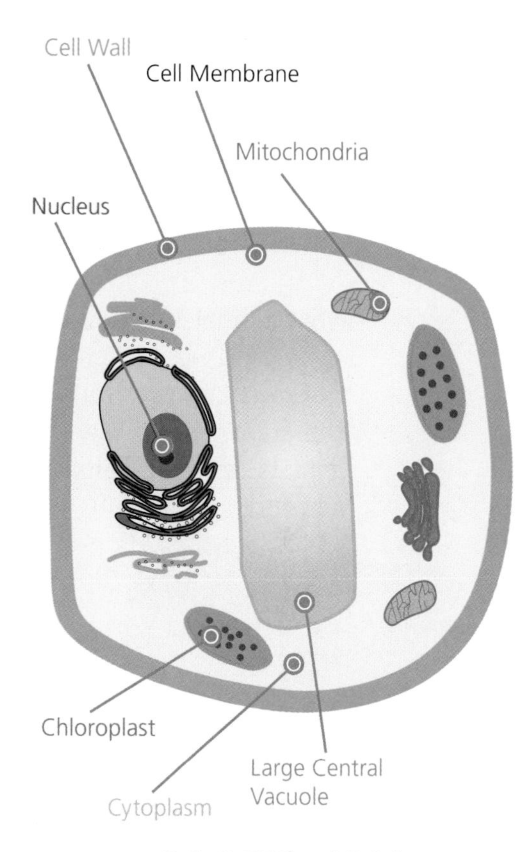

PLANT CELL

EXPLORE MORE

CHECK YOUR ANSWERS, PAGES 408–419!

CELL-U-LEARNING

Your school building is a lot like a plant or animal cell. The building has a strong outer structure. It provides services—water, electricity, and even food. The school has storage areas. Waste materials are carried away. In your science notebook, write a short paragraph comparing a plant or animal cell to your school.

Animal Tissue

Multicellular organisms have specialized cells. Specialized cells do only one job. Two or more specialized cells that work together are called a **tissue**.

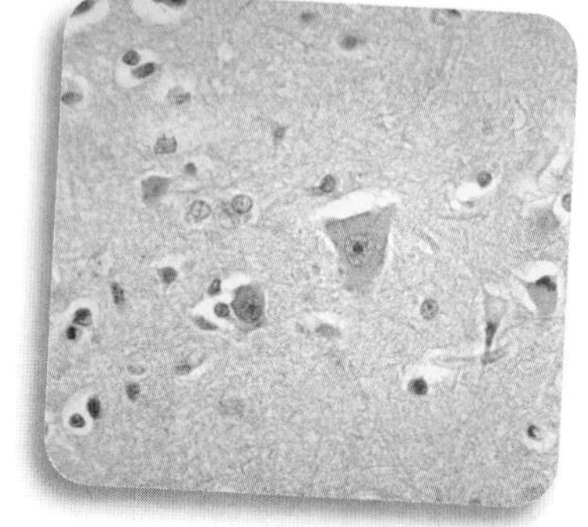

Most animals have four kinds of tissue.

Nervous tissue is found in nerves and the brain. It is made of nerve cells. Signals between cells are passed back and forth by nervous tissue.

Muscle tissue is made of muscle cells. Muscle tissue is found in muscles and in other organs. When muscle tissue contracts, it shrinks. When it relaxes, it expands. The body moves as muscle tissues contract and expand.

Connective (kuh NEK tiv) tissue holds body parts together. Tendons are tissues that connect bones and muscles. Bone is connective tissue that supports the body. Cartilage is connective tissue that cushions joints. Blood is connective tissue that moves food, oxygen and water through the body.

Sharks do not have bones. Cartilage holds their bodies together. Cartilage makes their bodies very flexible.

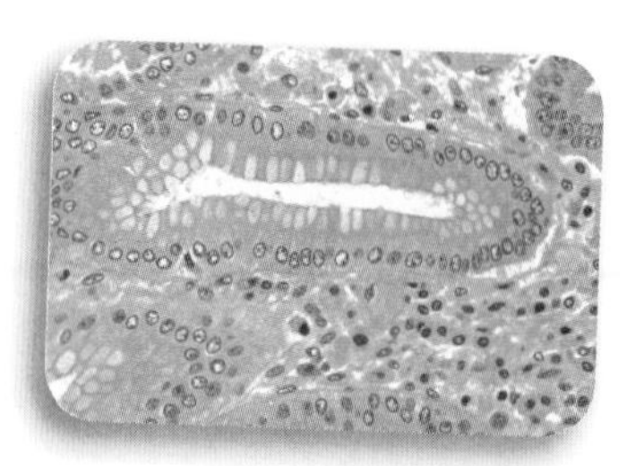

Epithelial (eh puh THEE lee uhl) tissue is thin and smooth. It protects the body. Skin is made of flat sheets of epithelial tissue. Epithelial tissue lines blood vessels, organs and other body parts.

Plant Tissue

Most plants have three kinds of tissue.

The outside of leaves, stems, and roots is covered by **dermal tissue.** The cells of dermal tissue are closely packed together. Tree bark is made of dermal tissue.

Vascular tissue moves nutrients and water in the plant. Vascular tissue helps hold the plant up. Not all plants have vascular tissue. Plants without vascular tissue cannot grow very tall. Moss does not have vascular tissue.

The third kind of plant tissue is **ground tissue.** Carbohydrates are stored in ground tissue. Plants use sunlight and carbon dioxide to change carbohydrates into energy in ground tissue.

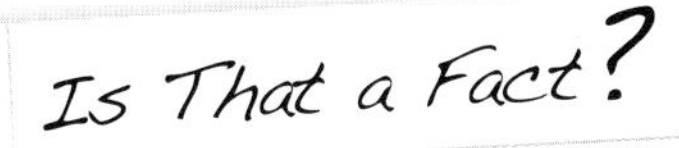

Let's Connect

Did you know that blood is actually a tissue? It's true! Blood may look like a liquid. But it is full of blood cells. Because red blood cells all do the same job, it is a tissue. It is a connective tissue because the blood connects other tissues to the respiratory, digestive, excretory, and endocrine systems.

Animal Organs

Two or more tissues that work together form an **organ.** Animals and plants have organs.

All animals do not have the same organs. For example, clams have hearts, but do not have brains. Most fish do not have lungs. They breathe through gills.

Most mammals, including humans, have the same kinds of organs. The brain, stomach, heart, lungs, and skin are examples of animal organs. These organs may look different from one animal to another, but they perform the same tasks.

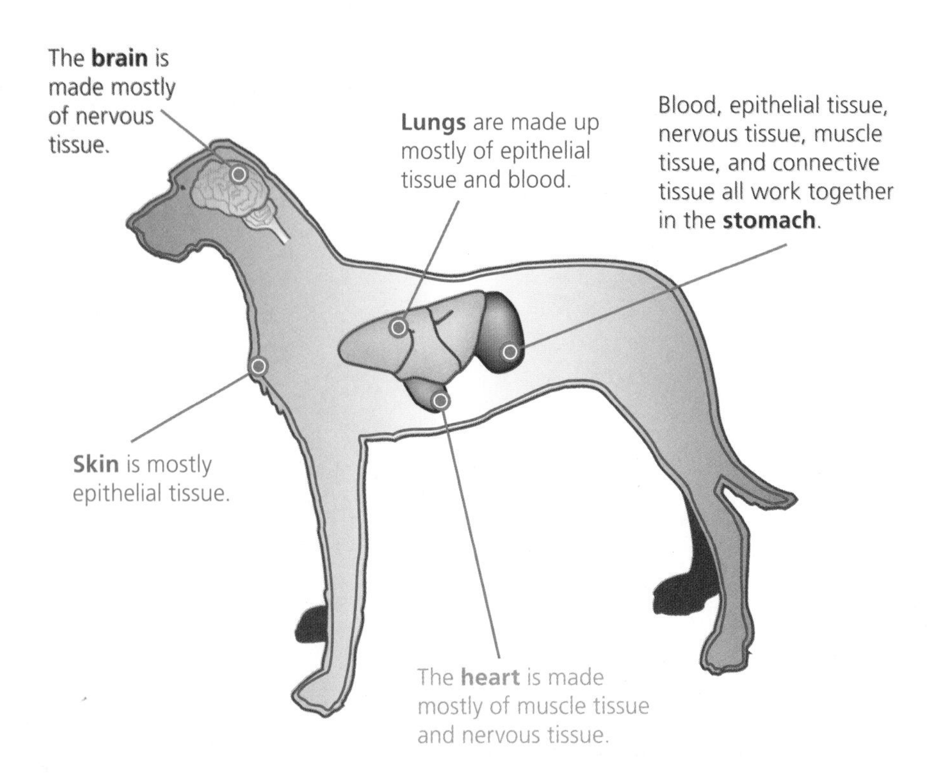

Plant Organs

Like animals, plants need water. They use energy. Plants get rid of wastes. They reproduce. Plants have organs that do these jobs. Plant organs are made of specialized tissues.

All plants do not have the same organs. Mosses do not have roots. They get moisture from the air. Most land plants have three organs. These are the roots, leaves, and stems.

Roots are usually underground. The roots of a plant hold it in place in the soil. Food for the plant is stored in the roots. The outside of the root is covered in dermal tissue. Roots take up water and minerals for the plant. Roots push nutrients and water upward from vascular tissue.

See also: page 116 Plants

Stems hold the leaves up into the sunlight. New leaves, branches and flowers are made on stems. Stems are surrounded by dermal tissue that helps support the plant. Water and minerals move from the roots to the leaves through vascular tissue in the stems. Some food is stored in the stem in ground tissue.

Plants make food in their leaves. Food is carried from the leaves to other parts of the plant through vascular tissue. Dermal tissue protects the leaf and helps it conserve water. Gases pass in and out of the plant through the leaves.

A plant has three organs: leaf, stem, and root.

Animal Organ Systems

Organs work together in **organ systems**. Animals and plants have organ systems.

Humans and other mammals have many organ systems. The organ systems are connected. They work together to keep the body healthy.

For example, the nervous system controls every other system. The brain, spinal cord, and nerves are part of the nervous system. Eyes and other sense organs are part of the nervous system.

The organs in the digestive system break down food. The food supplies nutrients to all of the other organ systems.

Animals do not all have the same organs and organ systems. Insects do not have lungs. A system of tubes brings oxygen to the body. Starfish do not have brains. Nerve nets act as the nervous system.

See also: page 90 The Human Body

The organ systems of a bird help it fly. The light skeletal system is connected to the lungs. Air pockets in the bones work with the respiratory system to make breathing more efficient.

Plant Organ Systems

Plant organ systems are different from animal organ systems. An animal organ system has different organs. For example, the digestive system includes a stomach, intestines, and several other organs. Plant organ systems are groups of the same type of organ. The leaf system of a plant is made up of all the leaves on the plant. The stem system is made up of all the stems and branches. The root system is made up of all the roots.

Plants grow new organs as they get older.

Plant organ systems grow as the plant grows. The plant grows new leaves when the plant needs more energy. More stems or branches may grow to support the extra leaves. Stems also can grow thicker to support extra weight. As the plant grows, it needs more nutrients and water. So it grows more roots as well. When animals grow, their organs grow larger, but they do not grow extra organs.

Which-Is-Which?

ORGAN

Explain how the four terms above are related to each other. Write your answers in your science notebook.

Organisms

What Are Organisms?

What do seahorses, horseflies, and woodland horsetails all have in common with horses? Well, yes—they all have the word "horse" in their name. But they are alike in another way: they are all ALIVE! Anything that is alive is called an **organism** (OR guh NIZ um).

Organisms are everywhere. They can be found growing in water and soil. Some grow on rocks. Some can even grow inside other organisms! Some organisms are attached to the sea floor. Others move around on land. Still others hitch a ride on the skin of larger organisms. Plants and animals are organisms. YOU are an organism too!

Did You Know?

The Processes of Life

What makes something "living" or "not living"? I looked it up because I thought it would be cool to program Kelvin to do everything an organism can do. I found out that there are seven basic things all organisms do. To be alive, you have to be able to do all seven of these!

Organisms can ...

- **move**
- **grow**
- **use air**
- **get nutrients**
- **remove wastes**
- **reproduce**
- **sense their surroundings**

So it turns out there are a few things that I can't program into Kelvin after all. Sorry, pal!

What Organisms Need

See also: page 78
Animals

What do you need to live? Food, water, air—pretty basic, right? They may be basic but they're pretty important. All organisms need the same things you need to survive.

Plants

All Organisms Need...

Energy

Energy is like the gas in a car. Without gas, a car won't run. Without energy, organisms can't run!

Plants get energy from the Sun. They use that energy to make their own food.

Animals get energy from the food they eat.

Nutrients

Organisms need nutrients to grow. Nutrients are like the bricks and boards of a house.

Plants take in most of their nutrients from the soil through their roots.

Animals get nutrients by breaking down food they eat. They get food from plants and other animals.

Water

Water is one of the most important nutrients. It gives cells their shape. Many processes that happen inside organisms must use water.

Plants take in water from the soil through their roots.

Animals get water by drinking it.

Air

Air contains gases that plants and animals need to live.

Plants use carbon dioxide from air to make food. They release oxygen into the air after they make the food.

Land animals get oxygen from air. Animals that live in water get oxygen from water.

Organisms Are Organized

Organisms come in all shapes and sizes. But all of them are made of cells. A **cell** is the smallest unit of life.

Green algae

Unicellular (yoo nih SELL yoo ler) organisms are made of only one cell. This single cell is the organism's entire body! All bacteria are unicellular. So are algae (al JEE), euglena (you GLEE nuh), and many other organisms.

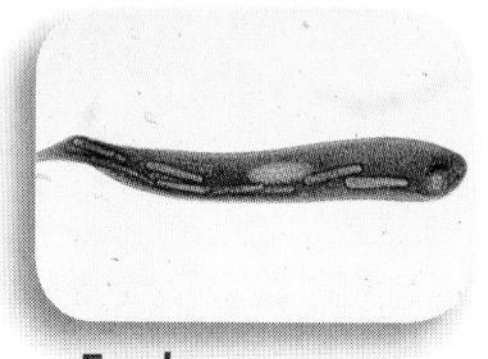

Euglena

Caladium

Sea star

Organisms that are made of many cells are called **multicellular** (mul tih SELL yoo ler). Plants and animals are multicellular organisms.

All cells are surrounded by a membrane. Inside the cell is a thick fluid that contains structures called **organelles** (or guh NELZ). There are many types of organelles, and each has a job to do. Some organelles store food. Others make energy or control wastes.

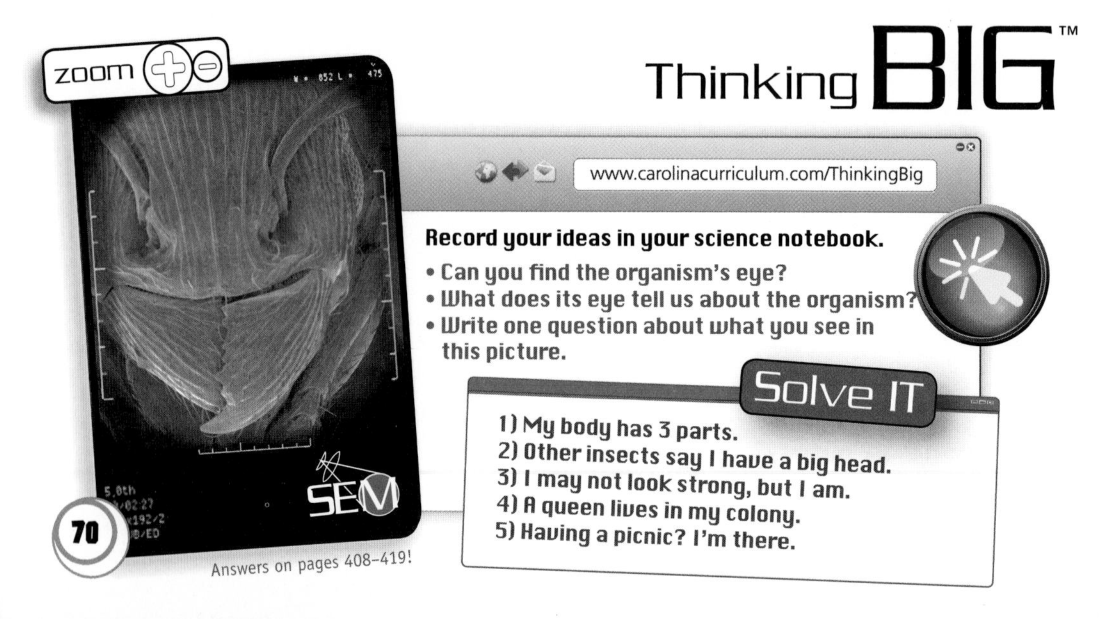

Answers on pages 408–419!

If you flex your arm, the muscle cells in your biceps contract at the same time. When you straighten your arm, the muscle cells all relax.

In a unicellular organism, one cell does all the work! But in a multicellular organism, a cell might have just one job. A cell that has a single task is called a **specialized cell** (SPEH shuh leyezd SELL).

A **tissue** is a group of cells that work together to do the same job. For instance, muscles are made of muscle tissue. Muscle tissue is made of muscle cells.

See also: page 58 Cells, Tissues, and Organs

An **organ** is a group of different tissues that work together to perform a task. Your stomach is an organ. It has different tissues that work together to help digest food. Some stomach tissues make chemicals that break down food. Muscle tissue in the stomach contracts to mix the food and chemicals.

An **organ system** is a group of organs that carry out a process. The stomach is part of the digestive system. This system also includes the intestines, the liver, and other organs. The organs in the digestive system work together to change the food you eat into nutrients your body can use.

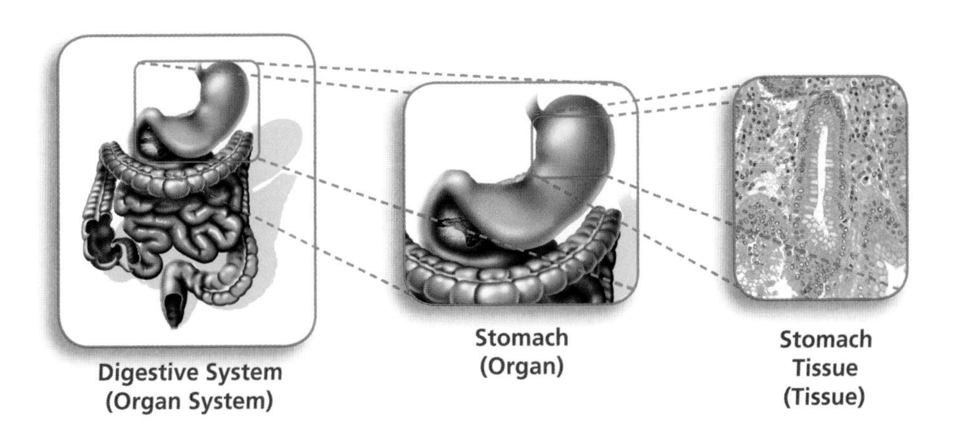

Digestive System (Organ System)

Stomach (Organ)

Stomach Tissue (Tissue)

Inherited Traits

All organisms can reproduce. A **species** (SPEE sheez) is a group of organisms that can mate with each other and produce offspring. When organisms reproduce, they pass certain traits to their offspring. A **trait** (trayt) is a feature, such as eye color or nose shape. If your eyebrows are the same shape as your mother's, you inherited that trait from her. Traits that are **inherited** are passed from parent to offspring, and from generation to generation.

Red maple leaf

Look at the shape of this red maple leaf. All red maple leaves have the same shape. Now look at the silver maple leaf. It looks different compared to the red maple leaf, right? The silver maple leaf inherited its shape from its parents. And the red maple leaf got its shape from *its* parents!

Silver maple leaf

History Makers

Gregor Mendel (1822–1884)

Gregor Mendel was a Moravian monk and scientist. He observed that most plants grown from seeds looked like their parents. Did parents pass on traits through their seeds? Mendel crossed plants that had certain traits and noted when these appeared in offspring. He found that plants did inherit some traits from their parents! Mendel wasn't the first to observe that offspring and parents look alike. But he was the first to test this in an experiment.

GO ONLINE

Thinking BIG™

Look online for a micrograph of a bee's leg.

http://www.carolinacurriculum.com/ThinkingBig

Adaptations

The saguaro's folds expand as the stem fills with water.

Some traits can help organisms survive by making it easier for them to get nutrients or other resources. A trait that helps an organism survive in its surroundings is called an **adaptation**.

It only rains a few days each year in the Sonoran Desert. Plants that grow there must store water between rainy days. The folds on the saguaro cactus spread out as the plant takes in water. This adaptation lets the cactus survive on stored water during long dry periods.

Cardinal

Beak size and shape is an important adaptation in birds. The cardinal's strong, thick beak is well shaped for cracking hard seeds. Robins have long, slender beaks that help them catch their favorite food: insects.

Robin

Some species have adaptations that change with the seasons. Look at the arctic fox. Its brown coat helps it hide in the dark vegetation in the summer. In winter, it grows a thicker white coat. This keeps the fox warm and helps it hide in the snow.

Animals

Plants

Nature's Disguise Artists

An octopus can change its body color faster than you can change your shoes! You know how you can mix red and blue paint to make purple? Octopuses have specialized skin cells with different colors called pigments. These cells mix their pigments to change the octopus's skin color to match its surroundings! An octopus also has special muscles that can make its skin look like rocks or sand. When the octopus swims to a new area, it can change its color and texture again!

THAT WOODPECKER LOOKS LIKE IT'S LISTENING TO SOMETHING!
SHH...IT'S TRYING TO HEAR GRUBS MOVING UNDER THE BARK.
WOW! THAT'S SO COOL!
IT USES ITS BEAK TO DRILL A HOLE IN THE BARK. THEN IT FISHES OUT THE GRUBS WITH ITS LONG TONGUE.
BEAK SHAPE IS LIKE A TOOL THAT HELPS BIRDS HANDLE CERTAIN TYPES OF FOODS. TOUCANS USE THEIR HEAVY BEAKS TO CRACK LARGE SEEDS AND NUTS.
I'M GOING TO SKETCH THESE IN MY SCIENCE NOTEBOOK.
HUMMINGBIRDS USE THEIR LONG BEAKS LIKE STRAWS TO DRINK THE NECTAR DEEP INSIDE FLOWERS.

Adaptations and Behavior

A **behavioral adaptation** is an activity that helps an organism survive in its surroundings. Some animals use behaviors to adapt to living in very hot or very cold areas. Behaviors can protect animals from predators. Behaviors also help predators capture prey.

Many desert animals rest under a shady bush or underground during the hottest hours each day. They become active in the cool evening after the Sun goes down.

Penguins stay warm by huddling. They form very large groups and stand very close together during the coldest months of the year.

There's safety in numbers! Predators usually won't attack a large group. That's why animals like these antelope stay together.

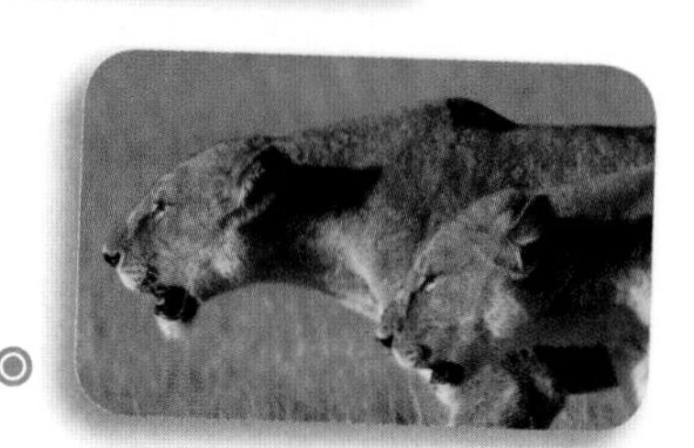

But that won't stop the lions. Sometimes one antelope falls behind or gets separated. Then the lions make their move.

GO ONLINE

To learn more about organisms, check out these Web sites!

- **OLogy**
  http://www.amnh.org/ology/
- **Cool Science for Curious Kids**
  http://www.hhmi.org/coolscience/forkids/
- **Spot the Arctic Hare**
  http://nature.ca/ukaliq/

Classifying Organisms

Can you count all the types of organisms on Earth? It might take a while. That's because there are quite a few—as many as 10 million, in fact! How do scientists keep track of all these organisms? They classify them. **Classification** (class ih fih KAY shun) is a system that groups organisms by how much they are alike. The organisms in each group are more like each other than they are like organisms in other groups.

You know you can classify organisms as unicellular or multicellular. You can classify multicellular organisms into two groups: plants or animals.

A good way to tell the difference between plants and animals is by looking at their cells. Plant cells and animal cells are alike in many ways. But there is one sure way to tell them apart.

Cell wall

Chloroplasts

Plant cells have cell walls. Many plant cells include organelles called chloroplasts. These structures help plants make their own food.

Animal cells do not have cell walls or chloroplasts. This is the main way to tell animal cells from plant cells.

History Makers

Carl Linnaeus (1707–1778)

Carl Linnaeus was a Swedish doctor and scientist who studied ways to classify plants. Early plant names were very long and confusing. Sometimes scientists even gave different names to the same plant! Linnaeus solved this problem. He organized plants into groups based on ways they were alike. He gave each type of plant a standard name based on this system. Then he used his system to classify all living things. Today his system is still used to name and classify organisms.

Dichotomous Keys

A **dichotomous key** (dy KOT o muss KEE) is a tool that helps identify an object by answering a set of questions. Each question has two answer choices. Each choice leads to another question with two more choices. The choices lead you step by step to identify each object.

See also: page 78 Animals

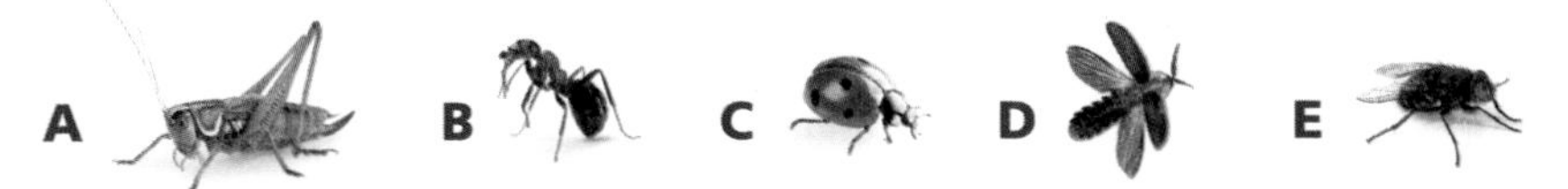

Let's identify these insects with the dichotomous key at the bottom of the page. The directions below will guide you through each clue.

See also: page 116 Plants

Clue 1 is about legs. Look at all five insects. Insect A has long back legs. The key says it must be a grasshopper. That's great! You've identified one insect. The other insects have legs of the same length. To identify these insects, you need more clues. Go to the next clue.

Clue 2 looks at whether the insect can fly or can't fly. Insect B doesn't have wings, so it can't fly. The key says it is an ant. The other insects have wings—they can fly. Go to the next clue.

Use the same process for clues 3 and 4. Check each insect as you read each clue. When you're done, you will have identified all five insects!

Dichotomous Key

1a. Back legs longer than other legs........... Grasshopper
1b. All legs are the same length .... Go to 2

2a. Can't fly.. Ant
2b. Can fly.. Go to 3

3a. Body does not have spots .... Go to 4
3b. Body has spots..Ladybug

4a. Can make light ....Firefly
4b. Can't make light...Housefly

Animals

What Is an Animal?

Did you know that you are like a sand dollar, a bumblebee, and a prairie dog? You may not look like these organisms, but you have a lot in common with them. That's because you are all animals!

Sand dollar

Prairie dog

Bumblebee

Scientists classify all organisms into very large groups called kingdoms. For example, all plants belong to the Kingdom Plantae. All animals belong to the Kingdom Animalia. You don't have to take a test to qualify for the Kingdom Animalia ... but there are a few "rules"!

Animals...

- have **cells** that do not have cell walls.
- **cannot** make their own food.
- **reproduce** by laying eggs or giving birth to live offspring.
- are **multicellular** organisms.
- **take in** oxygen and **release** carbon dioxide.
- can **move** at some stage of their lives.

Animal Bodies

Sea sponge

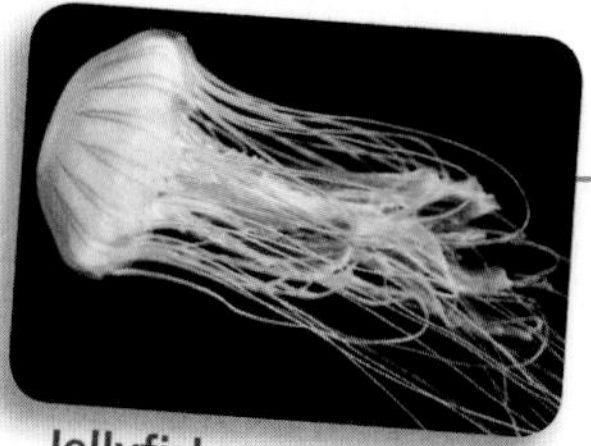

Jellyfish

Jellyfish and sponges are sometimes called simple animals because their bodies are not organized into organs or organ systems.

Animals are multicellular. Their bodies have many different types of cells. These cells are organized into tissues in most animals. Most larger animals have organs and organ systems.

Sponges have very simple bodies. They don't have tissues or organs. But they have specialized cells that work together. Some cells work together to move water through the sponge. They have structures called flagella (fluh JEL uh) that "sweep" water through the sponge's body.

Jellyfish don't have organs or organ systems. They do have tissues. These include muscle tissue and digestive tissues. A network of nerves helps jellyfish sense their surroundings. Then it sends information through their bodies.

Large animals need organs and organ systems to use nutrients and move them through their bodies. Organs systems give animals more choices about what to eat and where to live. This gives them more ways to live in the world.

Cells, Tissues, and Organs

Here's Looking At You!
Some jellyfish have simple eyes, but these are not true organs. Box jellyfish have 24 eyes! Most of these are very simple light detectors. But several pairs can detect colors and shapes. Okay, box jellyfish can't see things as well as we do. But even simple eyes are a great tool that helps these animals hunt for food.

Invertebrates and Vertebrates

Scientists divide animals into two main groups: animals that have a backbone and animals that do not. A backbone contains bones called **vertebrae** (VER tuh bray). Some people call the backbone a spine.

Invertebrates (in VER tuh brayts) don't have backbones. Their bodies are very soft. Most invertebrates have some type of hard material that covers and protects their bodies.

Snail

Mollusks have a hard mineral shell. This group includes clams, oysters, and snails.

Crab

Arthropods have an exoskeleton and jointed legs. An exoskeleton is lighter than a shell. Crabs, lobsters, and spiders are in this group.

Earthworm

Worms, squid, and jellyfish have no outer covering. They live in water or in damp areas to keep their bodies from drying out.

Sea lion

Sting ray

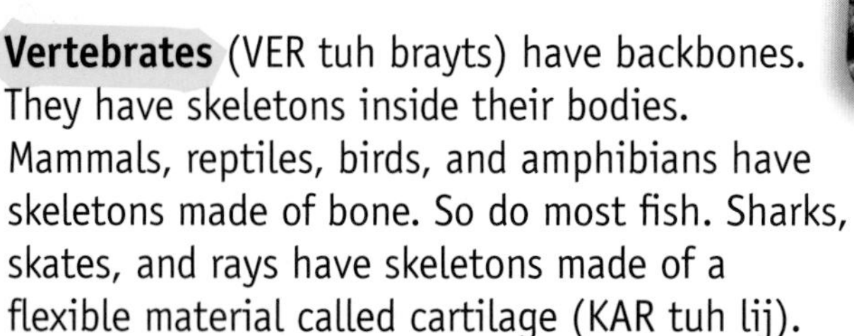

Vertebrates (VER tuh brayts) have backbones. They have skeletons inside their bodies. Mammals, reptiles, birds, and amphibians have skeletons made of bone. So do most fish. Sharks, skates, and rays have skeletons made of a flexible material called cartilage (KAR tuh lij).

Shape Changers

The skeleton inside some invertebrates is made of water! This is called a hydroskeleton. One of its jobs is helping the animal move. The hydroskeleton is made of tubes of water surrounded by muscle. To pull itself forward, an earthworm contracts the muscles around its hydroskeleton. Its body grows longer and thinner as the front of the body reaches forward. Then the worm relaxes the muscles. The body gets shorter and wider. The "tail" end of the worm pulls forward. This lets it catch up with the rest of its body!

Endotherms and Ectotherms

Some animals maintain a constant body temperature. Other animals take on the temperature of their surroundings. Which group do you belong to?

If you guessed the first group, you're right! Like all mammals, you are an endotherm. So are birds. An **endotherm** is an animal whose temperature stays the same regardless of its environment.

Blue jay in summer

Blue jay in winter

Birds fluff their feathers in cold weather. Fluffing traps warm air near their skin.

If an endotherm gets cold, its brain tells the muscles to shiver to make heat. Many mammals grow thick winter coats that trap warm air near the skin. Sea mammals and penguins have thick fat layers that stop heat loss.

When an endotherm gets too warm, its brain sends signals to cool down. Apes, monkeys, and humans cool down by sweating. Dogs and cats can't sweat. They pant to cool down. Elephants flap their ears to fan themselves.

Desert iguanas move into and out of the sunshine to control their body temperature.

An **ectotherm** is an animal whose body temperature depends on its surroundings. All animals except birds and mammals are ectotherms. Ectotherms control body temperature with behavior. They hang out in sunny areas to warm up. When they get too warm, they move to the shade.

continued

Both ectotherms and endotherms like to hang out in the sunshine. But ectotherms use this behavior to change their body temperature.

NUMBER CRUNCH

CHECK YOUR ANSWERS, PAGES 408–419!

You're Getting Warmer!

Plot the data on a graph. Use the graph to identify the endotherm and ectotherm.

Outside Temperature (°C)	Body Temperature (°C)	
	Animal 1	Animal 2
5	38.0	3.0
10	38.5	9.0
20	38.5	19.0
30	38.7	28.0

1. Draw a graph on graph paper. Label the horizontal axis "Outside Temperature (°C)." Label the vertical axis "Body Temperature (°C)."
2. Plot data for Animal 1 with a blue pencil. Connect the dots.
3. Plot data for Animal 2 with a red pencil. Connect the dots.
4. How are the lines for Animal 1 and Animal 2 different?
5. Which animal is an ectotherm? Which animal is an endotherm? Explain your answers.

Habitat and Niche

The area where an animal lives is called its **habitat** (HAB ih tat). A habitat provides everything that an organism needs for life. That includes the living things and nonliving things in its environment.

The shrubs and grasses in a desert habitat are important to the animals that live there. The plants provide shelter from the hot desert sunlight. They also provide food for desert rodents and rabbits, and hide them from predators. A **predator** is an animal that eats another animal.

The shrubs and rocks in this desert provide food and shelter for small animals such as this desert cottontail rabbit.

The role that an animal plays in its habitat is its **niche** (NEESH). A niche includes the ways the animal interacts with other organisms and its environment. Desert plants are the main food that a desert rabbit eats. It competes with other animals for those plants.

The rabbit itself is hunted and eaten by desert predators. These include coyote, foxes, and hawks. All of these factors make up the rabbit's niche.

GO ONLINE Thinking BIG™

Look online for a micrograph of a bee's wing at
http://www.carolinabio.com/ThinkingBig

Resources

When biologists study an animal's niche, they look at how the animal uses its resources. A **resource** is anything an animal needs to survive. Food, water, and places to live are basic resources for all animals.

See also: page 136 Ecosystems

organism	bat	lion	squirrel
resource	cave	pond	nut
resource type	shelter	water	food

A habitat must have enough resources for everything that lives there. But resources are limited. That means animals must compete for resources. Animals compete with members of their species. They also compete with other species. Two cheetahs might compete with each other for food. They might also have to compete with lions.

Only a certain number of organisms can live in a habitat. If the habitat gets too crowded it could run out of some resources. Some animals will be able to adapt to this change. Other animals will have to move to a new habitat. Animals that can't move or adapt will not survive.

Is That a Fact?

Over the Hump

Lots of people think that camels store water in their humps. But actually, a camel's hump is made of fat. The fat acts as an energy store. That means the camel can travel for long distances without eating. As for drinking ... well, a camel actually stores water in its blood.

Dromedary camels have one hump. They live in the deserts of North Africa and Southwest Asia.

Bactrian camels have two humps. They are found mainly in the deserts of Mongolia and Central Asia.

Adaptations and Behavior

Even a small change in resources or climate can have a large effect on a habitat and its communities. Some animals may not survive these changes. Others may have traits that help them survive when things change. These traits are called **adaptations**.

A **structural adaptation** is a feature of the animal's body that helps it survive. Body color and foot shape are structural adaptations.

continued

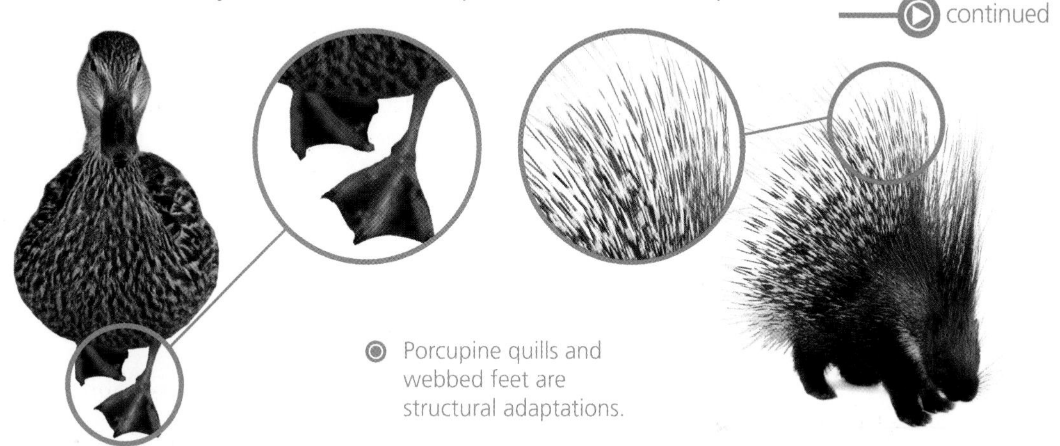

Porcupine quills and webbed feet are structural adaptations.

Answers on pages 408–419!

www.carolinacurriculum.com/ThinkingBig

Record your ideas in your science notebook.

- What state of matter is this? Why do you think this is so?
- Might this be manmade, or found in nature?
- Do you see any patterns? Draw what you see.
- Might this be strong? Smooth? Soft? Explain.

1) I am strong but light.
2) I keep a body warm.
3) No raincoat? I keep a body dry.
4) What I do best? Move high in the sky.
5) Wings on an airplane? Scientists got ideas from me.
6) What am I?

You already know that adaptations are passed from parent to offspring. But how do adaptations arise in the first place?

An adaptation starts as a form of a trait found in some population members before the habitat changes. Animals with the trait may be able to survive in the changed habitat. They will pass the trait to their offspring. The offspring will pass it to their offspring. Soon the trait may become common in the population.

Behavior is how an animal acts. Some behaviors are learned. Inherited behavior is passed from parent to offspring. Inherited behaviors that help animals survive in their environment are called **behavioral adaptations**.

Mimicry (MIM ick ree) is an adaptation that lets some animals avoid being eaten by other animals. The mimic is usually a harmless animal that looks like a dangerous animal or poisonous plant. This tricks the mimic's enemies into staying away from it.

History Makers

Ivan Pavlov (1849–1936)

Ivan Pavlov was a Russian scientist. He used dogs to study behavior. He knew that dogs' mouths watered when they saw food. He wondered what would happen if he rang a bell when he fed the dogs. After he tried this a few times, the dogs' mouths watered when the bell rang. This happened even if they didn't see food. Pavlov's work led to new ways to study animal behavior.

Some animals mimic plants. This lets them hide by "disguising" themselves as plants. A predator would have a hard time finding this dead-leaf butterfly in a leaf pile.

Animals that mimic other animals often have bright colors. These are a warning sign. Many poisonous animals have bright colors.

Coral snake Milk snake

Can you tell the difference between these snakes? The coral snake is the most poisonous snake in the United States. The milk snake is the mimic. It is not poisonous. But don't tell its predators! They avoid it because they think it's a coral snake.

Some animals move from one area to another each year. This is called **migration** (my GRAY shun). Some birds fly from cold northern regions to the Equator each winter to mate. They return north in the spring to build nests and hatch offspring.

Some animals hibernate when food is scarce. **Hibernation** is a state of deep rest. It lets an animal survive the winter without food. Hibernation can last for weeks or months. The animal's temperature drops during this time. Its heart rate and breathing slow down.

GO ONLINE

To learn more, check out these Web sites!

- **The Pavlov's Dog Game**
  http://nobelprize.org/educational_games
- **Amazing Animals**
  http://kids.nationalgeographic.com/Animals/
- **Animals**
  http://www.seaworld.org/index.asp
- **The Smithsonian Zoo**
  http://nationalzoo.si.edu/default.cfm
- **Animal Bytes**
  http://www.sandiegozoo.org/animalbytes/

LOOKING LIKE A STICK HELPS HIDE THE PRAYING MANTIS FROM THE INSECTS IT HUNTS. THE INSECTS THINK IT'S PART OF A TREE OR A PLANT. SO THEY IGNORE THE MANTIS. UNTIL IT'S TOO LATE ...
I THOUGHT THAT WAS A STICK!

I'M CAMOUFLAGED, TOO!

SHH! ANIMALS THAT USE CAMOUFLAGE DON'T MOVE UNTIL THEY'RE READY TO GRAB THEIR VICTIM. YOU CAN'T STAY STILL FOR 5 SECONDS, DUDE!

WHOA! THAT LEAF JUST BLINKED AT ME! OH HEY, THAT'S ... WHAT IS THAT?!
KATYDIDS ARE HARD TO SPOT—AND THEY LIKE IT THAT WAY! THEIR LEGS LOOK LIKE STEMS AND THEIR WINGS LOOK LIKE LEAVES. THAT HIDES THEM FROM ANIMALS THAT WANT TO EAT THEM.

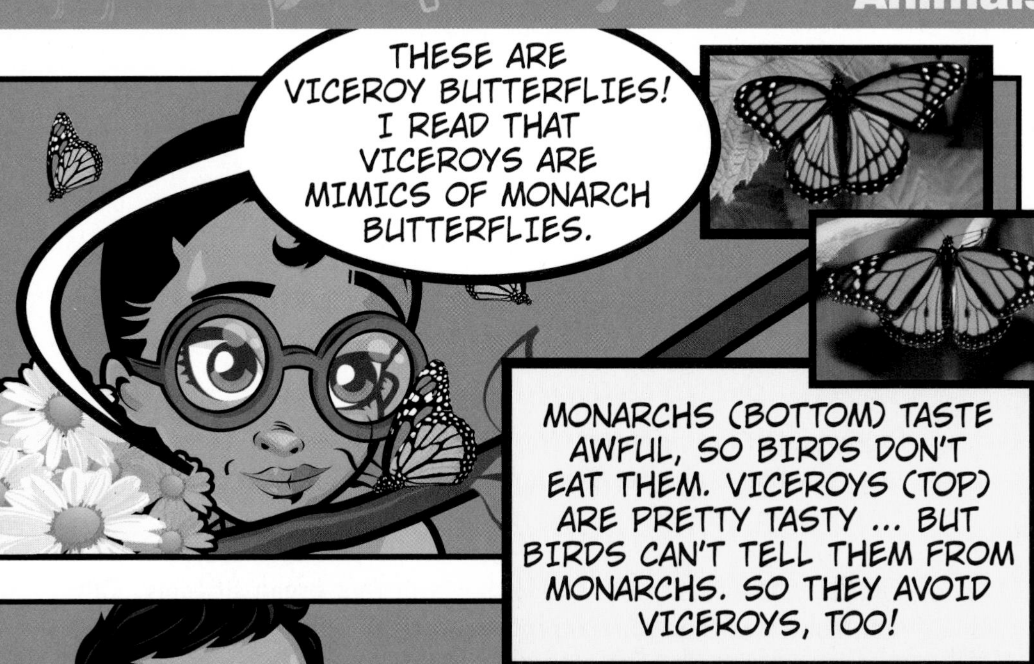
THESE ARE VICEROY BUTTERFLIES! I READ THAT VICEROYS ARE MIMICS OF MONARCH BUTTERFLIES.
MONARCHS (BOTTOM) TASTE AWFUL, SO BIRDS DON'T EAT THEM. VICEROYS (TOP) ARE PRETTY TASTY ... BUT BIRDS CAN'T TELL THEM FROM MONARCHS. SO THEY AVOID VICEROYS, TOO!
I'M GONNA WRITE DOWN THAT BEING A MIMIC AND USING CAMOUFLAGE ARE ADAPTATIONS THAT HELP ANIMALS HIDE!

IT'S YOUR TURN: WHAT'S THE DIFFERENCE BETWEEN BEING A MIMIC AND USING CAMOUFLAGE?

The Human Body

What Are Organ Systems?

Did you look in the mirror this morning? What did you see? You saw your hair and face. If the mirror was big enough, you may have seen other parts of your body.

Did you know that you also saw parts of four organ systems? Can you name them? If you said integumentary, digestive, nervous, and respiratory systems, you were right! All of these systems, and more, make up the human body.

Your skin and hair are parts of your integumentary system. Your mouth and teeth are part of your digestive system. Your eyes, nose, tongue, and skin are sense organs that are parts of your nervous system. Your nose is part of your respiratory system.

Organ Systems in the Human Body...

- **circulatory system**
- **digestive system**
- **nervous system**
- **muscular system**
- **respiratory system**
- **urinary system**
- **skeletal system**
- **integumentary system**

The cell is the smallest unit of life. Tissues are groups of cells that work together. A group of tissues that works together is an organ. Organs that carry out a process in the body are called an organ system.

All of your body's systems work together to keep your body working and healthy.

The Nervous System

The **nervous system** controls everything the body does. The brain, the spinal cord, and the nerves are parts of the nervous system.

Cerebrum The cerebrum (suh REE brum) is the largest part of the brain. Thought and reason take place in the cerebrum. This is the part of the brain that learns and plans. The cerebrum also controls the muscle movements that you want to make. Information from your sense organs is carried to the cerebrum.

Cerebellum The cerebellum (sair uh BELL um) controls balance. It also controls how the body moves and how muscles work together.

Brain Stem The brain stem controls important life functions like breathing, digestion, blood circulation, and heartbeat.

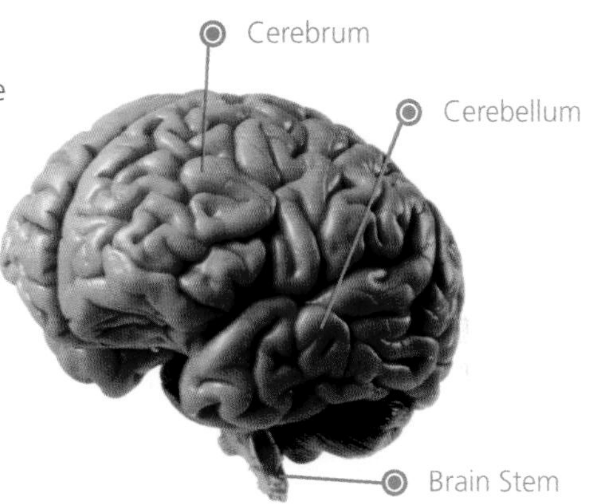

The **spinal cord** is a bundle of nerves that runs down the spinal column. It is the pathway for information between the brain and the rest of the body.

Messages are carried to and from the brain by nerves. There are nerves all through the body.

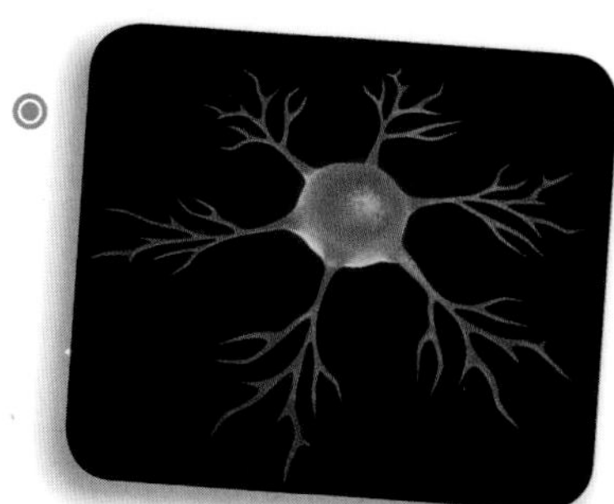

Special nerve cells called neurons (NUR onz) carry messages through the nervous system.

Your ears, eyes, nose, tongue, and skin are also part of your nervous system. These are called sense organs.

The Circulatory System

The **circulatory** (SUR kyoo luh tor ee) **system** supplies the body with nutrients and oxygen. It also carries carbon dioxide and wastes away from the tissues. The heart, blood vessels, and blood are parts of the circulatory system.

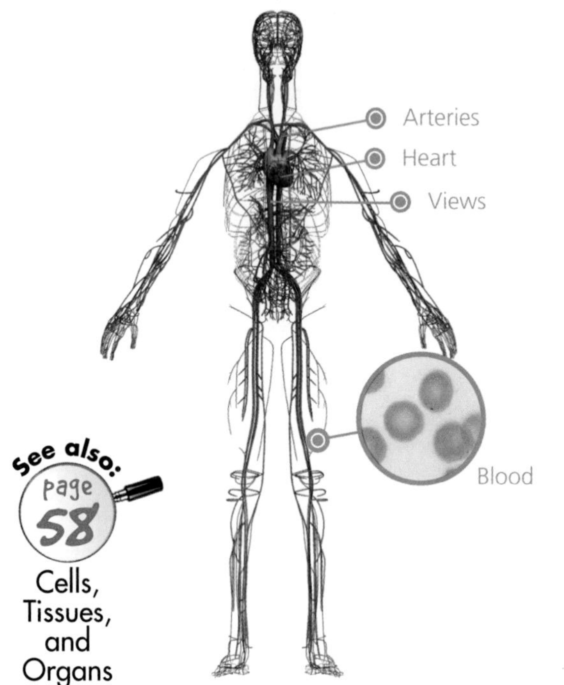

Heart The heart is an organ made of muscle tissue. The job of the heart is to pump blood through the blood vessels.

Arteries Arteries (AR tuh reez) are the largest blood vessels. They carry blood away from the heart.

Veins Veins (VAYNZ) carry blood to the heart. Capillaries (CAH puh lair eez) are the smallest blood vessels. They connect veins to arteries at the tissues.

Blood Blood is made of red blood cells, white blood cells, platelets, and plasma (PLAZ muh). Red blood cells supply oxygen to the body and carry carbon dioxide away from the tissues. White blood cells destroy germs. Platelets repair breaks in blood vessel walls. Plasma is the liquid part of blood.

See also: page 58 Cells, Tissues, and Organs

History Makers

William Harvey (1578–1657)

William Harvey was an English doctor and scientist. He was the first scientist to explain that blood actually circulates through the body. People thought that blood was simply held in the body, like water in a cup. Harvey saw that blood spurts from an artery as the heart beats. He also saw that pressing on a vein makes it swell on the side away from the heart.

The Respiratory System

Breathing is a part of respiration (res puh RAY shun). The organs of the **respiratory** (RES puh ruh tor ee) **system** make it possible to breathe. The respiratory system works closely with the circulatory system. Both systems help to supply the body with oxygen. Oxygen helps your body use the stored energy from the foods you eat. The nose, the larynx (LAIR inks) and pharynx (FAIR inks) in your throat, the trachea (TRAY kee uh), and the lungs are respiratory system organs.

Oxygen enters your lungs when you inhale, or breathe in. You get rid of carbon dioxide when you exhale, or breathe out. You inhale and exhale through your nose.

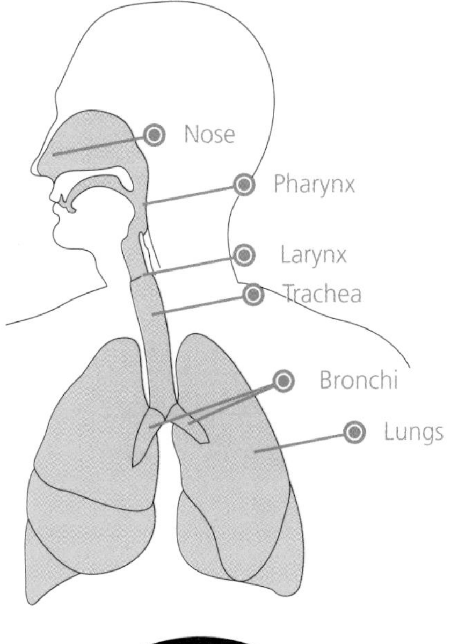

You might think that your lungs are like balloons that fill with air when you inhale and empty out when you exhale. But lungs are more like sponges. There are millions of air sacs called **alveoli** (al VEE oh lie) in your lungs. They look like tiny clusters of grapes. Each air sac sits in a web of capillaries. Blood cells that come through the capillaries next to an alveolus drop off carbon dioxide (CO_2) and pick up fresh oxygen (O_2).

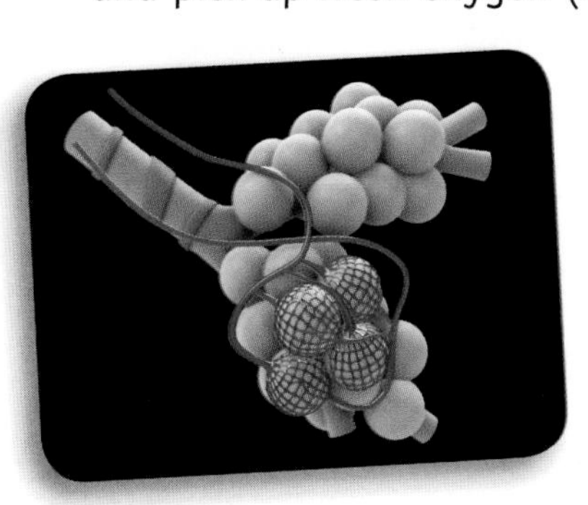

There are capillaries around each air sac. Oxygen moves into blood cells as the blood passes by the air sac. At the same time, carbon dioxide moves out of the blood.

The Urinary System

Wastes are removed from the blood by the **urinary** (YOOR in air ee) **system.** The system includes the kidneys, ureters (YOOR uh turz), urinary bladder, and urethra (yoo REE thruh).

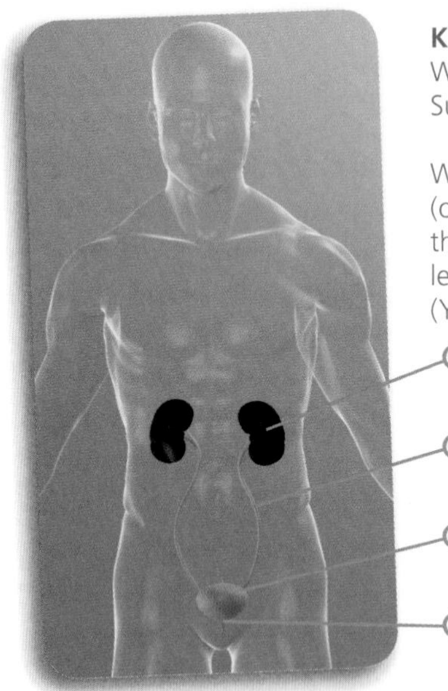

Kidneys The body has two kidneys. Kidneys filter blood. Water, salts, and other substances are forced out of the blood. Substances the body needs are pumped into the blood.

Water follows salts back into the blood through osmosis (oz MOH suhs). In osmosis, water moves from places where there is more water and less salt to places where there is less water and more salt. The remaining liquid is called urine (YOOR in).

Ureters Urine leaves the kidneys through the ureters. Ureters are tubes that connect the kidneys to the bladder.

Bladder Urine is stored in the bladder.

Urethra Urine leaves the body through a tube called the urethra.

Go Online

To learn more about the human body, check out these Web sites!

- **Heart Rate**
  http://www.californiasciencecenter.org/FunLab/FunLab.php
- **Organs and Organ Systems**
  http://kidshealth.org/kid/htbw/
- **Germs and Disease**
  http://www.scrubclub.org/home.aspx

The Digestive System

Food must be broken down into chemicals cells can use. This is the job of the **digestive** (dy JES tiv) **system.** The digestive tract includes the mouth, esophagus (ih SOF eh gess), stomach, small intestines (in TES tinz), large intestines, rectum, and anus. Some of these organs make acids and enzymes (EN zimes) that help digest food. Enzymes are special substances that speed up chemical reactions.

Mouth and **Salivary Glands** Teeth break food into smaller pieces. The tongue mixes the food with saliva (suh LY vuh). An enzyme in saliva begins breaking down starch.

Mouth and Salivary Glands

Esophagus

Stomach

Large Intestine

Small Intestine

Rectum

Anus

Esophagus The esophagus pushes the food into the stomach.

Stomach Food mixes with more enzymes and acid.

Small Intestine Food moves from the stomach into the small intestine. Enzymes break down food into chemicals.

Large Intestine Liquid waste moves into the large intestine where the water is absorbed.

Rectum Solid waste is stored in the rectum.

Anus Waste leaves the body through the anus.

Is That a Fact?

Going to Great Lengths

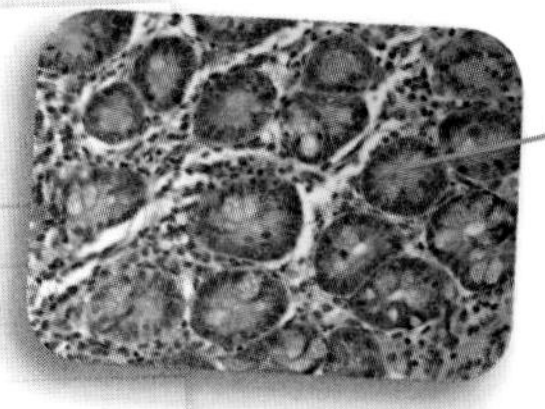

Villi

The small intestine is a long, narrow, muscular tube. An adult's small intestine is over six meters long. That's as long as a chain of 141 large paper clips! Most digestion happens in the small intestine. Inside, the intestine is lined with little bumps called villi (VIL eye). The villi give the intestines a very large surface area. The large surface gives the small intestine more room to absorb nutrients.

The loops on the towel give the towel more surface area. The towel can absorb more water.

The Muscular and Skeletal Systems

Your skeleton protects important organs like your brain and heart. The skeleton gives the body a shape. It protects soft internal organs. The **skeletal** (SKEL uh tul) **system** is made up of bones and cartilage (KAR tuh lij). Bones are hard and rigid. They contain calcium and phosphorus, two important minerals. Most of the calcium in your body is found in your bones.

Most bones have cartilage between the joints. This lets bones move without grinding together. Some cartilage becomes bone as the body grows. Other cartilage, like the cartilage in the nose, stays flexible.

See also: page 58 Cells, Tissues, and Organs

The **muscular** (MUS kyoo ur) **system** causes the parts of your body to move. Some muscle movement is voluntary. You can make the muscle do what you want it to do. Some muscle movement is involuntary. The muscle moves whether or not you want it to.

Skeletal muscle works with bones so the body can move.

Smooth muscle is found in organs and blood vessels.

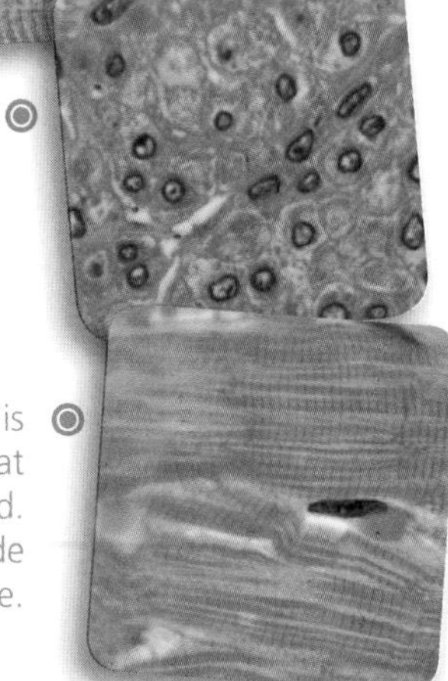

Your heart is a muscle that pumps blood. The heart is made of cardiac muscle.

Did You Know?

Who do you think has more bones, a baby or a teenager? Actually, it's the baby! Human infants have about 300 bones at birth! As we grow, some of these bones knit together. By the time we're all grown up, we have 206 bones in our skeletal system.

Human Growth and Development

Glands are part of the **endocrine** (EN doh krin) **system.** Glands make chemical messengers called **hormones** (HOR mohnz). Hormones travel through the bloodstream. Each hormone affects one type of cell. Hormones control how your body grows and develops.

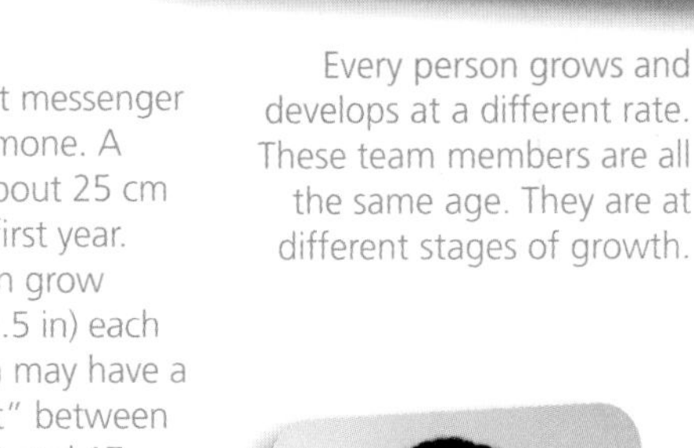

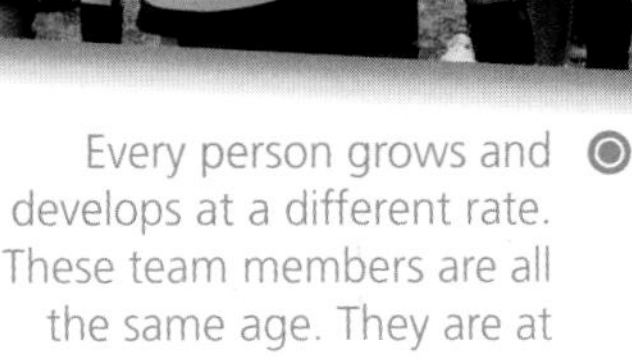

Every person grows and develops at a different rate. These team members are all the same age. They are at different stages of growth.

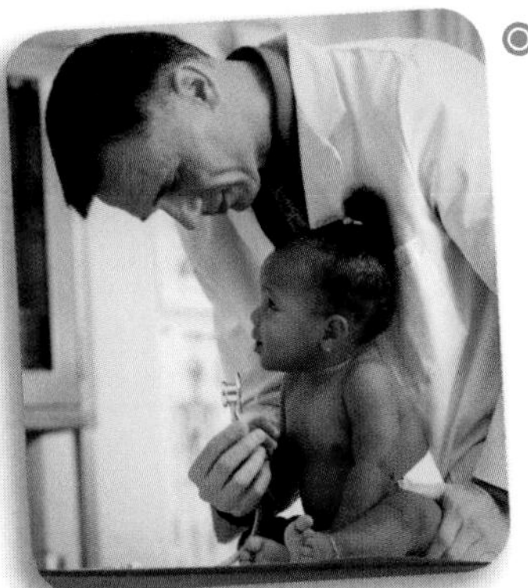

One important messenger is growth hormone. A baby grows about 25 cm (10 in) in the first year. Young children grow about 6 cm (2.5 in) each year. A person may have a "growth spurt" between the ages of 10 and 17. Some teens grow as much as 10 cm (almost 4 in) in a year!

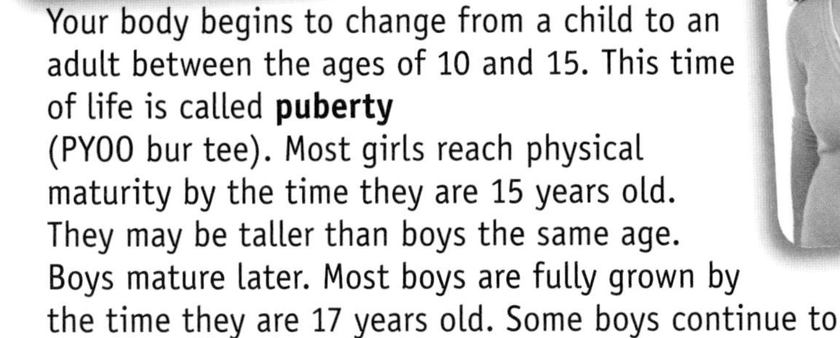

Your body begins to change from a child to an adult between the ages of 10 and 15. This time of life is called **puberty** (PYOO bur tee). Most girls reach physical maturity by the time they are 15 years old. They may be taller than boys the same age. Boys mature later. Most boys are fully grown by the time they are 17 years old. Some boys continue to grow into their early twenties.

Adults may feel some physical changes beginning at about age 30. Adults between the ages of 40 and 65 experience more changes. Skin begins to wrinkle. Hair may turn gray. Some people may lose hair.

Adults over the age of 65 are considered older adults. Many older adults are very active.

The Integumentary System

The **integumentary** (in TEG u MENT uh ree) **system** is made up of skin, hair, and nails. Skin is the body's largest organ. Skin is the first line of defense against disease. It also helps control body temperature. Nerve endings in skin bring the body important information about the outside world.

Sweat glands help the body stay cool and help remove some wastes.

The outer layer of skin cells makes the body waterproof.

Many cells in the integumentary system are actually dead! As new cells are formed, older cells are pushed upward. The outer layer of skin, called the epidermis (eh pih DUR mis), is made up mostly of dead cells. Hair follicles (FOL ih kulz) are alive, but hair itself is not alive. The nails on fingers and toes grow from the nail roots. The nail roots have living cells, but fingernails and toenails are not alive.

Hair is not alive. That's why it does not hurt when your hair is cut.

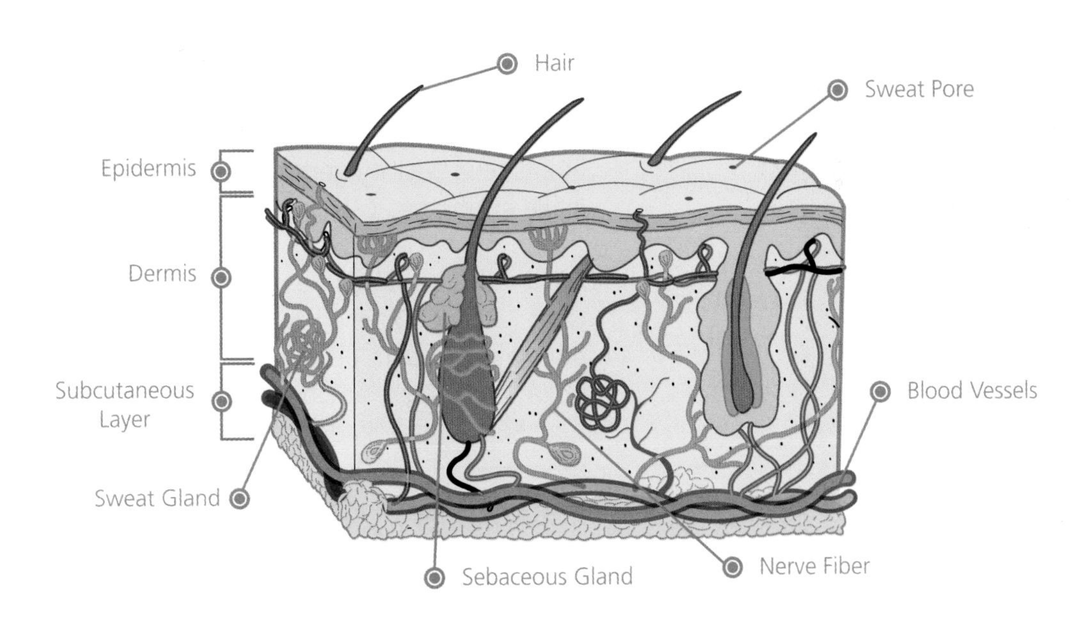

Germs and the Immune System

Germs cause infections and diseases. Bacteria, viruses, protozoa, and fungi are germs. The immune system fights germs. Parts of every organ system are parts of the immune system.

Skin keeps germs from getting inside your body. Some germs get into your body through your nose, eyes, or mouth. Mucus carries them to your stomach where they are digested. Enzymes in saliva help keep your teeth clean. They also fight infections in the mouth. Tears wash away smoke, dust, and allergens.

Salmonella typhimurium are bacteria that cause salmonellosis. Salmonella can be spread in raw or undercooked foods.

When your body makes too much mucus, you have to blow your nose.

Germs inside your body are handled by the **lymphatic** (lim FAT ik) **system**. White blood cells destroy germs. They also help make antibodies (AN tee bah deez).

Antibodies recognize a disease every time you are exposed to it. You will not get the same disease more than once if you have antibodies. Vaccines (vac SEENZ) are made from dead or weakened viruses or bacteria. Your body makes antibodies when you are vaccinated.

Tobacco smoke and air pollution damage lungs and cause heart disease.

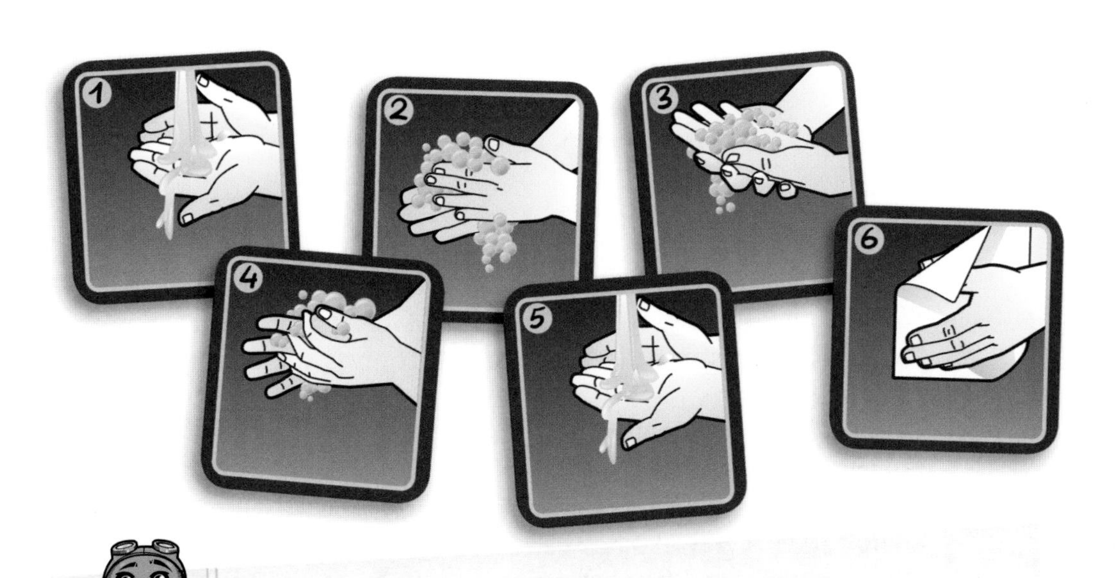

Make the Connection

How to Wash Your Hands

Washing your hands is the most important thing you can do to help prevent diseases and infections. Germs try to get into your body through the skin. Germs on your hands can get inside your body if you eat without washing your hands.

1. Wet your hands with warm running water.
2. Use soap!
3. Rub your hands together to loosen dirt and germs.
4. Wash the front and back of your hands and around the nails. Wash for 20 seconds. Sing the chorus of your favorite song—twice!
5. Rinse your hands well.
6. If possible, dry your hands with a paper towel. Use the towel to turn the water off.

Food Energy

What Is Food Energy?

Have you ever wondered why you feel better after you eat a meal? You can run faster, play harder, and think better. Food does more than make your stomach stop growling. It gives you **energy.**

In science, energy is the ability to do work. Cars need the energy in fuel to move. Living things need energy so that their bodies can do the work of living. Your brain uses energy when you think. Your heart needs energy to beat.

The chemical energy from this girl's after-school snack gives her the fuel that her muscles will need to sink a basket!

Energy can take different forms, such as light energy and heat energy. Energy can also change forms. For example, when you turn on your flashlight, chemical energy in the flashlight's batteries changes into light energy.

Food is a form of **chemical energy.** When you eat, the chemical energy in food is converted to other forms of energy that your body can use to do work.

Plants change light energy from the Sun to chemical energy when they make food during photosynthesis. They use the chemical energy in that food as fuel for their life processes.

Nutrients

See also: page 116
Plants

To learn about food energy, let's look at food. What's in it? And what really happens to it when you eat?

Organisms

Food is made up of chemicals called **nutrients** (NOO tree ents). These are materials the body needs to grow and repair itself. There are six types of nutrients in food:

- water
- carbohydrates
- proteins
- fats
- vitamins
- minerals

Nutrients are the building blocks that make up our bodies. That's why you need to take in a certain amount of each type of nutrient every day to stay healthy.

This turkey sandwich has the six types of nutrients that your body needs to grow!

Water is the most important nutrient of all. About 70 percent of your body is water. It gives your cells their shape. It keeps your tissues from drying out. Most of the processes in your body cannot take place without water.

Drinking water helps you cool down and replaces water you lose through sweating.

Your body loses water when you sweat and through excretion. You need to replace this water every day. Everything you eat and drink contains some water. Most of the water in the turkey sandwich is in the lettuce and tomato. But drinking water is the best way to make sure you get enough of this nutrient each day.

Cells, Tissues, and Organs

Carbohydrates

The whole-grain bread in this turkey sandwich has complex carbohydrates—both starch and fiber.

Cars need fuel to run. Your body needs "fuel" to work, too. You get fuel from carbohydrates. **Carbohydrates** are nutrients that give your cells energy. There are two types of carbohydrates: simple and complex.

Fruit and candy both have simple sugars. But fruit also has vitamins and other important nutrients. Candy doesn't.

Simple carbohydrates are also called sugars. Table sugar, honey, fruit, and milk all contain sugars. The basic building block of all carbohydrates is a sugar called **glucose**. It is the sugar that plants make in **photosynthesis**. And it is the sugar your cells burn for energy during **cellular respiration**.

Your body breaks down simple carbohydrates very quickly. This gives you a quick burst of energy. But it doesn't last very long.

Simple carbohydrate = glucose + glucose

Complex carbohydrates are also called **starches**. They contain many glucose units linked in a long chain. Cereal grains such as wheat, rye, and oats, and vegetables such as potatoes, corn, and beans all contain starch.

It takes your body a long time to break down starch into glucose units. That means that the energy you get from starches lasts a long time.

Complex carbohydrate = glucose + glucose + glucose + glucose + glucose + glucose + glucose + glucose ...

Pastas, breads, and cereal made from whole grains are good sources of complex carbohydrates.

Potatoes, corn, beans, and peas are loaded with complex carbohydrates.

Fiber is a complex carbohydrate that gives plants their structure. Humans can't digest fiber. So it just passes through the intestines and is removed as waste.

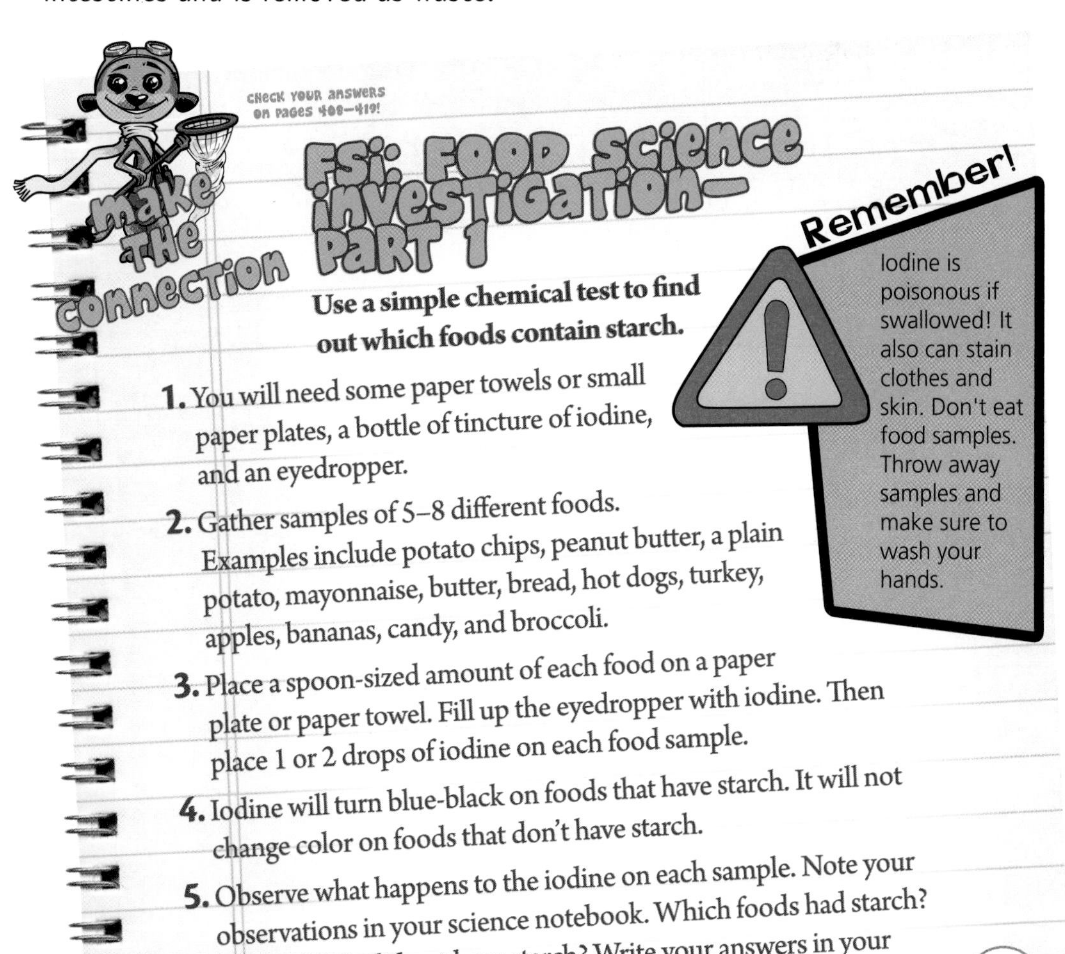

Make the Connection

Check your answers on pages 408–419!

FSI: Food Science Investigation—Part 1

Use a simple chemical test to find out which foods contain starch.

1. You will need some paper towels or small paper plates, a bottle of tincture of iodine, and an eyedropper.
2. Gather samples of 5–8 different foods. Examples include potato chips, peanut butter, a plain potato, mayonnaise, butter, bread, hot dogs, turkey, apples, bananas, candy, and broccoli.
3. Place a spoon-sized amount of each food on a paper plate or paper towel. Fill up the eyedropper with iodine. Then place 1 or 2 drops of iodine on each food sample.
4. Iodine will turn blue-black on foods that have starch. It will not change color on foods that don't have starch.
5. Observe what happens to the iodine on each sample. Note your observations in your science notebook. Which foods had starch? Which foods did not have starch? Write your answers in your notebook.

Remember!

Iodine is poisonous if swallowed! It also can stain clothes and skin. Don't eat food samples. Throw away samples and make sure to wash your hands.

Proteins

Most of the protein in a turkey sandwich is in the turkey.

Proteins are nutrients that have many missions. Your skin and muscles—in fact, most of your tissues—are made up mainly of protein. Hemoglobin is a protein that carries oxygen in your red blood cells. Enzymes are proteins that help cell processes take place.

But what exactly is a protein?

A **protein** is a very large molecule made up of smaller units called **amino acids**. When you eat, your digestive system breaks down proteins into amino acids. Your cells use these amino acids to make new proteins.

Protein = amino acid + amino acid + amino acid + amino acid + amino acid + amino acid + amino acid ...

Go Online

To learn more, check out these Web sites!

- **Body and Mind**
  http://www.bam.gov/
- **SmallSteps**
  http://www.smallstep.gov/kids/html/index.html
- **Fish Kids**
  http://www.epa.gov/fishadvisories/kids/

True Blue
Scientists can use chemical tests to find out which nutrients are in foods. When you place a drop of a chemical called Coomassie blue on a food sample, it will turn blue if the food contains protein. The darker the color, the more protein is in the sample.

A lunch of bean soup and whole-grain bread gives you enough protein, energy, and fiber to last all afternoon.

See also: page 58
Cells, Tissues, and Organs

There are 20 amino acids that organisms need to make proteins. Your cells can make some of these amino acids. But there are a few that your body cannot make. Scientists call these essential amino acids. You must get these amino acids from your food.

For good sources of protein, check out chicken, fish, meat, milk products, eggs, beans, peas, and nuts. And don't forget turkey!

The Great Pyramid
A great place to learn about nutrients and food is MyPyramid. No, it's not my pyramid! It's a Web site! And it's all about nutrients and what to eat. And it's got some pretty cool games, too! The pyramid is like a chart that is divided into six colored parts. Each color represents a major food group. To see the pyramid, go online to www.mypyramid.gov/kids. Check it out!

Did You Know?

Fats

Most of the fat in this turkey sandwich is in the mayonnaise.

Fat in food has a bad reputation. Too much fat can cause serious health problems. But fats are very important to your body. The trick is to eat just enough fat for good health.

Fats are nutrients that store energy. A fat molecule is made up of many smaller units. Let's call these "fat building blocks"—fbbs, for short.

Fat = fbb + fbb + fbb + fbb + fbb + fbb + fbb + fbb + fbb + fbb + fbb + fbb ...

Your digestive system breaks down fats in food into fbbs. Your body uses these building blocks to make new fats. Some of these new fats end up in cell membranes. Others help make hormones or transport nutrients.

Most of the fat in your body is stored in fat cells. Your body converts this stored fat into glucose when you need extra energy. Some vitamins are stored in body fats. Fat helps insulate you from the cold.

GO ONLINE

Milk Matters

http://www.nichd.nih.gov/milk/kids/kidsteens.cfm

Low-fat dairy products are a great way to get the nutrients you need from milk without extra fat.

Fats come from both animal and plant sources. Animal fats aren't as healthy as fats from plants. So aim for plant-based sources of fat when you can. And remember: a little fat goes a long way toward staying healthy!

PLANT AND ANIMAL SOURCES OF FAT

Plant Sources:
- nuts
- seeds
- olive oil or corn oil
- olives
- avocados

Animal Sources:
- meat, poultry, and fish
- milk
- butter
- cheese
- ice cream

CHECK YOUR ANSWERS ON PAGES 408–419!

FSI: FOOD SCIENCE INVESTIGATION—PART 2

Predict which foods contain fat. Then test your predictions with a simple test.

1. Cut a brown paper bag into 5–8 squares.
2. Gather samples of 5–8 different foods. Some examples include potato chips, peanut butter, a plain potato, mayonnaise, butter, bread, hot dogs, turkey, apples, bananas, chocolate, and broccoli.
3. Place a spoon-sized amount of each food on a paper square. Leave samples overnight.
4. Predict which foods will have the most fat and which will have the least fat. Write your predictions in your science notebook.
5. Remove the food samples and examine the paper squares. Foods that have fats will leave grease spots on the paper. Which foods had the most fat? Which foods have the least fat? Write your answers in your science notebook.

Vitamins and Minerals

Your body needs only tiny amounts of most vitamins and minerals. But those tiny amounts pack quite a wallop inside your body! Most of the work in your cells would come to a complete stop without the help of these important nutrients.

The lettuce, tomatoes, and whole-grain bread have vitamins and minerals.

Vitamins (VI tuh minz) are chemical compounds. They control many processes in your body.

All About Vitamins

VITAMIN	WHAT IT DOES	WHERE TO GET IT
A	keeps eyes healthy; builds healthy bones and skin	orange fruits and veggies, dark green veggies like spinach and kale
D	builds strong bones and teeth	fortified milk, liver, fish, cereals
B group	helps body use energy; helps heart, nerves, and blood work	whole grains, fish, eggs, leafy green veggies, milk products, beans, peas
C	keeps gums and muscles healthy; fights infections	oranges, tomatoes, broccoli, strawberries, green peppers
E	protects eyes, liver, skin, and lungs	whole grains, egg yolks, nuts, leafy green veggies
K	helps stop bleeding	dairy products, broccoli, leafy green veggies

Minerals are elements. Your body uses them to build tissues and for other tasks. Calcium is a very important mineral. It builds strong bones and teeth. It also helps your muscles work. Magnesium and potassium help your muscles work. And potassium helps your heart keep a steady beat. Other minerals your body needs include iron, zinc, and fluorine (FLOR een).

How can you know you're getting enough vitamins and minerals each day? Here's a tip: the colors in fruits and vegetables are connected to vitamins and minerals. So make sure there's always a mix of colors on your plate at every meal!

Milk products are loaded with calcium. Fortified cereal, broccoli, nuts, and avocados are good sources too!

A colorful mix of fruits and veggies will supply you with the vitamins and minerals you need to stay healthy.

Fortified Foods

Not getting enough vitamins can make you really sick. Too little vitamin C can cause a disease called scurvy that makes your gums bleed. Anemia is caused by too little iron. People with anemia become very weak. Many people got these diseases long ago when it was hard to get fruits and vegetables and whole grains. Many foods today are fortified with vitamins and minerals. This helps make sure everyone gets the nutrients they need each day.

Healthy Eating

How can you get the right nutrients every day? Go for the mix! Eat foods from each major food group. The foods in each group have more than one nutrient—so eating something from each group covers all your bases!

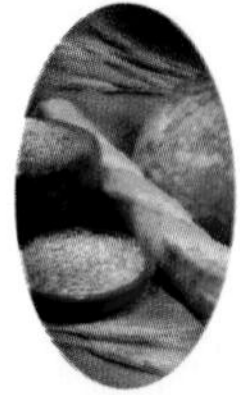

WHOLE GRAINS | VEGETABLES | FRUITS | MILK PRODUCTS | MEAT AND BEANS | FATS AND OILS

TIP: Remember, a little goes a long way when it comes to fats and oils!

Eat a good breakfast. Doughnuts taste good, but they're not a good breakfast choice. They're loaded with sugar and fat, but not much else. A bowl of whole-grain cereal with fruit and low-fat milk has protein, vitamins, fiber, and complex carbohydrates. That's a good mix! And it will keep you full all morning!

Choose colorful foods. Lots of color means lots of vitamins and minerals. Red, orange, yellow, and green veggies are full of vitamin A. Dark green veggies have lots of iron and calcium. Red grapes are loaded with vitamins C and E. Aim for two or more colors on your plate and in your lunch each day!

Eat healthy snacks. We all need a snack or two to keep us going. Just go for the mix! Try a stick of string cheese and whole grain crackers, or some low-fat dip and raw veggies. They're better choices than chips and a cola!

Pizza is a healthy snack when you top it with veggies and low-fat cheese.

Activity burns up calories and keeps your body strong.

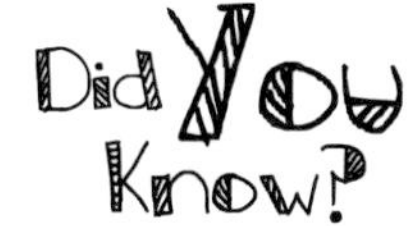

Measuring Food Energy

You've probably heard the word "calorie" before, but do you know what it means? A **calorie** is simply a unit of measurement. Calories measure the amount of energy in a certain amount of food. This tells you how much energy your body will get if you eat or drink that food. For example, a cup of low-fat milk has about 80 calories. That means it will give your body about 80 calories of energy.

Your body needs a certain number of calories each day. These calories give your cells energy to do work. Most students your age need about 1600–1800 calories a day. If you eat more calories than your body uses, then your body will store the extra calories as fat. And that can lead to health problems.

Reading Labels

How can you find out about the nutrients and energy in different foods? Nutrition labels are a good place to start. A **nutrition label** is a lot like the table of contents in a book. It tells you which nutrients are in the food. Nutrition labels are usually found on the back or side of a package.

1. Serving size and number of servings in the package
2. The number of calories in each serving, and how many calories come from fat
3. The amounts of important nutrients in each serving
4. The % Daily Values. Experts know how much of each nutrient a person should have each day. The percent daily value tells you how this food can help you reach that goal.
5. These values remind you how much of each nutrient a person should eat each day. These are based on a 2000-calorie-a-day diet.

Nutrition Facts

1 Serving Size 1 cup (228g)
Servings Per Container 2

2 **Amount Per Serving**
Calories 250 Calories from Fat 110

	% Daily Value*
Total Fat 12g	18%
Saturated Fat 3g	15%
Trans Fat 3g	
Cholesterol 30mg	10%
Sodium 470mg	20%
Potassium 700mg	20%
Total Carbohydrate 31g	10%
Dietary Fiber 0g	0%
Sugars 5g	
Protein 5g	
Vitamin A	4%
Vitamin C	2%
Calcium	20%
Iron	4%

(3, 4)

5 * Percent Daily Values are based on a 2,000 calorie diet. Your Daily Values may be higher or lower depending on your calorie needs.

	Calories:	2,000	2,500
Total fat	Less than	65g	80g
Sat fat	Less than	20g	25g
Cholesterol	Less than	300mg	300mg
Sodium	Less than	2,400mg	2,400mg
Total Carbohydrate		300g	375g
Dietary Fiber		25g	30g

You can use nutrition labels to decide if a food is healthy for you. Try comparing labels of different brands of the same food to see if there are any major differences in nutrients. What you find may surprise you!

Go Online

To learn more, check out these Web sites!

- **Healthy Eating**
  http://www.mypyramid.gov/kids/kids_game.html
- **Fruits and Vegetables**
  http://www.fruitsandveggiesmorematters.org/
- **Staying Healthy**
  http://kidshealth.org/kid/stay_healthy/index.html

Check your answers on pages 408–419!

FSI: Food Science Investigation—Part 3

Compare nutrition facts labels from a sugary cereal and a whole grain cereal. Use this information to determine which cereal is a better choice for a healthy breakfast.

1. Make sure the serving sizes of both cereals are equal.
2. Compare the number of calories per serving without milk. Which cereal has more calories?
3. Which cereal has a greater amount of fat?
4. Which cereal has more sodium?
5. Which one has the greater amount of fiber?
6. Which one has more sugar?
7. Based on your observations, which cereal is better for you? Why?

Plants

What Is a Plant?

Would you say that a plant has leaves and roots? Would you say that plants have flowers and stems? Would you say that all plants grow in soil?

Some plants do have these traits. But not all plants have leaves and roots. Not all plants have flowers and stems. Not all plants grow in soil. All **plants** ARE members of the kingdom **Plantae** (PLAN tay).

See also: page 58 Cells, Tissues, and Organs

Members of the kingdom Plantae may be unicellular or multicellular. Like animal cells, all plant cells have a nucleus. Unlike animal cells, plant cells have cell walls. Other organisms also have cell walls. Only plants have cell walls that contain **cellulose**. Members of Plantae use sunlight, water, and carbon dioxide to make food. The process is called **photosynthesis** (foh toh SIN thuh sis). Photosynthesis happens in an organelle called a **chloroplast** (KLOR uh plast). Animals do not have chloroplasts in their cells.

These plants look very different. But they all are members of the kingdom Plantae.

What Plants Need

Like animals, plants need water. Water gives plant cells their shape. It is an important part of photosynthesis. Water also helps move nutrients into and through plants. Most plants absorb water through their roots.

Plants need carbon dioxide and oxygen. They use carbon dioxide and water to make carbohydrates during photosynthesis. Plants use oxygen to release energy from carbohydrates.

Plants make all the other organic molecules they need. To make these molecules, they need minerals. Minerals are plant nutrients. Most plants get minerals from soil. A few plants do not grow in soil. Green algae take in minerals that are dissolved in water. Some plants are **parasites** (PAIR uh sites). They get their nutrients from other plants. A few plants even get minerals from the air.

Most plants get their minerals from soil. Spanish moss gets its minerals from the air!

Plants need energy. Plants get their energy from sunlight. They capture the energy with their chloroplasts.

Unlike animals, plants cannot move around to find supplies. A plant's environment must have everything it needs for it to grow and reproduce.

Plant Tissues, Organs, and Organ Systems

Most plants have **dermal**, **vascular** (VAS kyoo lar), and **ground tissues**. Dermal tissue is like skin. It covers the outside of the plant. Vascular tissue carries water and nutrients through the plant. Photosynthesis takes place in ground tissue.

The organs and organ systems of a plant are its stems, leaves, and roots.

Potatoes and tulip bulbs are underground stems. The stems store food for the plant.

Stem The stem connects the leaves to the roots. It supports the leaves. Tubes in the stem carry water and nutrients from the roots to the leaves and flowers.

Special vascular tissues in the stem carry water and nutrients to the rest of the plant.

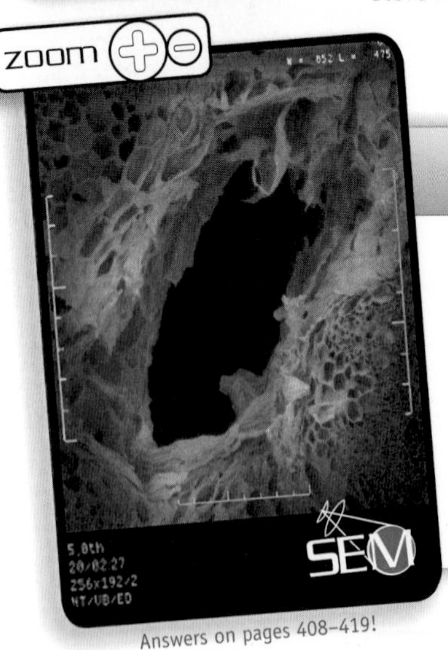

Answers on pages 408–419!

Thinking BIG™

www.carolinacurriculum.com/ThinkingBig

Record your ideas in your science notebook.

- What shapes do you see?
- What parts do you see that are the same? How so?
- How might those parts be used?
- Why do you think so?

Solve IT

1) I'm a connector.
2) I can be long or short.
3) I have tubes like drinking straws but no soda pop.
4) I carry water and nutrients to leaves and flowers.

Leaf The outer layer of a leaf keeps the leaf from drying out. Carbon dioxide, oxygen, and water move into and out of the leaf. Photosynthesis occurs in the middle of the leaf.

As days grow shorter in the fall, some plants stop producing chlorophyll. Leaves are shed to protect the plant and save energy.

The spines on the saguaro cactus are a kind of leaf.

Pine needles are leaves.

Chlorophyll in leaves makes them green.

Roots Roots absorb water and nutrients from the soil. Roots hold the plant up. They keep plants from blowing over when the wind pushes on their leaves and branches. Roots anchor the plant's weight deep in the ground and sometimes over a large area.

Food is stored in the roots. Some plants store sugars there as well.

This large beech tree needs long, thick roots for support.

These beet roots were grown to be made into sugar. Beets are used to make table sugar and to sweeten other products.

Photosynthesis and Respiration

Plants make their own food. This process is called **photosynthesis.** In photosynthesis, plants use sunlight, carbon dioxide, and water to make **carbohydrates**. Carbohydrates store food energy. Sugars and starches are carbohydrates.

Water is carried into the plant through the roots. The water travels to the leaves through the stem.

Plant cells have a special organelle to capture energy from the Sun. It is called a **chloroplast**. A chemical called **chlorophyll** is in the chloroplast.

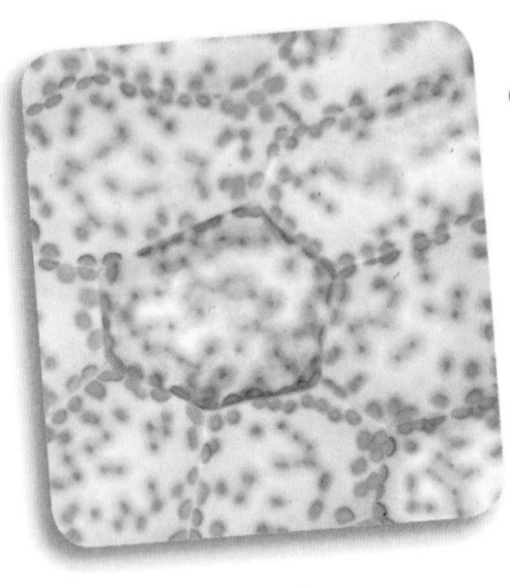

The tiny green organelles in these plant cells are chloroplasts.

Chlorophyll gives the plant its green color. Chlorophyll also captures energy from sunlight. Leaf cells use this energy to change carbon dioxide and water into sugar.

Is That a Fact?

It All Adds Up

Here is the chemical equation for photosynthesis:

Carbon Dioxide + Water + Sunlight (Energy) → Sugars + Oxygen

Here is the chemical equation for cellular respiration:

Sugars + Oxygen → Carbon Dioxide + Water + Energy

Look at the two equations. Did you notice that respiration is the reverse process of photosynthesis?

Stem tissue carries sugars to the roots for storage. Why does the plant store sugars? Remember that the Sun only shines during part of each day. The plant still needs energy at night. Many plants lose their leaves in the autumn. They need energy to make new leaves in the spring.

Plant cells release energy from sugar during **cellular respiration**. Plant cells burn sugar with oxygen to release energy. This produces carbon dioxide and water.

Go Online

To learn more about plants, check out these Web sites!

- **Plant Parts, Seeds, and Soil**
  http://urbanext.illinois.edu/gpe/
- **Seed Germination and Water Needs**
  http://www2.bgfl.org/bgfl2/custom/resources_ftp/client_ftp/ks2/science/plants_pt2/growth.htm
- **Plant Life Cycle**
  http://www.ngfl-cymru.org.uk/vtc/plant_life_cycles/eng/Introduct/mains.htm

Land Plant Life Cycles

Most land plants begin their lives as seeds. Inside the seed coat is a young plant and a food supply. For a while, the young plant lives off the stored food. As the seed absorbs water, it begins to swell. The seed coat cracks open. When the seed splits, the plant begins to grow. This is called **germination** (jer muh NAY shun).

The root pushes downward. It anchors the plant to the ground. A shoot pushes upward through the soil. The shoot, or sprout, will become stems and leaves. As the first leaves develop on the sprout, chlorophyll makes the plant green.

The young plant is now a **seedling.** Plants need light and water. The seedling uses air, water, and sunlight to produce its own energy.

When the plant reaches a certain height, it stops growing bigger. Then it spends energy making seeds.

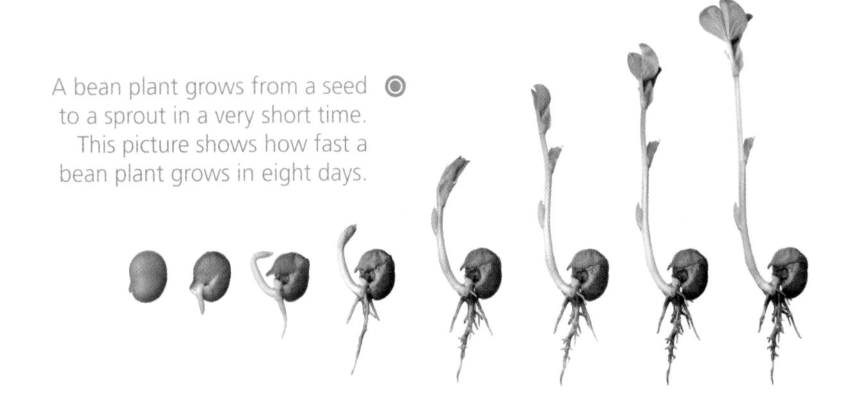

A bean plant grows from a seed to a sprout in a very short time. This picture shows how fast a bean plant grows in eight days.

Once the seeds are formed, the plant begins to die. Plants that go through their entire life cycle in one growing season are called **annual** plants.

Some plants live for more than one season. They are called **perennial** plants. Perennial plants lose their flowers and leaves in the autumn and winter. They grow again in the spring from their roots. Perennial plants can live for many years.

The silk on this corn plant is the flower. Corn is an annual plant. It starts from seed in the early spring. Seeds are harvested in late summer. Then we eat the seeds!

This plant is a perennial. It produces flowers and seeds for many years. The plant grows each year from the same root.

Plants put most of their energy into making seeds. Seeds are made in the flowers. When the seeds are formed, the rest of the plant dies.

How Land Plants Reproduce

Like animals, plants can reproduce sexually. This means that they make eggs and sperm. The sperm fertilizes the egg to make an embryo. The embryo is usually in a seed.

See also: page 128 Cycles

Some plants do not make seeds. Mosses, ferns and some other plants produce **spores** (SPORZ). Spores have all of the information a plant needs to make a new plant.

Spores have formed on this fern leaf.

Mosses do not grow from seeds.

Some plants make seeds, but they do not have flowers or fruit. In many of these plants, seeds form in cones. Male cones produce **pollen** (PAH luhn). The pollen contains the sperm. The female cones produce **eggs**. Wind blows pollen to the female cone and fertilizes the egg inside the cone.

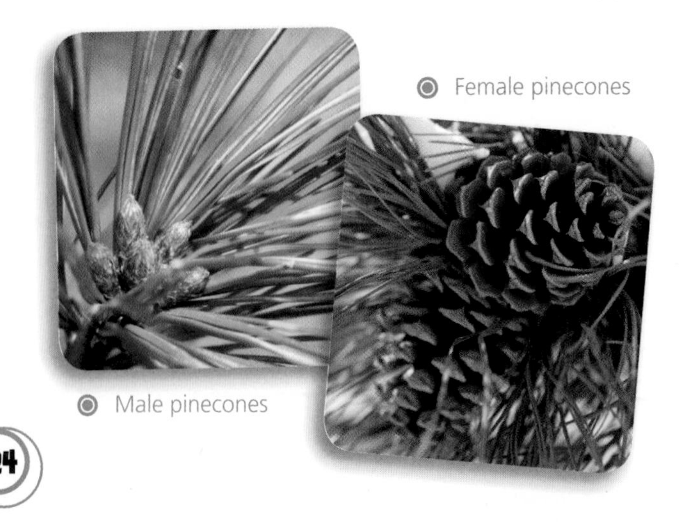

Female pinecones

Male pinecones

Many land plants use flowers to produce seeds. The male parts of the flower produce pollen. The female parts of the flower produce eggs.

Many plants rely on wind to spread pollen. Plants produce a lot more pollen than they need to reproduce. All the extra pollen causes runny noses and sneezing in people with allergies.

Usually, plants depend on animals to carry pollen. Bees, birds, bats, and butterflies are just some of the animals that spread pollen. Pollen is sticky. Bees and other animals come to the flower to get food. Pollen rubs on their legs or other body parts. When the animal visits another plant, it leaves some of the pollen behind.

The flower **petals** attract pollinators. **Sepals** protect the flower bud.

1. A flowering plant has male and female parts. The male part is called the **stamen**. Stamens make sperm in grains of pollen.

2. The female part of the flower is the **pistil**. Pollen sticks to the stigma at the top of the pistil. When pollen moves from the **stamen** to the pistil it is called **pollination** (PAH luh NAY shuhn).

3. The pollen fertilizes eggs in the **ovary**.

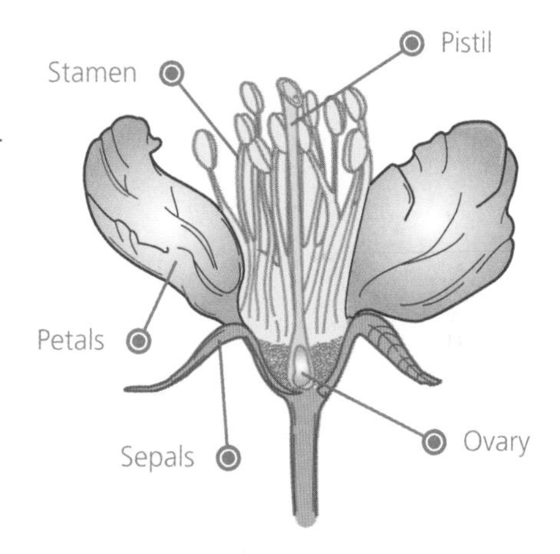

Plants do not always make new plants from seeds. Some plants send out shoots from their roots that grow into new plants. Others form little plants called **plantlets**. Still others grow new plants from their leaves or stems.

This plant reproduces by making plantlets.

Plant Inheritance and Adaptations

See also: page 136
Ecosystems

There are plants in almost every part of Earth. Plants live in the Arctic and in the desert. They live in places where it rains almost all of the time. They live in places that have very little sunlight. Plants grow in between rocks. They grow under the ocean water. Some plants even grow on other plants!

Plants have inherited many behaviors and adaptations that help them survive and reproduce in different environments.

Cattleya orchids come from rain forests in South America. This orchid is growing on the branch of a tree. Orchids get water and nutrients from the air.

Purple saxifrage is found in arctic areas. Many arctic plants are very small. They grow close to the ground. This helps them survive harsh winds, cold temperatures, and a lack of nutrients. Most arctic plants have shallow roots. They push their roots between rocks or into frozen ground.

Hummingbirds are attracted to the red color of this flower.

See also: page 68
Organisms

Many flowers have bright colors and special shapes. These colors and shapes attract certain insects or birds that will pollinate the flower. Other plants use fragrance to attract pollinators.

Desert cacti store water in their tissues. Their leaves have become sharp needles that conserve water and protect the plant. Cranberry bushes like having wet roots, so they grow in bogs. Their fruits float so their seeds can be scattered more easily. Many trees in cold climates shed their leaves in the autumn. They grow new leaves each spring.

Cones of lodgepole pines only open when heated. The young trees grew from seeds released by the 1988 Yellowstone National Park wildfire.

Plants Respond to Their Environment

Phototropism Many plants are able to respond to their environment. Plants need sunlight to make food. If sunlight comes from only one direction, the leaves of a plant will turn toward the light.

Hydrotropism Plants need water. If plants do not have enough water where they are planted, their roots will move toward moisture.

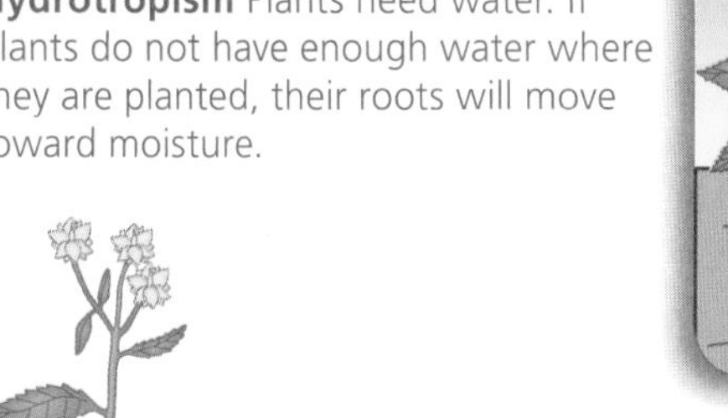

Gravitropism Plant roots get nutrients from the soil. Plant roots grow downward. The force of gravity pulls them down into the soil. Plant stems grow toward the sunlight, away from the pull of gravity.

Cycles

What Is A Cycle?

This morning you woke up. Tonight you'll go to bed. You'll do these things again tomorrow, and again the day after that. Getting up each morning and going to bed each night are events in your daily cycle. A **cycle** is a series of events that repeat over and over on a regular basis.

Many changes in nature occur in cycles. The Sun rises and sets each day as part of the Earth's daily cycle. The Earth also has yearly cycles. The Moon has daily, monthly, and yearly cycles, too.

Sunrise and sunset occur every day because of the daily cycle of Earth's rotation on its axis.

Many natural materials have cycles. Plants take carbon dioxide from the air and release oxygen to the air during photosynthesis. Organisms take oxygen from the air and release carbon dioxide to the air during cell respiration. Water and nitrogen also cycle between Earth, the atmosphere, and living things.

Organisms also have cycles. Some cycles, such as migration, are based on behaviors. A **life cycle** is the way an organism grows and reproduces. Some organisms have specialized life cycles. But one cycle takes place in all living things: the cell cycle.

The Cell Cycle

Cells pass through a number of stages during their life cycle.

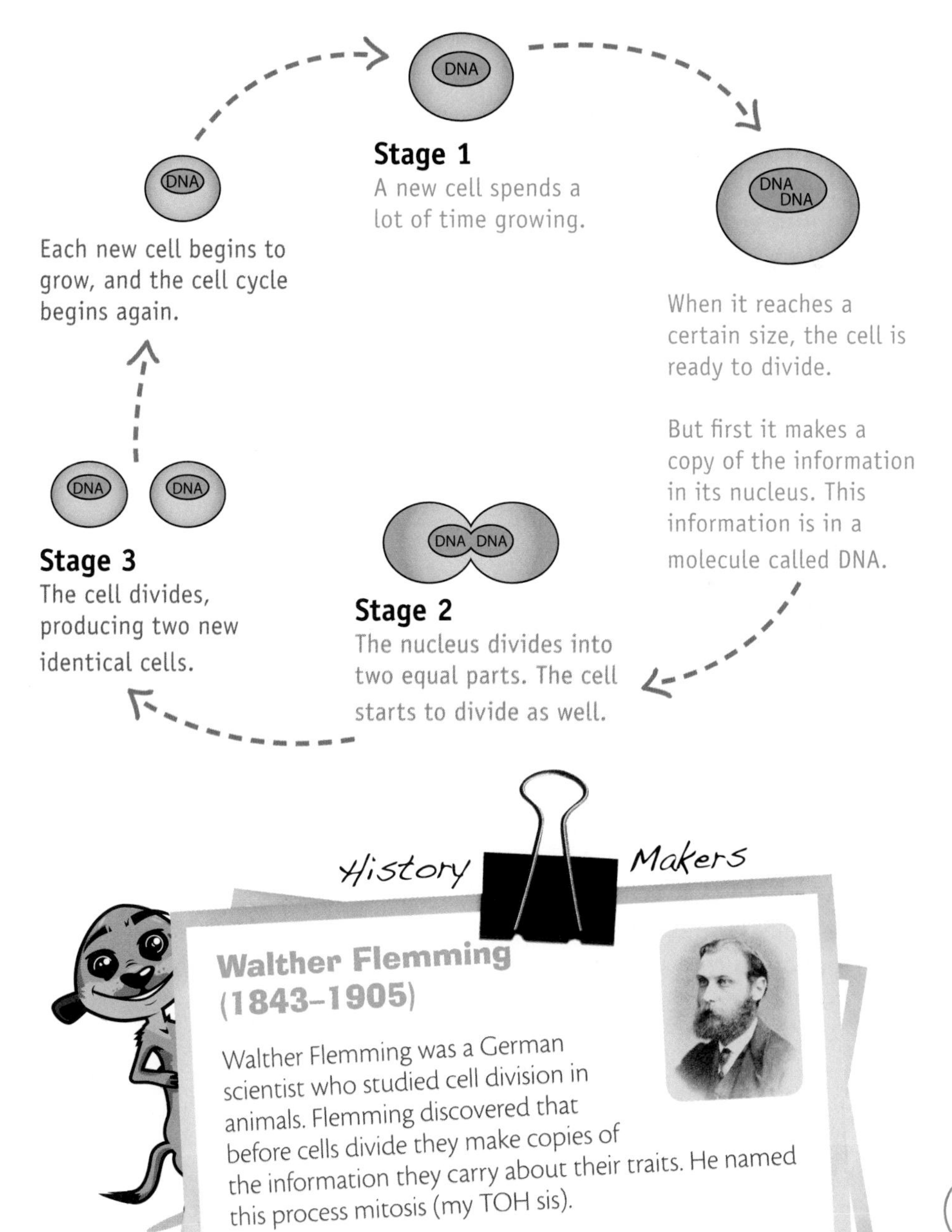

History Makers

Walther Flemming (1843–1905)

Walther Flemming was a German scientist who studied cell division in animals. Flemming discovered that before cells divide they make copies of the information they carry about their traits. He named this process mitosis (my TOH sis).

The Life Cycles of Seed Plants

The life cycle of most plants starts with a **seed**. The seed contains a tiny plant. It also contains food that will nourish the plant until it is ready to make its own food. There are two types of seed-bearing plants: nonflowering plants and flowering plants.

Nonflowering plants and flowering plants go through similar stages early in their life cycles. Both produce seeds that take root in soil. They produce a stem that grows upward above the ground and produces leaves. The stem grows into a young plant called a seedling. The seedling grows into an adult plant.

Nonflowering plants don't produce flowers during their life cycle. They produce cones. The adult nonflowering plant produces male cones and female cones. Pollen from male cones fertilizes the eggs in female cones. The fertilized eggs develop into seeds inside the female cone. The cone opens and releases the seeds, which are carried on the wind to other areas.

The Life Cycle of a Pine Tree

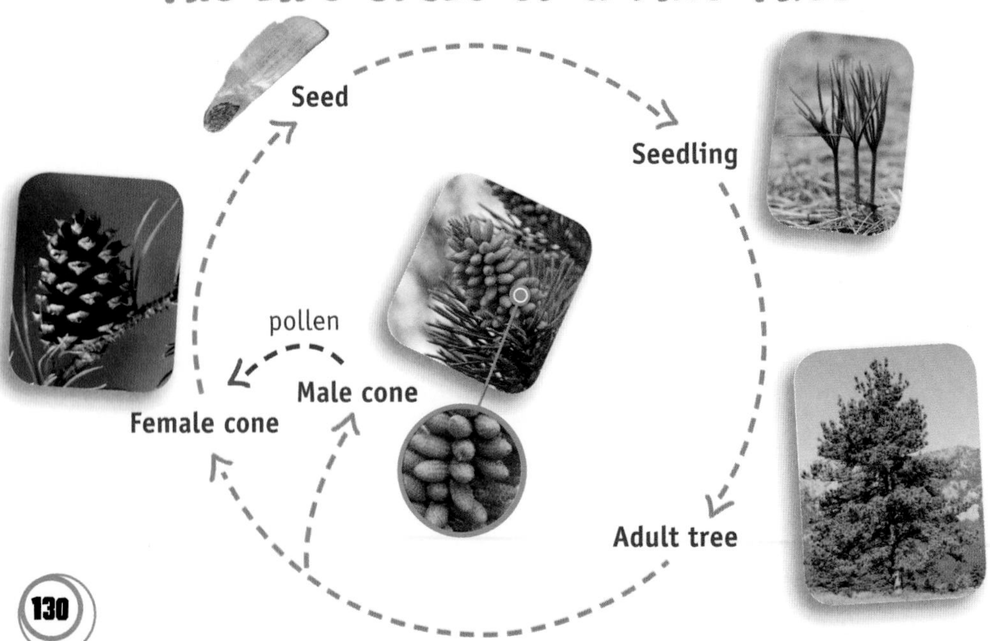

Flowering plants produce flowers and fruit. The adult flowering plant produces a flower that has a male part and a female part. The male part is the **stamen**. The female part is the **ovary**. Pollen from the stamen fertilizes the eggs in the ovary. The fertilized eggs develop into seeds inside the ovary. The **fruit** develops around the seeds. Fruit is an adaptation that attracts animals. When animals eat the fruit, they pass the seeds with their wastes. This way, seeds can sprout and grow up away from the parent plants.

See also: page 116 Plants

The Life Cycle of an Apple Tree

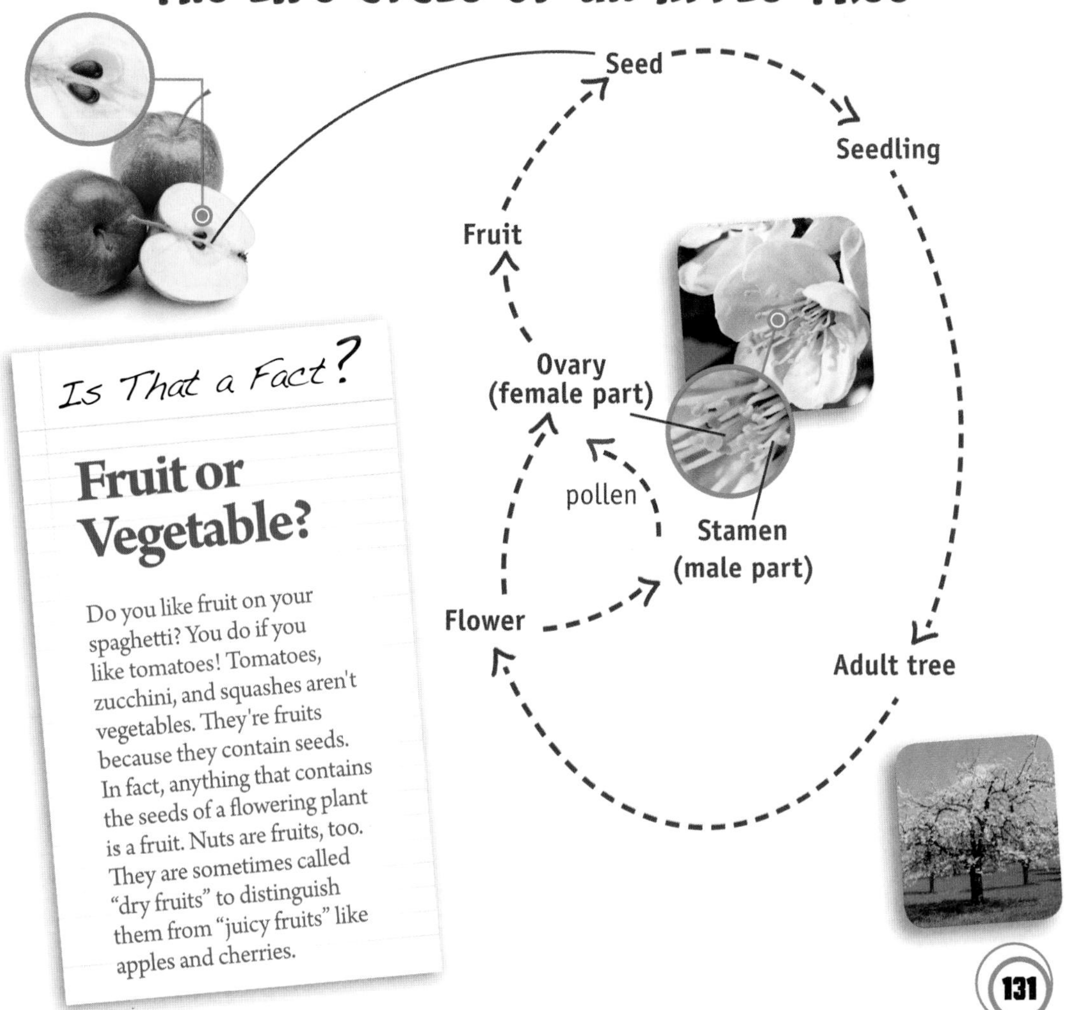

Is That a Fact?

Fruit or Vegetable?

Do you like fruit on your spaghetti? You do if you like tomatoes! Tomatoes, zucchini, and squashes aren't vegetables. They're fruits because they contain seeds. In fact, anything that contains the seeds of a flowering plant is a fruit. Nuts are fruits, too. They are sometimes called "dry fruits" to distinguish them from "juicy fruits" like apples and cherries.

Incomplete Metamorphosis

If you've ever seen a caterpillar and a butterfly, you've probably wondered how they could be the same species. Many animals change forms as they grow. The series of changes is called **metamorphosis** (MET uh MOR fuh sis). There are two types of metamorphosis: incomplete metamorphosis and complete metamorphosis.

See also: page 116
Plants

Incomplete metamorphosis has three stages: egg, nymph (NIMF), and adult. The nymph looks a lot like the adult. Damselflies, dragonflies, and many other invertebrates have this life cycle. **Invertebrates** are animals that do not have backbones.

The Life Cycle of a Damselfly

Egg
The female lays her eggs in water.

Nymph
Nymphs look like tiny adults. But their wings are not fully grown.

Adult
Once the flies become adults, they are ready to produce more offspring!

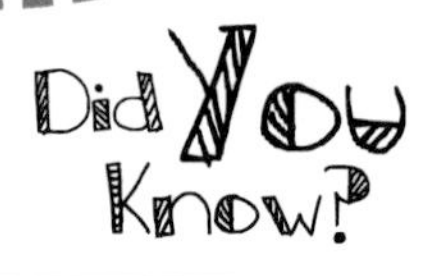

Nymphs and larvae shed their **exoskeletons** as they grow. This is called **molting**. Some nymphs and larvae molt several times as they grow. When they molt, they actually crawl out of their exoskeleton. This mealworm larva is about to leave its old exoskeleton behind!

An **exoskeleton** is a hard covering on the outside of many invertebrates.

Complete Metamorphosis

Complete metamorphosis has four stages: egg, **larva**, **pupa** (PYOO puh), and adult. In this life cycle, the larva and pupa stages look very different from the adult stage. Butterflies, beetles, and mealworms are examples of invertebrates with this life cycle.

The Life Cycle of a Monarch Butterfly

Egg
The female butterfly lays hundreds of eggs on milkweed leaves

Larva (caterpillar)
Each egg hatches to produce a larva. The larva grows and molts a few times before it enters the next stage.

Pupa
The pupa forms a hard case outside of its body called a chrysalis. Inside, the pupa's body is slowly changing.

Adult
When the adult butterfly emerges from the chrysalis, it looks very different from its larva stage!

Which Is Which?

NYMPH LARVA PUPA ADULT

Metamorphosis Which life stage is found only in incomplete metamorphosis? Which stages occur only during complete metamorphosis? Do any stages occur in both types of metamorphosis?

Vertebrate Metamorphosis

Amphibians (am FIB ee uns) are the only vertebrates that go through metamorphosis. **Vertebrates** are animals that have backbones. Frogs, toads, and salamanders are amphibians.

Amphibian larvae are called tadpoles. They live in water and breathe through gills. As the tadpole changes into a young frog, or froglet, its lungs develop. When the froglet can breathe through its lungs, it is ready for life on land.

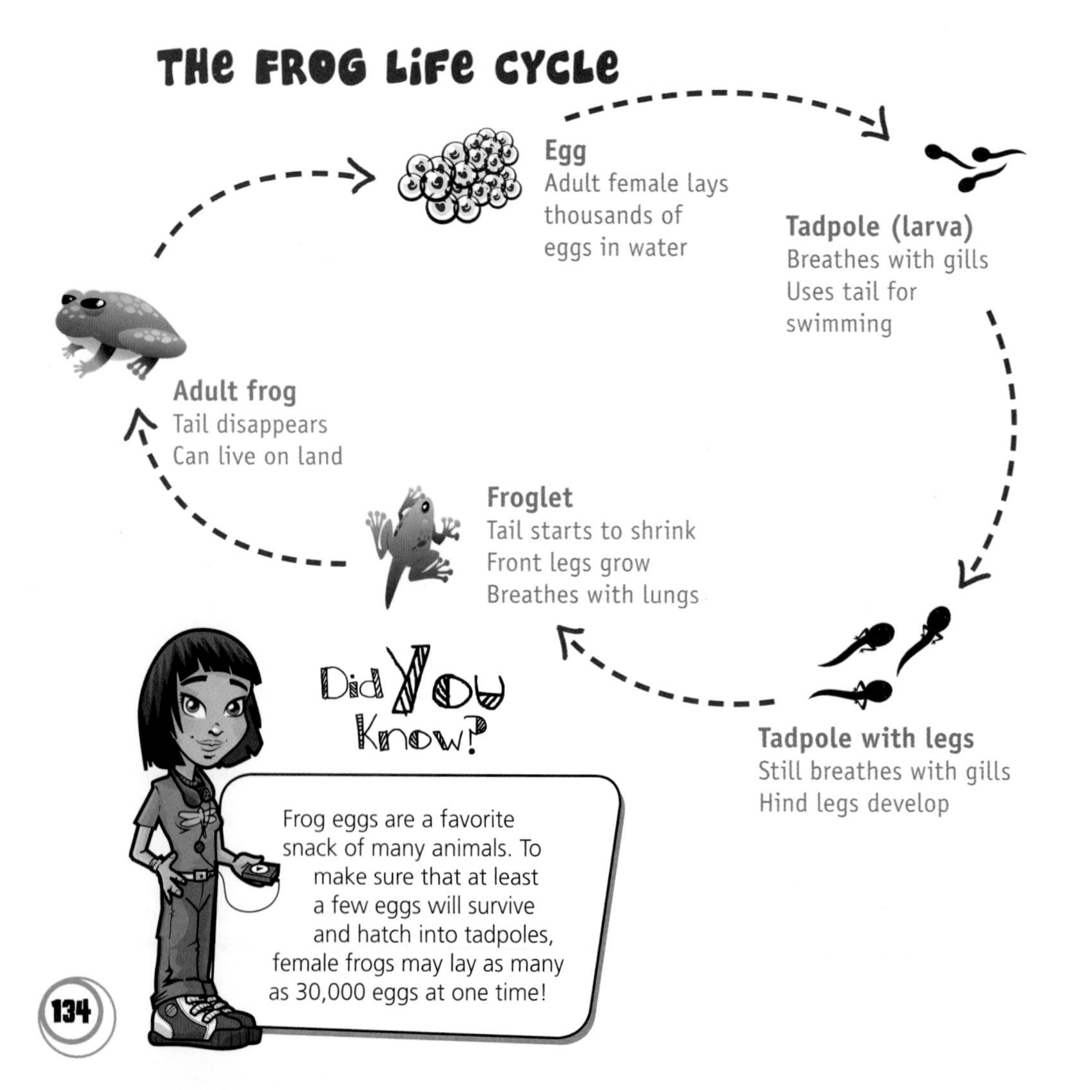

Earth's Cycles

The Earth has many cycles. The day/night cycle occurs because the Earth **rotates** on its **axis**. An object that rotates spins completely around on its axis. It takes 24 hours for one complete rotation around Earth's axis. During that time Earth passes through one day and one night.

Animals

The Earth's yearly cycle occurs because the Earth **revolves**, or orbits, around the Sun. When an object revolves, it moves around another object. It takes the Earth just over 365 days to revolve once around the Sun.

The Solar System

Seasons occur in cycles because the Earth always tilts in the same direction as it revolves around the Sun. The axis tilts toward the Sun in the summer. It tilts away from the Sun in the winter. This means that the two hemispheres have opposite seasons. It is summer in the northern hemisphere when it is winter in the southern hemisphere.

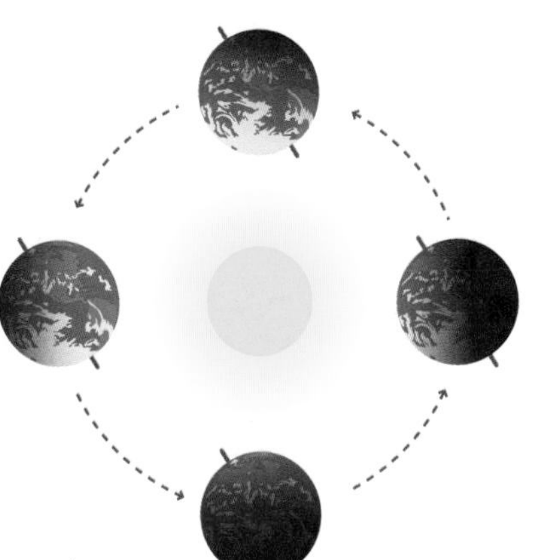

Ecosystems

The Moon has cycles too. It rotates on its axis once every 28.5 Earth days. This is also how long it takes the Moon to revolve once around the Earth.

GO ONLINE

@

To learn more, check out these Web sites!

- **Frogs**
  http://seagrant.wisc.edu/frogs/
- **The Children's Butterfly Site**
  http://kidsbutterfly.org
- **Get Growing**
  http://www.realtrees4kids.org/threefive/getgrowing.htm

Ecosystems

What Is an Ecosystem?

Somewhere in a forest, two squirrels are chasing each other around an oak tree. The squirrels are part of a population. A **population** is a group of organisms of the same species that live in the same place at the same time. The populations in an area form a **community**. The squirrels and oaks are part of the forest community.

Nonliving things like soil and water affect the members of a community. Together, the community and the nonliving things in it make up an **ecosystem**.

The organisms in an ecosystem are connected to each other and to their surroundings. A change in one part of the ecosystem can affect everything that lives there.

The living and nonliving parts of an ecosystem are connected.

Make the Connection: Knock Knock... Who's There?

Identify the living and nonliving parts of a local ecosystem.

1. Choose an ecosystem near your school or your home to study.
2. Draw two columns in your science notebook. Label one "Living Things." Label the other "Nonliving Things."

Carbon and Oxygen Cycles

The flow of carbon dioxide and oxygen through an ecosystem is important for everything that lives there.

Plants take in carbon dioxide from the air through their leaves. They use this carbon dioxide to make their own food. This process also produces oxygen. Plants release this oxygen into the air through their leaves.

Animals get oxygen from the air when they inhale, or breathe in. Their bodies use this oxygen to release energy from the food they eat. This process also produces carbon dioxide. Animals release this carbon dioxide into the air when they exhale, or breathe out.

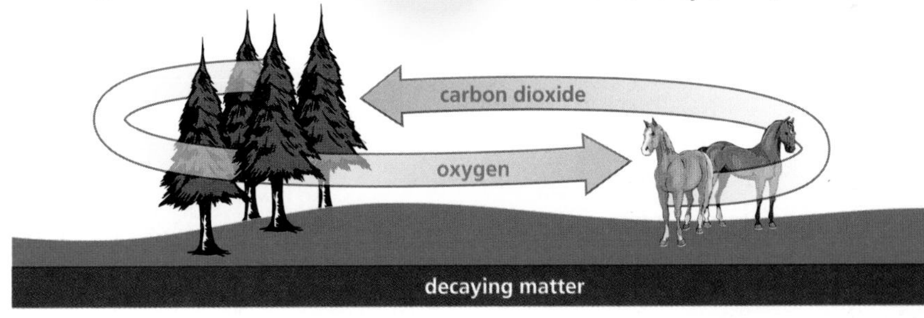

See also: page 116 Plants

Older plants that aren't eaten will die and begin to rot. The horses pass undigested plant matter as waste materials. Decomposers such as flies and worms get energy by breaking down wastes and rotting matter. Then they release carbon dioxide into the air.

3. Walk slowly through the ecosystem. Which parts are living? Which parts are nonliving? List these in your notebook.
4. Make sketches of organisms you don't recognize. Give them temporary names like "Bird A" and "Bird B."
5. Write a short summary describing the living and nonliving parts of your ecosystem. Save this for a later activity.

Types of Organisms in Ecosystems

Each organism in an ecosystem has a role to play. An organism's role depends on how it gets energy. There are three main roles in ecosystems: producers, consumers, and decomposers.

Most plants are producers.

Producers make their own food. They use carbon dioxide, water, and energy from sunlight to make food. Plants and algae are producers. So are some bacteria.

Consumers cannot make their own food. They get energy by eating other organisms. All animals are consumers. So are some single-celled organisms.

Whales and birds are consumers.

The amanita mushroom is a very poisonous decomposer.

Decomposers get energy by feeding on decaying organisms and wastes. Mushrooms and other fungi are decomposers. So are some insects, worms, and bacteria.

WHAT'S MY JOB?

Identify the roles of the organisms in a local ecosystem.

1. Go back to the local ecosystem you visited earlier. Be sure to bring your notes and summary from that visit.
2. Draw three columns in your science notebook. Label these as "Producer," "Consumer," or "Decomposer."

Types of Consumers

There are three types of consumers.

Snowshoe hare

Herbivores (HER bih vorz) eat only plants and plant products. Some herbivores eat grasses and young plants. Others eat bushes or plant roots. Still others eat only seeds or fruits. Rabbits and hares, deer, cows, and horses are herbivores. So are some insects and some birds. So are elephants and gorillas.

Omnivores (OM nih vorz) are animals that eat plants and other animals. Raccoons and skunks are omnivores. They eat insects and small mammals. They also eat berries and nuts. Bears, foxes, and wolves are omnivores too. So are humans, chimpanzees, and many monkeys.

Red fox

Carnivores (CAR nih vorz) only eat other animals. Owls and hawks are carnivores. So are spiders and many snakes. All of the "big cats"—lions, tigers, cheetahs, jaguars, leopards, and cougars—are carnivores. So are the small domestic cats we keep as pets!

Great horned owl

3. Observe the organisms in the ecosystem.
4. Identify each organism as a producer, consumer, or decomposer and mark this in your notebook.
5. Observe how the organisms interact. Be sure to take notes about what you see.
6. Write a summary of your observations.

Food Chains

When an organism eats another organism, food energy moves from the “food” organism to the “eater” organism. This path of energy through an ecosystem is called a **food chain.**

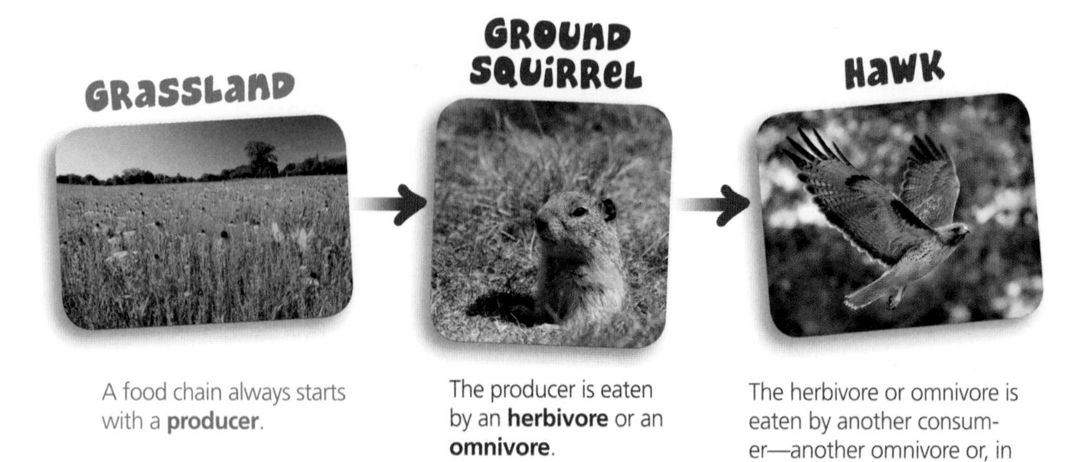

A food chain always starts with a **producer**.

The producer is eaten by an **herbivore** or an **omnivore**.

The herbivore or omnivore is eaten by another consumer—another omnivore or, in this case, a **carnivore**.

Some food chains have one or two more consumer levels. And of course, an ecosystem has more than one food chain! Grasses, ground squirrels, and hawks form one grassland food chain. Here are some others.

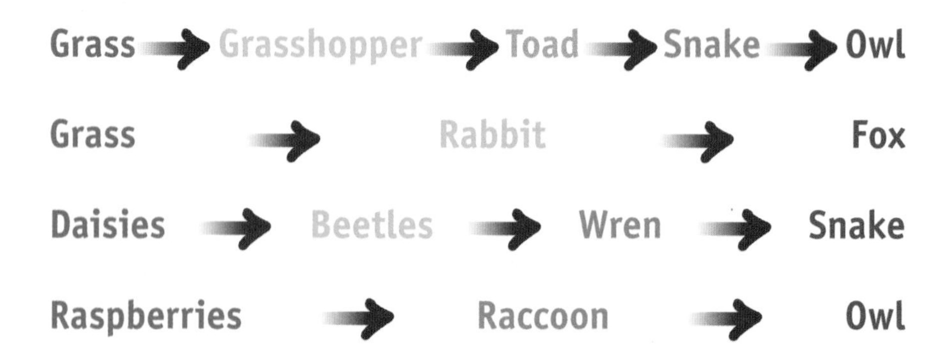

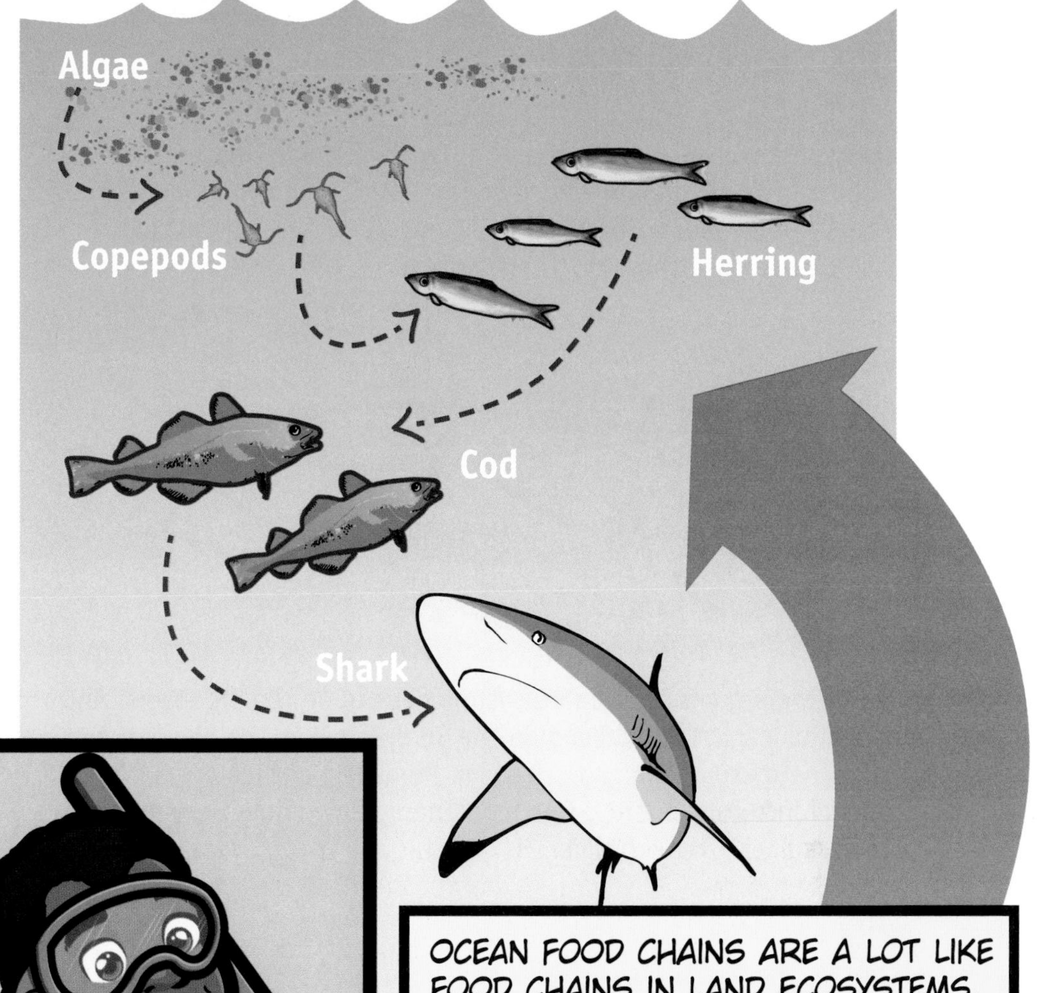

OCEAN FOOD CHAINS ARE A LOT LIKE FOOD CHAINS IN LAND ECOSYSTEMS. BUT SOMETIMES AN OCEAN FOOD CHAIN LOOKS UPSIDE DOWN! OCEAN PRODUCERS ARE ALWAYS NEAR THE WATER SURFACE, WHERE THERE'S PLENTY OF SUN. OCEAN HERBIVORES, LIKE THESE COPEPODS (KOH PEH PODS), STAY NEAR THE PRODUCERS. THE OTHER ORGANISMS ARE NEARBY, TOO. IN FACT, MORE THAN 90% OF OCEAN LIFE HANGS OUT WITHIN 180 METERS (600 FEET) OF THE OCEAN SURFACE!

Food Webs

Most organisms eat more than one type of food. That means they appear in more than one food chain. Overlapping food chains form a food web. A **food web** contains all of the paths that energy can take through an ecosystem.

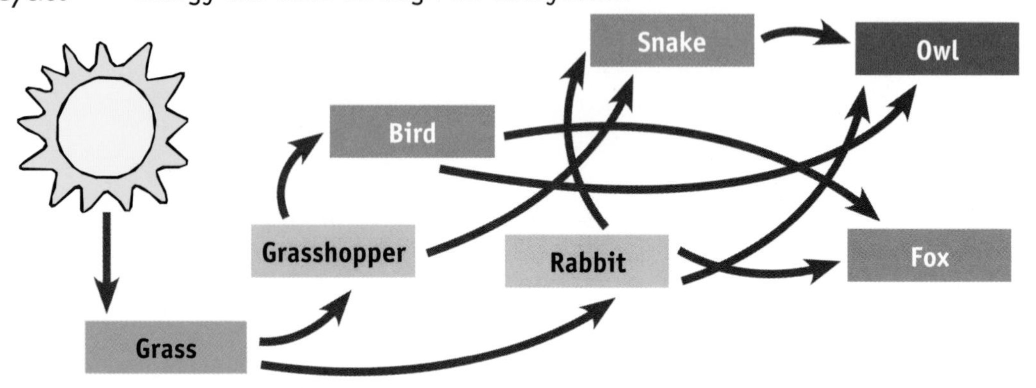

All of the organisms in an ecosystem depend on the producers. And the producers? They depend on the Sun! After all, the Sun is the true energy source in an ecosystem. Producers change the Sun's energy into food energy. This food energy flows from the producers to consumer after consumer.

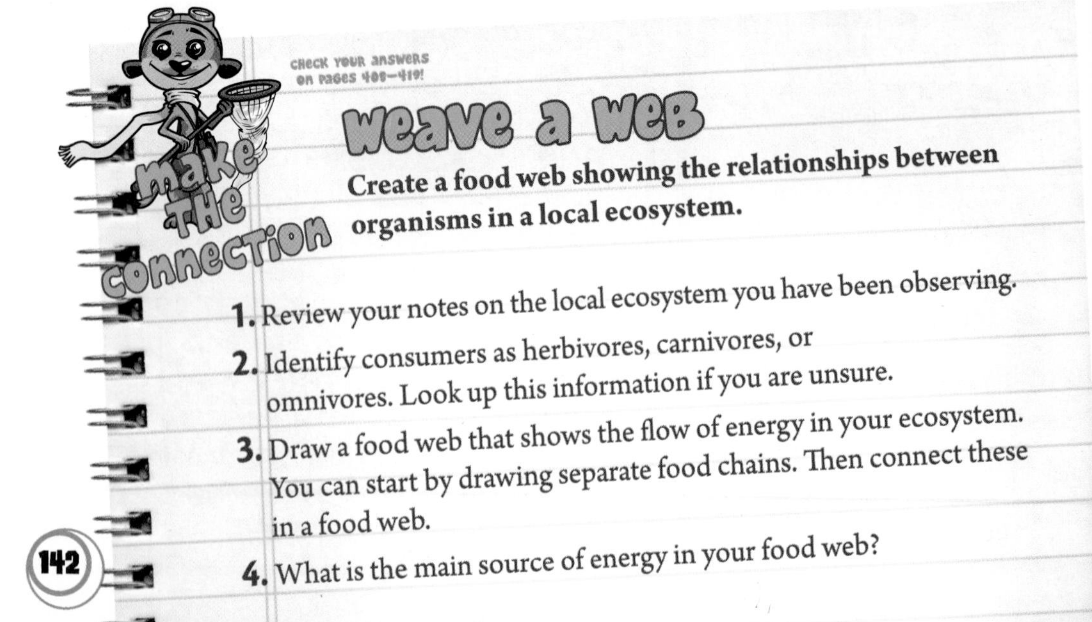

Weave a Web

Create a food web showing the relationships between organisms in a local ecosystem.

1. Review your notes on the local ecosystem you have been observing.
2. Identify consumers as herbivores, carnivores, or omnivores. Look up this information if you are unsure.
3. Draw a food web that shows the flow of energy in your ecosystem. You can start by drawing separate food chains. Then connect these in a food web.
4. What is the main source of energy in your food web?

Energy Pyramids

Food chains show the direction of energy flow between organisms. An **energy pyramid** shows how much energy passes from one organism to another in a food chain. Each link in a food chain is a level in the energy pyramid.

Organisms on each level in the pyramid get energy from organisms on the level below them. The organism uses up most of this energy and stores the rest in its body. This stored energy then passes to the next level. For example, herbivores eat producers. The herbivore uses about 90 percent of the energy it gets. It stores the rest in its body. Then this passes to the carnivore that eats the herbivore.

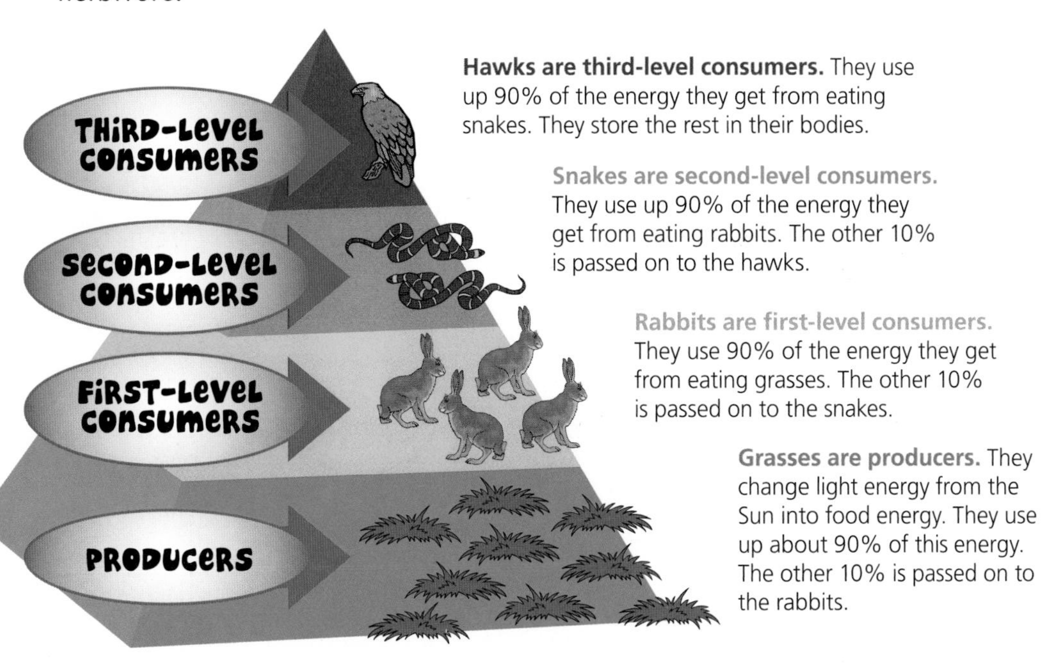

Hawks are third-level consumers. They use up 90% of the energy they get from eating snakes. They store the rest in their bodies.

Snakes are second-level consumers. They use up 90% of the energy they get from eating rabbits. The other 10% is passed on to the hawks.

Rabbits are first-level consumers. They use 90% of the energy they get from eating grasses. The other 10% is passed on to the snakes.

Grasses are producers. They change light energy from the Sun into food energy. They use up about 90% of this energy. The other 10% is passed on to the rabbits.

It takes thousands of producers to support all of the organisms in one food chain. By the time energy reaches the top level, there's very little left. That's why most food chains have just three or four levels. In most natural systems, there isn't enough energy left to support more levels at the top.

Ecosystem Balance

A balanced ecosystem has enough resources to support the organisms that live there. But ecosystems don't always stay balanced. Changes happen in nature just as they do in our lives. These changes can increase or decrease population sizes.

The organisms in a food chain are connected to each other even when they are separated by one or more energy levels. For example, here's a typical grassland food chain:

Grass **Rabbits** **Owls**

The producers (grass) and the carnivores (owls) are connected even though they are separated by the herbivores (rabbits).

When the ecosystem is balanced, there is enough grass to support all of the rabbits. And there are enough rabbits to support all of the owls.

Let's see what happens if one day there are fewer rabbits:

Fewer rabbits = less food for owls = fewer owls

Fewer rabbits = less grass is eaten = more grass (A LOT more)

The rabbits now have more food and fewer predators.

More grass + fewer owls = more rabbits. (A LOT more)

More rabbits = more grass is eaten = less grass. (A LOT less)

The owls now have more food. And no predators.

More rabbits = more food for owls = more owls (A LOT more)

Less grass + more owls = fewer rabbits (A LOT less)

The owls now have less food. A lot less.

Fewer rabbits = fewer owls

With very little to eat, the owls are in trouble. Some owls move away. Some starve.
Some eat other prey.
The owl population becomes very small.

There's very little grass to support the rabbits.
Some rabbits move away. Some starve. Some eat other plants.
Now there are very few rabbits and very few owls.

Fewer rabbits means the grass population can recover.
More grass will lead to more rabbits.
More rabbits will lead to more owls.
Soon, the producer-herbivore-carnivore populations will reach normal levels.

The ecosystem is balanced again!

Ecosystem Relationships

All of the organisms in an ecosystem interact with each other. There are three main types of interactions: competition, predator-prey, and symbiosis.

Competition occurs when organisms need to use the same resource. All organisms need food, water, and space. Plants and other producers also need light. But there is only a certain amount of these resources in any ecosystem. Organisms must compete with each other to get the resources they need. In this way, competition helps control population sizes.

Zebra, antelope, elephants, and many other animals compete for water in this community. The plants compete for water, too.

Predator-prey interactions control population sizes, too. A **predator** (PRED uh tor) is an animal that eats another animal. A **prey** (PRAY) animal is the one that is eaten. Without predators, prey populations would grow too large. The prey would run out of food. Many would starve. By eating prey, predators help keep the ecosystem in balance.

Spiders use webs to trap prey.

Salmon are a favorite prey of grizzly bears.

Some organisms depend on another organism in order to survive. This is called a **symbiosis** (SIM bee OH sis). There are three types of symbiotic interactions.

Types of Symbiosis	Species 1	Species 2	Examples
Mutualism	Benefits	Benefits	• Cleaner shrimp and moray eels • Bees and flowers
Commensalism	Benefits	Not affected	• Spanish moss and trees • Remora fish and sharks
Parasitism	Benefits	Harmed	• Ticks and mammals • Sea lampreys and fish

A **mutualism** (MYOO tchoo uhl iz im) is a symbiosis where both organisms benefit. Cleaner shrimp get food by eating harmful organisms that live on the moray eel's skin. This keeps the eel healthy and gives the shrimp a meal.

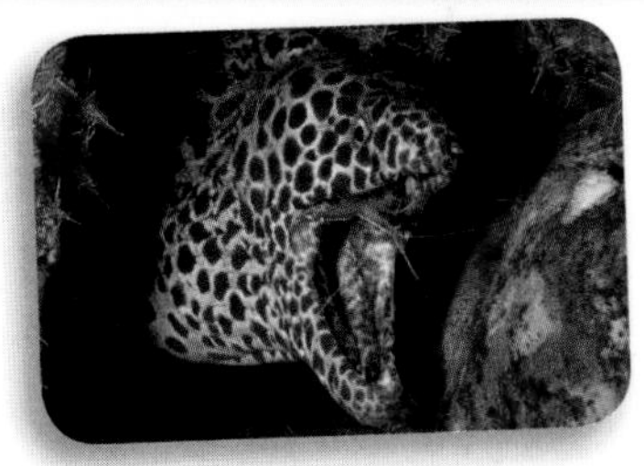

Cleaner shrimp and moray eels have a relationship where everyone wins.

Spanish moss is an epiphyte that grows on oak and cypress trees.

In **commensalism**, one organism benefits. The other is not affected. Epiphytes are plants that grow on top of other plants. The epiphytes have tiny roots that hang in the air and trap rain water. The epiphyte benefits because it doesn't have to compete for space on the forest floor. The tree isn't affected by the epiphyte.

In **parasitism**, one organism is helped and one is harmed. The organism that is helped is a **parasite**. The host is the organism that is harmed. Parasites live in or on the **host**. Parasites survive by absorbing nutrients from the host. This weakens the host and makes it sick.

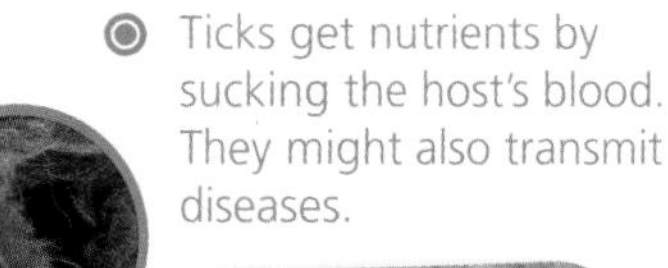

Ticks get nutrients by sucking the host's blood. They might also transmit diseases.

Ecosystem Disturbances

All of the parts of an ecosystem are linked. This keeps the ecosystem balanced. But some events can disturb this balance. They can destroy habitats and resources. Some events kill organisms. Other events force populations to move away.

Many natural events can disturb an ecosystem. Changes in weather can have a large effect on an ecosystem. Lightning strikes cause wildfires. Tornadoes destroy trees. Storms cause floods and erode beaches.

Too much water causes flooding. Floods kill plants and drown animals.

Droughts (DROWTS) happen when there is too little rain for a long period. Without water, plants and animals will die.

Is That a Fact?

Fill Me In!

Some ecosystems formed in the craters of old volcanoes! Take Yellowstone National Park. The park we see today is inside the top of an old volcano! The volcano erupted 640,000 years ago and left a giant crater. Then the crater filled in with living things and resources over time. Today it's a whole ecosystem!

Volcanoes can cause many problems when they erupt. Lava and mudflows wipe out habitats and kill animals. Clouds of ash and dust can block out the Sun. If they block too much Sun, the climate is affected.

Humans activities also disturb ecosystems. Humans build roads and buildings that cause soil erosion and destroy habitats. Wildfires destroy large areas of forest. Oil spills contaminate soil. They poison ground and surface water. Oil spills also kill animals. Birds coated with oil cannot fly or hunt for food.

Humans cut down large sections of rainforest to grow crops every year. This habitat will never be able to restore itself.

Humans also introduce invasive species into ecosystems. These organisms do not have natural predators in the new habitat. Their populations grows out of control without predators. They crowd out native species and disrupt food webs.

The lily leaf beetle is a European bug that has damaged many gardens in its new habitat: the eastern United States.

To learn more, check out these Web sites!

- **Kids' Planet**
  http://www.kidsplanet.org
- **EcoKids**
  http://www.ecokids.ca
- **Sea and Sky**
  http://www.seasky.org

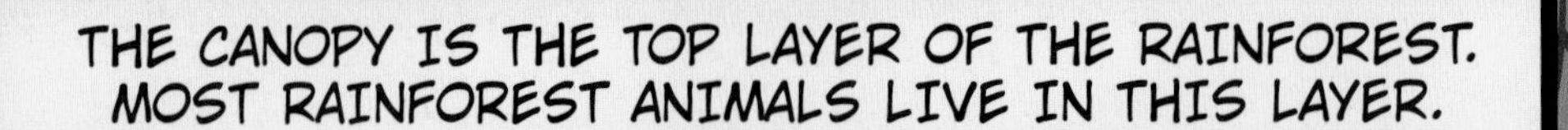

THE CANOPY IS THE TOP LAYER OF THE RAINFOREST. MOST RAINFOREST ANIMALS LIVE IN THIS LAYER.

A LOT OF THE MONKEYS HERE IN THE AMAZON HANG BY THEIR TAILS WHILE THEY EAT. THIS SPIDER MONKEY WAS SO BUSY EATING FRUIT THAT HE ALMOST DIDN'T SEE ME HERE!

THE UNDERSTORY IS REALLY DARK, SO ... HEY! IS THAT AN ANACONDA?! UH, CHASE...

MOST OF THE WORLD'S INSECTS LIVE IN RAINFORESTS. AND MOST OF THE RAINFOREST'S INSECTS LIVE HERE IN THE UNDERSTORY ... AAAAAANND I DON'T THINK I'M GONNA MOVE 'TIL THAT ANACONDA MOVES ON!

RAINFOREST SOIL HAS VERY FEW NUTRIENTS. MANY RAINFOREST TREES ADAPT TO THIS BY GROWING SHALLOW ROOTS.

I'M GLAD I BROUGHT MY FLASHLIGHT. IT'S REALLY DARK ON THE RAINFOREST FLOOR. NOT MANY ANIMALS AROUND HERE. EXCEPT FOR TAPIRS ... AND THAT GIANT ANTEATER. UH OH -- WHERE'S ROCKET?

HEY TOMAS,
CHECK IT OUT: I CAN USE THIS LIANA TO SWING TO THE NEXT TREE. LIKE TARZAN!!
A LIANA IS A THICK VINE THAT GROWS IN MANY RAINFORESTS. LIANAS WIND AROUND TREES AND GROW UP TOWARD THE SUNSHINE THAT PEEKS THROUGH THE CANOPY.
THE TREES AREN'T TOO CROWDED TOGETHER HERE IN THE UNDERSTORY. SO THERE'S ROOM FOR PREDATORS LIKE THE JAGUAR TO STALK PREY ... HEY, WHERE'S ROCKET?
?

IF YOU GO OUTSIDE AND TOUCH THE GROUND – THAT'S THE BIGGEST THING YOU CAN TOUCH! THE EARTH!

Earth Science

What is Earth science? **Earth science** is the study of Earth and its place in the universe. Our planet Earth provides a home for animals, plants, and people. Earth science teaches us about volcanoes and oceans and shooting stars. It teaches us about fossils and snow and sand. Can you touch the clouds? What makes volcanoes erupt? We can find these things out when we study Earth science.

Earth and Its Landforms

What Spheres Make Up Earth's System?

Rock, water, air, and life all act together in Earth's **system**. Each part of the Earth System needs energy to carry out its processes. Energy comes from the Sun and from deep inside Earth.

The **atmosphere** (AT muh sfeer) is the mixture of gases surrounding Earth. The air you breathe is part of the atmosphere.

See also: page 38 Systems

The **hydrosphere** (HY druh sfeer) includes all of Earth's liquid water and ice, water in soil and rock, and water vapor in the air. About three-fourths of Earth's surface is covered by water.

The **geosphere** (JEE oh sfeer) is the rock and soil on Earth's surface and the layers under the surface. The **lithosphere** (LITH uh sfeer) is the top part of the geosphere. Earth's **crust** is the top of the lithosphere. The crust includes the land under the oceans.

The **biosphere** (BY oh sfeer) is made up of all living things. The biosphere includes life in the atmosphere, the hydrosphere, and the geosphere. Humans are part of the biosphere.

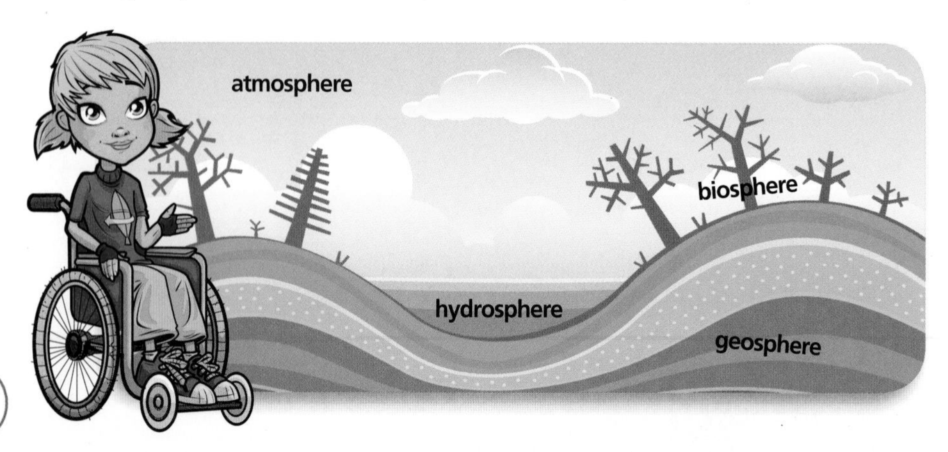

Earth's Geological Features

A **landform** is a natural geological (jee uh LAH ji cul) feature on Earth's surface. The surface of Earth constantly shifts and alters its features.

Mountains are Earth's highest landforms. Earth's highest mountains are on the ocean bottom.

Mid-ocean ridges are mountains on the ocean floor.

The **continental shelf** is the land around the edge of a continent, under ocean water.

A **plateau** is a large area of flat ground. A plateau is much higher than the land around it.

A **plain** is a large, flat area of ground. A plain is much lower than the land around it.

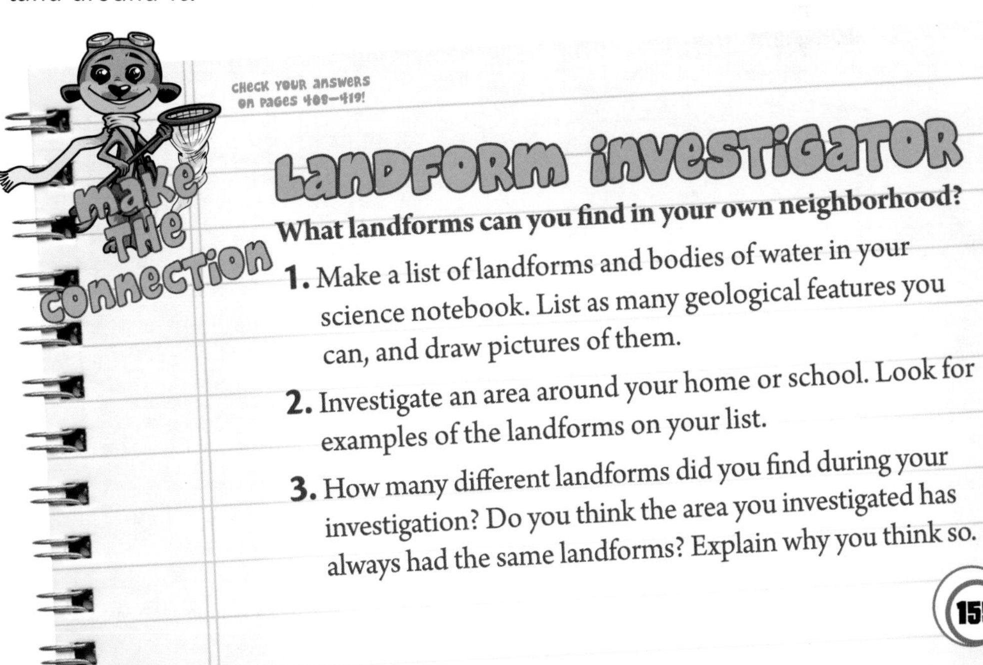

Check your answers on pages 408–419!

Landform Investigator

What landforms can you find in your own neighborhood?

1. Make a list of landforms and bodies of water in your science notebook. List as many geological features you can, and draw pictures of them.
2. Investigate an area around your home or school. Look for examples of the landforms on your list.
3. How many different landforms did you find during your investigation? Do you think the area you investigated has always had the same landforms? Explain why you think so.

Change by Weathering

Weathering (WEH ther ing) breaks rock into smaller pieces. Weathering can make rocks break apart more easily.

Wind, water, ice, changes in temperature, and living things cause **physical weathering**. When rock is broken down by physical weathering, it is still rock.

See also: page 166 Minerals and Rocks

Wind As the wind blows small particles against rock, it acts like sandpaper and wears away small pieces of rock.

Water Rock is worn down by moving water. These rocks were made smooth by flowing river water.

Ice Water seeps into cracks in rock and then freezes. When the ice expands, the rock breaks apart.

See also: page 186 Soil

Heating and Cooling When rock heats up, it grows. When rock cools down, it shrinks. The stress makes rock break down or peel apart.

Animals Animals that dig, tunnel, or burrow move soil and break down loose rock.

Plants Plants take root in cracks in rock. As the plants grow, the roots spread the rock apart.

Rock that goes through **chemical weathering** can change into other things. Chemical **reactions** cause chemical weathering.

Water Minerals in rock can dissolve in water. Gases in water react with rock to form other kinds of rock.

Weather, Seasons, and Climate

Oxygen Oxygen reacts with iron to form rust.

Carbon Dioxide Carbon dioxide and water make acid. The acid reacts with minerals to form a kind of salt called a carbonate.

Acids Acids from plants and rainwater wear rock away.

Stalactites look like icicles. They hang from the roofs of caves. Water carries dissolved carbon dioxide and minerals through cracks in rock. When the water drips through the roof of a cave, it meets the air. The minerals in the water become solid. Over time, the minerals build up, forming a stalactite.

Quick Question

You may not realize that you see examples of weathering every day. Which of the following is NOT an example of weathering? Write your answer in your science notebook.

- A railroad bridge is coated with rust.
- A tree is cut down and burned.
- Ivy roots crumble the mortar of a brick wall.
- Winter freezing and spring thawing make potholes in a road.

Change by Erosion

Erosion (uh ROH zhun) moves soft or loose weathered rock from one place to another. Erosion is most often caused by water, ice, wind, or gravity.

Erosion by Water Heavy rains wash loose soil away and leave deep gullies. Tiny pieces of rock, sand, and soil, called **sediment** (SED uh ment), are carried away in rivers and streams.

Erosion by Ice Glaciers (GLAY sherz) are huge sheets of ice that move very slowly.

Erosion by Wind Wind carries away loose soil and leaves rock behind.

Erosion by Gravity Gravity causes loose rocks and soil to roll downhill as **rockfalls** or **landslides**. Heavy rains cause **mudflows**.

Erosion Makes New Landforms
A **delta** forms as sediment builds where a river flows into an ocean or sea.

Sediment deposited by waves or currents forms a **beach**.

Sand carried by wind can build up. When a lot of sand is deposited in one place, it is called a **dune**.

A **valley** is formed when rivers or glaciers wear away rock and soil.

A **canyon** is a deep and narrow valley. The Grand Canyon was formed by the Colorado River as it wore away rock.

COME ON KARI–
YOU'RE AS SLOW
AS A GLACIER.
YOU KNOW
WILL, SOME
GLACIERS
AREN'T THAT
SLOW.

SOME
GLACIERS
ARE MOVING
AS FAST AS
45 METERS
EACH DAY!
WHEN GLACIERS REACH A STEEP SLOPE THEY SLIDE FASTER. AS GLACIERS MELT, THE GROUND GETS SLIPPERY. THE GLACIER SLIDES FASTER. MELTING GLACIERS LOSE WEIGHT. LIGHTER GLACIERS MOVE FASTER.

HOW ABOUT THIS?
TRY TO MOVE AS FAST
AS A GLACIER!

Earth's Water Resources

All life on Earth needs water. Animals, including humans, need water for their bodies to work properly. Plants need water to produce food.

When we look at Earth from space, we see that Earth is covered by water.

Water covers 75 percent of Earth's surface. Almost all of Earth's water is saltwater in oceans and seas. Only about 3 percent of Earth's water is freshwater. Just a tiny bit of Earth's freshwater is found in lakes, streams, rivers, and swamps. Most of Earth's freshwater is ice in the form of icecaps and glaciers. About 30 percent of Earth's freshwater is stored under our feet as groundwater.

See also: page 252 Matter

See also: page 128 Cycles

Water can be a solid, a liquid, or a gas. Very cold liquid water can become a solid. Warm liquid water can become a gas. Water in the form of a gas is invisible.

Snow is water in solid form. Ice is another form of solid water.

The Water Cycle

The Water Cycle

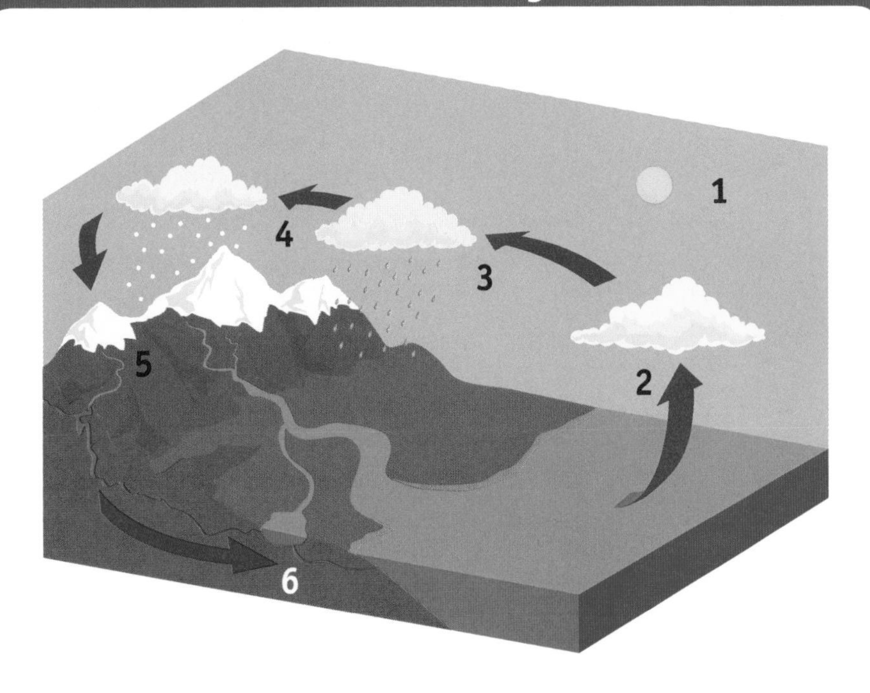

1. The Sun's energy drives the water cycle. The Sun heats liquid water and changes it into a gas called water vapor. This change is called **evaporation** (ee vap o RAY shuhn).

2. Most water that evaporates is ocean water. Freshwater, and water from plants and animals, also evaporates.

3. Water vapor moves up into cooler air and turns into drops of water. This change is called **condensation** (con den SA shuhn). Water drops become attached to small pieces of dust in the air. Clouds form this way.

4. Inside the clouds the small drops join to form large, heavy drops. The drops fall to Earth as rain, snow, sleet, or hail, called **precipitation** (pree sip uh TAY shuhn).

5. Most precipitation ends up in the oceans and becomes saltwater. Precipitation falls on ice caps, lakes, streams, rivers, and swamps.

6. Water soaks into soil and becomes part of Earth's groundwater. In time, surface water and groundwater become part of the oceans.

How Mountains Form

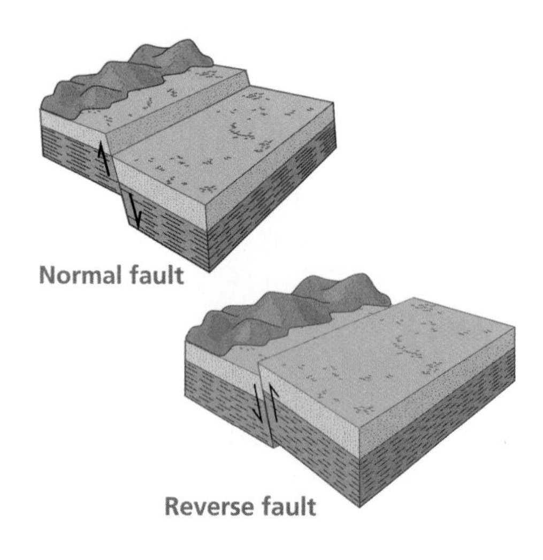
Normal fault

Reverse fault

Fault Mountains A **fault** (FAWLT) is a crack in the rocks of Earth's surface. A **normal fault** forms when rocks pull apart. One side of the land drops down. A **reverse fault** forms when rocks are pushed together. One side of the land is pushed up.

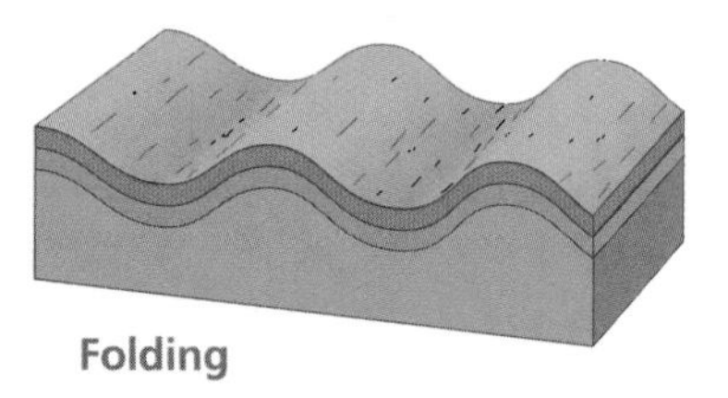
Folding

Folded Mountains When layers of rock are pushed into each other from both ends, they are squeezed into waves called **folds.**

Volcano

Volcanic Mountains Melted rock from deep inside the ground is called **magma.** Volcanoes form when great pressure forces magma up and out. Once the molten rock is out on Earth's surface, we call it **lava.**

Eroded Mountains Wind, rain, and ice have eroded softer portions of Earth's surface, leaving the harder rock behind.

Eroded mountain

We Investigate Earth

Scientists use special instruments to find out what is going on under ocean water and inside or above Earth.

Sonar Sonar (sound navigation ranging) works by sending a short pulse of sound called an acoustic signal. Scientists use sonar to find out how deep oceans are and to help them map the ocean bottom.

Radar Radar (radio detecting and ranging) works by sending out radio waves that are bounced back by an object. Radar is used to find out how fast and in which direction an object is moving. Radar is also used to detect nonmoving objects underground.

Seismograph A **seismograph** (SIZE moh graf) detects seismic waves. A seismic wave is energy that travels through the earth from an earthquake.

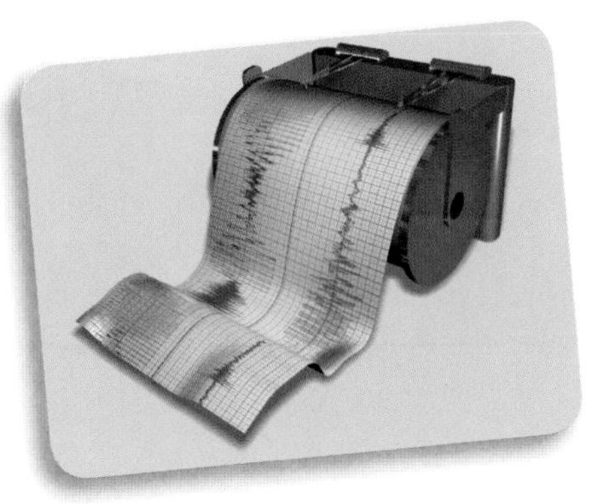

This seismograph records seismic waves on paper.

Weather Balloons Instruments to measure temperature, humidity, and atmospheric pressure are carried into the amosphere by **weather balloons**.

Go Online

To learn more about investigating landforms:

- **Earthquakes**
  http://earthquake.usgs.gov/learning/kids/
- **Geology**
  http://www.historyforkids.org/scienceforkids/geology/
- **Mountains**
  http://www.mountain.org/education/
- **The Water Cycle**
  http://ga.water.usgs.gov/edu/watercycleplacemat.html

WE HAVE OUR MAP—WE'RE ON OUR WAY TO MEET OUR GUIDE.
THE GUIDE IS BRINGING SPECIAL EQUIPMENT SO RAIN CAN CLIMB TOO.
STAY ON THE PATH! DON'T TRAMPLE THE PLANTS.
WE CAN'T ROCK CLIMB HERE WHEN THE PEREGRINE FALCONS ARE NESTING. THEY NEED PEACE AND QUIET UNTIL THEIR CHICKS ARE HATCHED.

WEAR THE RIGHT GEAR WHEN YOU CLIMB. ALWAYS WEAR A HELMET. MAKE SURE YOUR EQUIPMENT IS IN GOOD SHAPE. FOLLOW ALL OF THE PARK'S SAFETY RULES! DON'T CLIMB ALONE!

I'M GLAD I'M WEARING MY HELMET. HERE COMES A LOOSE ROCK!

IF YOU PACK SOMETHING IN, PACK IT OUT WHEN YOU GO. LEAVE ONLY FOOTPRINTS!

Minerals and Rocks

What Are Minerals and Rocks?

Take a look around you. Do you see things made of stone or brick? Is there anything made of ceramic or metal? These objects are made of Earth materials. **Earth materials** are natural materials that are found in the Earth and that are not living. Minerals and rocks are the most common Earth materials.

Did You Know?

Mica is a mineral that forms in thin, flat plates. Mica can be green, brown, black, or even clear. Some pieces of mica are so thin that you can see through them. Colonists in early America sometimes used large, thin pieces of mica to make window glass.

A **mineral** is a solid object. It has a certain crystal and chemical structure. It does not contain organic matter, which means that it is not living. Gold, silver, and copper are examples of minerals.

These are two minerals. The purple mineral is called amethyst. The clear mineral is called quartz. Minerals like these can be used to make jewelry.

A **rock** is a solid object that contains two or more minerals. Sometimes rocks contain fossils of living things. Granite, shale, and slate are examples of rocks.

Soil

These rocks were found along a beach. Look closely and you will notice that the rocks have different colors and patterns. This is because they are made up of different minerals. They are smooth because the motion of the waves in the water has worn them down.

There are many types of minerals and rocks, and each type is different from the others. Some are dull, while others are brightly colored. Some are rough and others are smooth. Some are hollow, while others are solid.

Arrowheads are rocks that were carved by early humans and used for hunting and fighting. One end of the arrowhead is pointed or rounded. The other end is notched so it can be fastened to a stick. Take a close look at these arrowheads. How are they similar? How are they different?

A **geologist** is a scientist who studies minerals and rocks for a living. Other people study them just because they find them interesting. They go on hikes to find and collect minerals and rocks. These people are nicknamed "rock hounds." Can you guess why?

Properties of Minerals

Have you ever seen a ring made of topaz? Have you ever put salt on your food, or smelled water that has sulfur in it? Then you've seen, tasted, or smelled minerals.

Sulfur is a yellow mineral. It is used to make products like rubber, gunpowder, and plastic wrap. This worker collects sulfur in Indonesia.

There are about 3,800 types of minerals. The most common are called "rock-forming minerals" because they combine to form rocks. Feldspar, quartz, and mica are some rock-forming minerals.

Suppose you went hiking and found an interesting mineral. How would you know what kind of mineral it was? You can identify a mineral by examining the properties of that mineral. Some properties of minerals are streak, luster, and cleavage.

To determine a mineral's **streak**, you rub a piece of it against a porcelain plate. The streak does not have to be the same color as the mineral itself. For example, hematite is a black mineral that leaves a red streak. **Luster** describes the way in which light reflects off a mineral's surface. You can think of this in terms of how much a mineral sparkles. Some types of luster are dull, greasy, silky, vitreous, and brilliant. A mineral can have one or more types of luster. If you split a crystalline mineral, it will divide easily along certain planes. This property is known as a mineral's **cleavage.** It is determined by the mineral's atomic structure.

GO ONLINE

To learn more, check out this Web site!

- **Mineral Matters**
  http://sdnhm.org/kids/minerals/

A mineral's **hardness** is determined by how easily it can be scratched by another mineral. In 1812, Friedrich Mohs created a scale to rank the hardness of ten common minerals. He found that diamonds are the hardest minerals, because no mineral can scratch a diamond except another diamond. He also found that talc is the softest mineral because any other mineral can scratch talc.

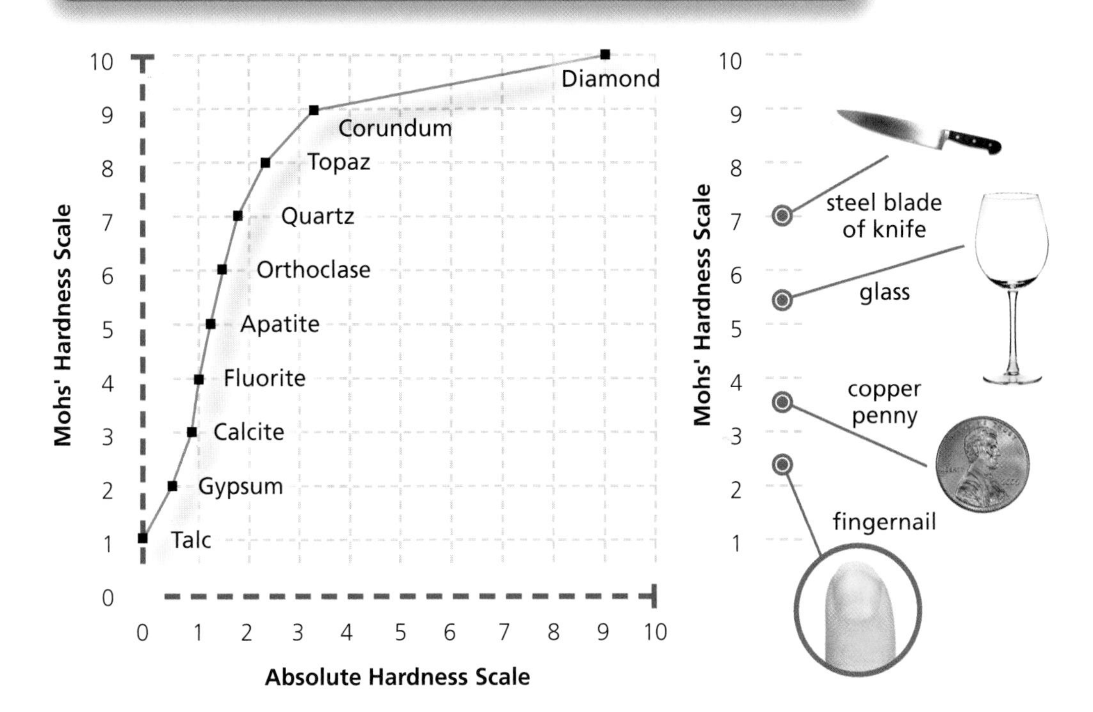

A scientist tests an unknown mineral by scratching it against others on the list. For example, if the mineral can be scratched by topaz but not by quartz, then it has a hardness between 7 and 8. Notice that diamond is far to the right on the chart. This means that, in terms of absolute hardness, diamond is about four times as hard as corundum. Gypsum is about the same hardness as a fingernail. Orthoclase is not as hard as quartz, but it is harder than glass. What other comparisons can you make from the chart?

The Rock Cycle

You may remember that different cycles occur in nature. These include the life cycle, the water cycle, and the rock cycle. A **cycle** is a series of events that happen in the same order again and again. Cycles can be short or long. The life cycle of a butterfly lasts only a few months. The rock cycle lasts millions of years.

There are three main types of rocks and each is formed in a different way. Let's take a closer look at each type of rock.

Igneous (IG nee us) **rock** is formed from lava that erupts from a volcano. The temperature deep below Earth's surface can reach 1,100°C (2,120°F) or more. At that temperature, rocks are hot enough to melt and mix together. The melted rock is called **magma**. A volcanic eruption often ejects magma above the ground. Magma that reaches Earth's surface is called **lava**. This lava then cools and hardens into igneous rock. When lava cools quickly, it forms volcanic rocks. **Volcanic** rocks have small mineral grains. When magma cools slowly, it forms plutonic rocks. **Plutonic** rocks have large mineral grains.

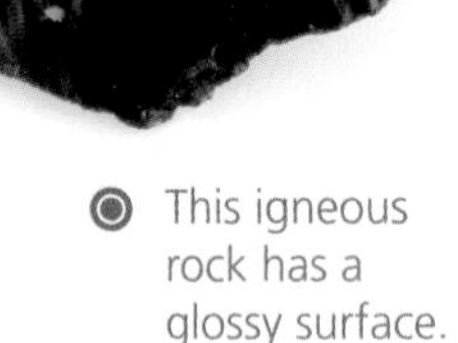

This igneous rock has a glossy surface.

Do you see the different layers in this sedimentary rock?

Sedimentary (sed uh MEN tuh ree) **rock** is made up of layers of sediment. Sediment contains minerals, rocks, and fossils that have been squeezed together over millions of years. If you look closely at this kind of rock, you might see the different layers of sediment that cemented together over millions of years.

What do you notice about this metamorphic rock?

When heat and pressure below Earth's surface melt and squeeze igneous or sedimentary rock for a very long time, a new type of rock forms. This new rock is called **metamorphic** (meh tuh MORE fik) **rock**. The word metamorphic means "change of form." Heat and pressure are powerful forces. Over time, heat and pressure can change igneous or sedimentary rock into metamorphic rock.

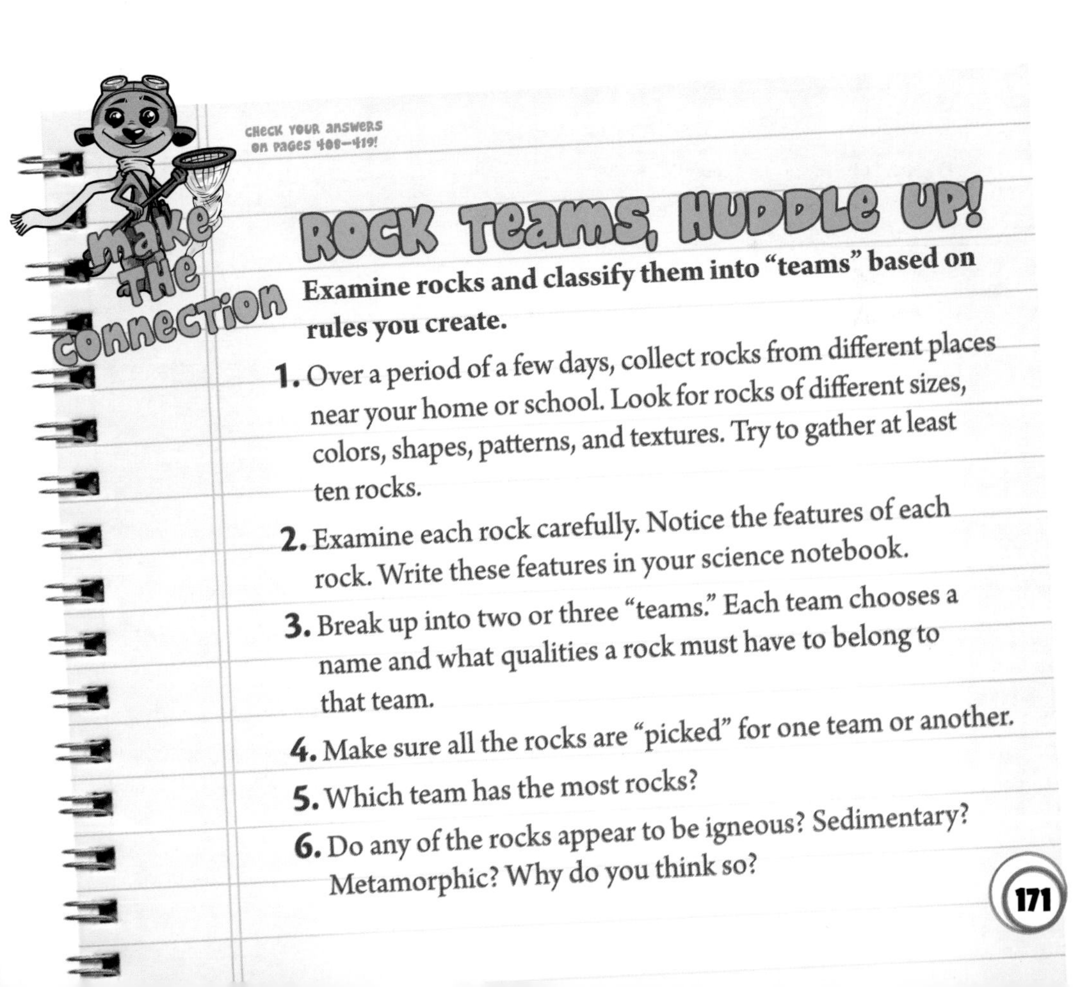

Check your answers on pages 408–419!

Rock Teams, Huddle Up!

Examine rocks and classify them into "teams" based on rules you create.

1. Over a period of a few days, collect rocks from different places near your home or school. Look for rocks of different sizes, colors, shapes, patterns, and textures. Try to gather at least ten rocks.
2. Examine each rock carefully. Notice the features of each rock. Write these features in your science notebook.
3. Break up into two or three "teams." Each team chooses a name and what qualities a rock must have to belong to that team.
4. Make sure all the rocks are "picked" for one team or another.
5. Which team has the most rocks?
6. Do any of the rocks appear to be igneous? Sedimentary? Metamorphic? Why do you think so?

The Rock Cycle

1. Rocks below Earth's surface melt to form magma. Magma reaches Earth's surface and becomes lava.
2. Lava cools and hardens into new rock.
3. Weathering and erosion break rocks apart. Weathering forms sediment. Erosion moves sediment to new places.
4. Compaction and cementation squash sediment together. When the water between the particles evaporates, the new rock is cemented together.
5. Heat and pressure melt and squeeze minerals and existing rocks into new rocks.

Finding Minerals and Rocks

You can hardly go anywhere without finding minerals and rocks. They can be found on the side of a road. They can be spotted in gardens and backyards. They are found on mountains and in rivers. Look closely at a pile of dirt dug up by an excavator. Most likely you'll see lots of rocks.

Some minerals and rocks are easy to find, such as granite (igneous), limestone (sedimentary), and quartzite (metamorphic). Common minerals include feldspar, dolomite, and hematite.

Certain minerals are rare and hard to find. For this reason, they can be very valuable. Diamonds and gold are rare and cost a lot of money to mine and refine. As a result, they cost more than other minerals, such as zircon or iron.

pro·nun·ci·a·tion

laccolith

(LAK-uh-lith)

A laccolith consists of magma that pushed up through Earth long ago. The magma cooled and hardened to create a bulge just under the surface of Earth. The land around the bulge later eroded to form the laccolith.

Earth and Its Landforms

Did You Know?

This is Devils Tower National Monument in Wyoming. Devils Tower is made of igneous rock. How it formed remains a mystery. Some scientists say that it is the remains of a laccolith.

Uses of Minerals and Rocks

People have used minerals and rocks in many ways throughout history. Early humans fashioned clubs from stone. They chipped arrowheads from minerals and rocks. Hard rocks were used to grind grain. Some people lived in caves that were formed of rock.

Although technology has advanced since the days of our early ancestors, we still have many uses for minerals and rocks.

Uses of Minerals and Rocks

MINERAL/ROCK	HOW WE USE IT
feldspar	glass; ceramics; enamelware; soaps
gold	dentistry; medicine; jewelry; art; currency
gypsum	wallboard; lawn fertilizer; chalk; cement
silica	computer chips; glass; ceramics; water filtration
slate	roofs; chalkboards; paving stones
sulfur	fertilizer; film; tires; detergents

These specially shaped cinderblocks form a section of a bridge.

Buildings and bridges are built of stone. Gravel is used on driveways and roads. Pumice in hand soap helps scrub away grease and dirt. Even our high-tech items like computers and cell phones depend on chips made from the mineral silica.

Metals such as iron, silver, and aluminum can be extracted from ore. Ore is rock that contains metals that are used to build cars, airplanes, and machines. Aluminum foil is, not surprisingly, made from aluminum. Talc is used to make baby powder. Diamonds are used in cutting tools. Copper can be fashioned into wires and sculptures. The list goes on and on! Can you think of others?

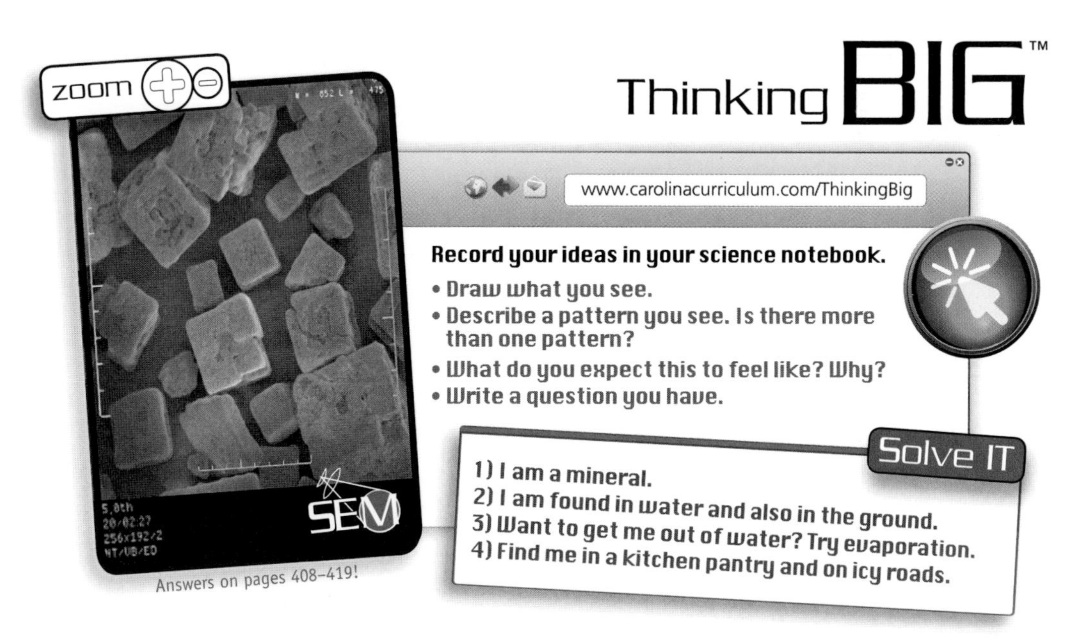

Fossils

What Are Fossils?

Did you know that the average person will live about 78 years? This may seem like a long time. To a butterfly, which may live only two weeks, it is a long time! Yet if you compare the human lifespan to the age of Earth, it is not a long time at all. Scientists believe that our planet is about 4.5 billion years old. During this time Earth has gone through many changes.

How do scientists learn about the early days of Earth? One way is by studying fossils. **Fossils** are the traces or remains of living things that died a long time ago. Many fossils come from tiny sea animals. Some fossils are the remains of plants. Still others come from large animals such as dinosaurs. The oldest fossils scientists have found are about 3.5 billion years old.

Wow, How Long Ago Was That?

You will need a meter stick or tape measure, a marker, a pen or pencil, sticky notes or labels, tape, and about 50 meters of yarn.

1. Pull the yarn down a long hall. Use tape or weights to hold down each end of the yarn. Make a mark on the yarn every 5 meters. Label the measurements.
2. Make a time line. Every 10 meters of yarn represents 1 billion years. Every half-meter of yarn represents 0.5 billion years. Label events. Write *BYA* for "billions of years ago." Write *MYA* for "millions of years ago." Write *YA* for "years ago." Events:
   - About 4.5 BYA Earth formed
   - About 3.5 BYA single-celled life appears
   - About 0.5 BYA many land plants appear
   - About 245 MYA dinosaurs appear
   - About 65 MYA mass extinction of dinosaurs happens
   - About 100,000 YA first humans appear

5 BYA

4.5 BYA Earth Formed

0 m

5 m

How Fossils Are Made

Usually, when something dies it does not leave any evidence that it ever existed. The organism will **decompose** (dee kuhm POHZ), or rot away. Animals may eat the animal or plant. But what if the dead organism is not eaten and does not rot away? Sometimes organisms are buried so quickly that they do not have time to decompose. These once-living things may become fossils.

Most fossils get their start in mud or sand. An animal may die in or near a river, lake, or ocean. Soft parts, such as flesh, skin, and organs, may rot. The hard parts of the animal, such as bones, teeth, or shell, are covered by clay or sand, called **sediment**. Over time, the sediment turns to **sedimentary rock**. The hard parts of the organism break down over time and leave an imprint behind. Minerals may replace the bones. Over millions of years, the movement of Earth's crust brings the rock close to the surface. When wind and rain wear rock away, the fossils are revealed.

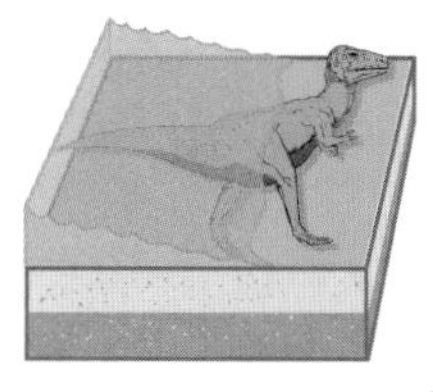

An animal dies. Soft body parts rot away or are eaten.

Hard body parts are covered with a layer of sediment.

More layers of sediment build up. The pressure turns the sediment layers to rock.

Minerals from groundwater or other sources replace the bone.

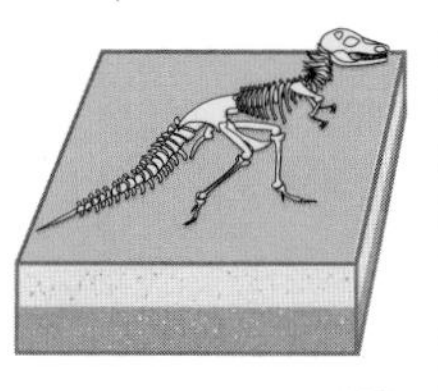

Rock surrounding the mineralized bone rises to Earth's surface. The rock is worn away and the fossil is exposed.

Types of Fossils

A **true form fossil** is formed from actual plant or animal parts. Usually, minerals replace plant or animal tissue. Dinosaur bones may look like real bones, but they are minerals. Petrified wood looks like trunks and branches of ancient trees, but the actual wood has been replaced by quartz. Some preserved animal parts may not become minerals. Insects may be trapped in sticky tree sap. Hardened tree resin is called **amber**. Animals and plants may be preserved in sticky pools of asphalt. Organisms may be buried in frozen soil or ice.

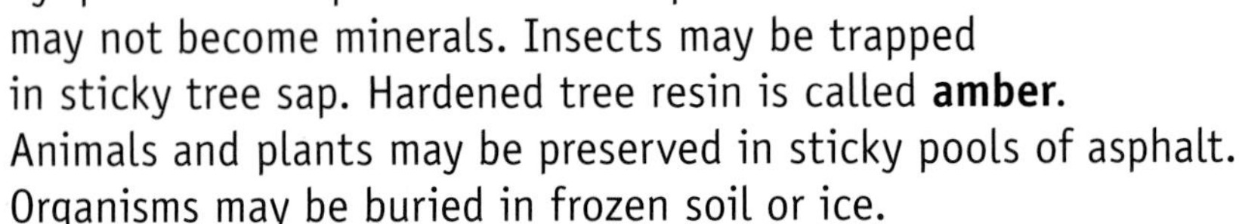

A **mold fossil** is made when the remains of an animal or plant leave an impression in soft sediment. The mark is preserved by pressure and heat. A **cast fossil** forms when sediment fills in the impression made by a mold.

This mold fossil shows the shape and texture of an ancient sea creature called an ammonite (AH muh nite). Ammonites lived millions of years ago. This fossil is molded in sandstone.

Is That a Fact?

Petrified Poo

To find out what was in the dinosaur diet, scientists study a kind of fossil called a coprolite. A **coprolite** (KAH pruh lite) is a trace fossil formed from feces, or dung. Coprolites as large as 40 cm (16 in.) in diameter have been found. Scientists believe these mineral piles of poo were left by large dinosaurs called **sauropods** (SOR uh podz).

3.5 BYA
Single-celled life appears

15 m

Trace fossils are proof of animal life in the past. Creatures have left behind burrows, footprints, and waste products. We can learn a lot about how ancient animals lived and what they ate by studying trace fossils.

Earth and Its Landforms

Models

This trace fossil is a footprint left by a dinosaur. It was found at what is now Dinosaur Valley State Park in Texas.

CHECK YOUR ANSWERS ON PAGES 408–419!

MAKE THE CONNECTION

YOUR OWN STATE'S FOSSIL!

States have their own birds, trees, flowers, and even insects. Did you know that almost every state has its own state fossil?

1. What is the official fossil for your state? What kind of organism created your state's fossil? If you state does not have an official fossil, investigate one of the fossils found in your state.
2. How long ago did the organism live?
3. What were the environmental conditions in the area when the organism was alive?
4. How have the conditions changed?
5. What kind of fossil is your state's fossil? A true form fossil? A mold or cast? A trace fossil?
6. Draw a picture of the fossil.

Finding Fossils

The winged griffin was a mythical creature of ancient Greek and Roman lore.

Ancient travelers told tales of strange bones found in the Gobi desert in Asia. People believed that the bones belonged to winged creatures called griffins. Now scientists believe those bones may have been fossils.

Today, scientists called **paleontologists** (pay lee uhn TALL uh jihsts) look for fossils. They look in sedimentary rock. They look at sites where fossils were found in the past. Scientists may also search for fossils in rocks that are as old as the fossils they want to find. A scientist looking for fossils of dinosaurs would look at rocks that formed during the time dinosaurs lived and died.

Paleontologists use string to mark off the area where they dig up fossils. It's part of good recordkeeping.

Mary Anning (1799–1847)

Mary Anning was born in 1799 in Lyme Regis in England. When she was about ten years old, she and her brother discovered a strange fossil. It belonged to an ancient sea creature called an **ichthyosaur** (ICK thee uh sore). It was the first fossil of its kind found in England. Mary grew up to be a famous and respected fossil hunter.

20 m 3 BYA

2.5 BYA 25 m

Using Fossils To Find the Age of Rocks

Fossils are found in layers of sedimentary rock. Older rock is in the bottom layers. Younger rock is in the upper layers. Scientists use **index fossils** to help find out the age of rock layers. They can also use index fossils to find the age of other fossils.

Four things must be true for a fossil to be an index fossil. The fossil must be well preserved and easy to recognize. It must be found over a wide area. The species that created the fossil must have lived during a short period of geologic time. The fossil must be plentiful.

Index fossils are used to find the **relative age** of rock layers. If scientists find an index fossil in a layer of rock, they know that rocks or fossils in the layer above are younger than the rocks or fossils in the layer below.

Index fossils are used to find out **absolute age**. Since scientists know the age of an index fossil, they know that the rock layer where it is found is as old as the fossil.

Using Index Fossils To Date Rocks

Index fossils are always found in the same layer of rock. If you find an index fossil in sedimentary rock at two different locations—even miles apart—you will know the rocks are the same age.

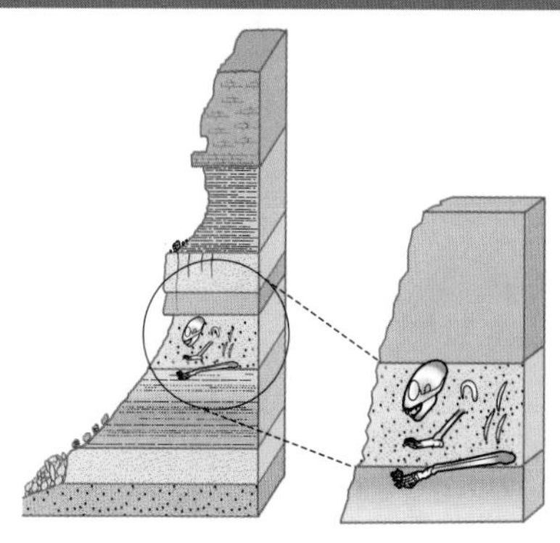

Fossil Evidence of Past Life

Life has existed on Earth for about 3.5 billion years. Earth and its environments have gone through many changes. Mountains have been formed and worn down. Oceans have appeared and disappeared. Land masses have moved. Life forms have appeared, changed, and disappeared. Scientists study fossils to learn about these changes.

Fern fossils look a lot like ferns today. Ferns have been around for more than 360 million years—longer than any flowering plant. Ferns can adapt to survive in both hot and cold environments.

Animals

This fossil may look like a "roly-poly bug," but it is actually a trilobite.

Trilobites (TRY luh bites) were ancient ocean creatures. Scientists know this because their fossils have always been found in areas where oceans once were. When trilobites moved, their tiny legs stirred up bits of food particles for them to eat.

Ecosystems

Petrified wood found in the arid grasslands of Arizona's Painted Desert is about 225 million years old. The tree trunks do not have growth rings. This means the trees grew year-round. There were no seasons. It was always hot, humid, and wet.

Comparing Life in the Past to Life Today

How do scientists know what ancient organisms looked like? They look at fossils. Scientists make models or paint pictures based on the fossils they see. Some animals and plants that lived long ago have descendents alive today. The modern plants and animals look a lot like their prehistoric ancestors.

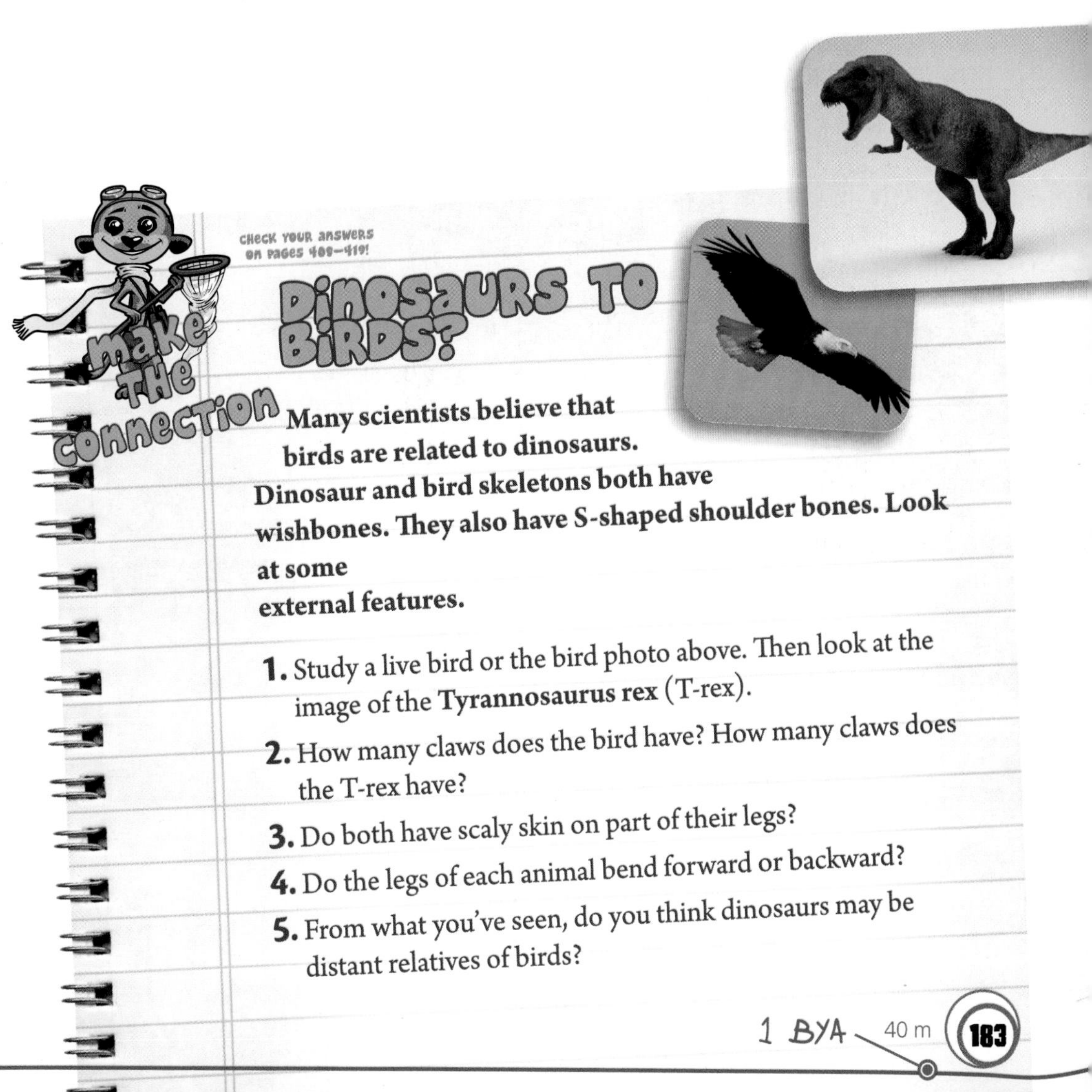

CHECK YOUR ANSWERS ON PAGES 409–419!

Make the Connection

Dinosaurs to Birds?

Many scientists believe that birds are related to dinosaurs. Dinosaur and bird skeletons both have wishbones. They also have S-shaped shoulder bones. Look at some external features.

1. Study a live bird or the bird photo above. Then look at the image of the **Tyrannosaurus rex** (T-rex).
2. How many claws does the bird have? How many claws does the T-rex have?
3. Do both have scaly skin on part of their legs?
4. Do the legs of each animal bend forward or backward?
5. From what you've seen, do you think dinosaurs may be distant relatives of birds?

Fossil Fuels

Are you ready to take a little field trip on a bus? Pack up your camera. Then pack up a fossil or two. Why? Because the gas you need for the bus comes from fossil fuels. **Fossil fuels** were formed from organisms that lived millions of years ago. Coal, petroleum, and natural gas are fossil fuels.

See also: page 166 Minerals and Rocks

Coal is a solid fossil fuel that formed from prehistoric land plants. It is made mostly of carbon. Humans have used coal as fuel for thousands of years.

Coal is a type of sedimentary rock. About 250 million years ago, trees in swamps fell into the water. Bacteria changed the trees into **peat**. The peat was covered by mud and sediment. Heat and pressure finally turned the peat into coal.

Petroleum was formed by countless tiny organisms that lived and died in ancient seas. About 84 percent of petroleum is carbon.

Natural gas also was formed from the remains of small marine organisms. Natural gas is mainly **methane**. Methane is a gas made up of carbon and hydrogen. Petroleum and natural gas are often found together.

0.5 BYA
Land plants appear

45 m

300–400 million years ago

Petroleum began forming even before dinosaurs existed. Tiny sea creatures died and fell to the ocean floor. Layers of sediment covered them.

50–100 million years ago

The remains were buried deeper. Heat and pressure changed the carbon in the sea creatures into crude oil and natural gas.

Alternative Energy

This oil rig drills deep into sedimentary rock to find oil and gas. Sediment that formed the rock covered dead sea creatures over 300 million years ago.

Fossil fuels are **nonrenewable resources**. They will eventually run out. Scientists and industries are working to help us use less fossil fuel. Cars and appliances now use less energy than they have in the past. The coal industry is working on ways to make coal burn cleaner and produce fewer pollutants.

Scientists and engineers are working on ways to use more **renewable resources**, such as wind and solar power.

How Long Ago?
About how many years passed from the time the first life appeared and the time that the first dinosaurs appeared?

Soil

What Is Soil?

See also: page 68 Organisms

Without soil, few plants would grow. With few plants, animals would struggle to survive. Life on Earth would change greatly. Living things might even die out completely.

Worms contribute to the soil when they decay, and their waste products add organic matter to the soil.

Living things leave waste products behind. When they die, they **decay**, or break down. **Soil** is a mixture of waste and the decayed remains of plants and animals, tiny pieces of weathered rock, and mineral particles. Soil also contains water and air. It take thousands of years to make 2 cm of soil.

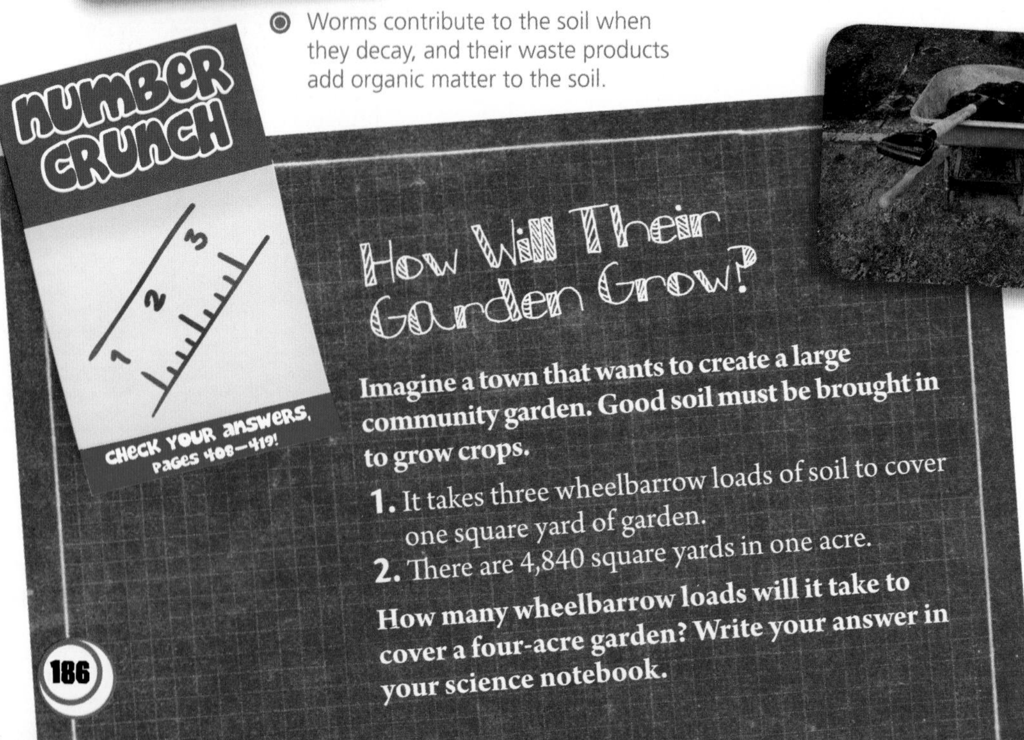

NUMBER CRUNCH

CHECK YOUR ANSWERS, PAGES 408–419!

How Will Their Garden Grow?

Imagine a town that wants to create a large community garden. Good soil must be brought in to grow crops.

1. It takes three wheelbarrow loads of soil to cover one square yard of garden.
2. There are 4,840 square yards in one acre.

How many wheelbarrow loads will it take to cover a four-acre garden? Write your answer in your science notebook.

Layers of Soil

Soil often forms layers, called **horizons**. The top layers are usually full of organic material. The middle and bottom layers have a lot of minerals. The appearance and texture of soil is different in each layer.

Minerals and Rocks

Layers of Soil

The O-horizon is mostly leaf litter and humus. **Humus** is made up of dead plants and animals, and decayed animal wastes.

The A-horizon is also called **topsoil**. It is a mix of humus and weathered rock particles.

The E-horizon does not have many nutrients. As water drips through soil, nutrients are washed to the layer below.

The B-horizon collects dissolved minerals and fine clay from the layers above it. This horizon is also called **subsoil**.

The C-horizon is mostly broken-up rock.

The R-horizon has no soil. It is made up of bedrock. **Bedrock** is solid rock that is not broken or weathered.

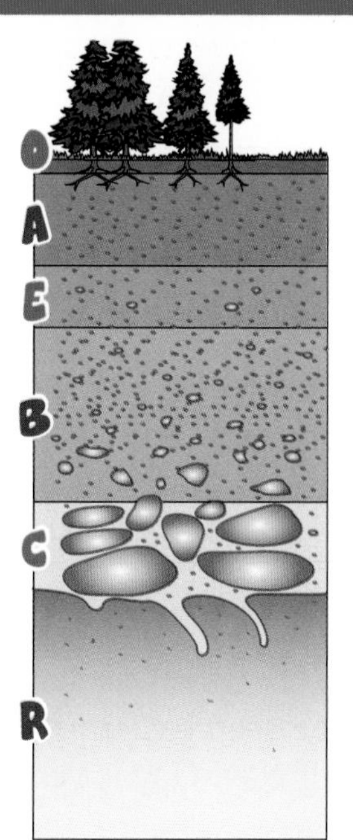

Which Horizon Is Which?

FIND ANSWERS ON 408–419

Put these soil horizons in order from top to bottom. Then name one important feature of each horizon. Write your answers in your science notebook.

Types of Soils

Different places on Earth have different kinds of soil. There are thousands of types of soils in the United States alone!

Soils are often grouped based on the size of the particles in the soil. If a particle of sand was the size of a basketball, then a particle of silt would be the size of a baseball and a particle of clay would be the size of a golf ball.

Sand Particles of sand are very large. Water runs right through these particles, so sandy soil is very dry. Sandy soil does not have much organic matter. It is made up mostly of weathered rock and minerals.

Silt Particles of silt are smaller than those of sand and larger than those of clay. Silt is made up of small organic particles and minerals. Silt feels smooth and powdery when you rub it between your fingers.

Clay Clay is made up of pieces of rock so small that they cannot be seen easily. There are not many air spaces between particles of clay. Clay is very heavy and sticks together when it gets wet.

Did You Know?

The Crowded Shovel

How many people live on Earth? Right now, there are almost 6.8 billion people! Yet in just one shovelful of rich soil, there are more organisms than there are humans on our planet. These organisms include animals we can see, like worms and grubs. There are also billions of living things that are too small to see without a microscope.

Properties of Soil

In most places, particles of different sizes are mixed together to make up soil. Different kinds of soils have different properties. Texture and structure are two important properties of soil.

Soil texture is how the soil feels. Texture depends on the proportion of each kind of soil particle found in the soil. For example, soil with a lot of sand in it feels gritty and rough. Soil with a lot of silt in it feels soft and smooth. Soil with a lot of clay feels heavy, smooth, and sticky.

The best soil texture for growing most crops is loam. **Loam** is made up of about equal parts of humus, sand, silt, and clay. Loam holds water and nutrients well, and also allows water to circulate.

Soil structure is the shape soil takes. Soil structure affects how water will drain and how much air will get into the soil. Soil structure is described as granular, blocky, platy (PLAY tee), columnar (CALL um nar), prismatic, single-grained, or massive.

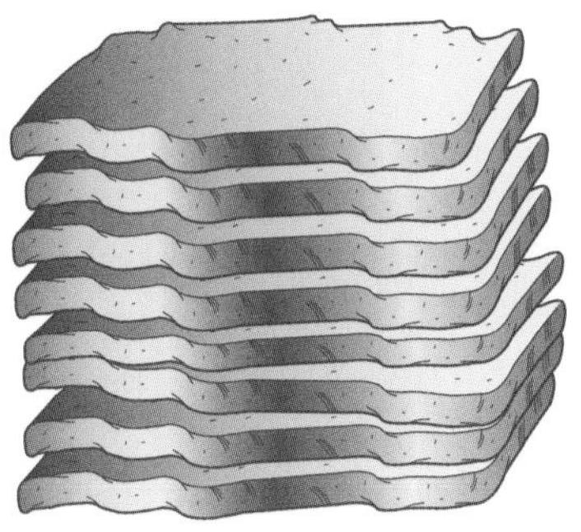

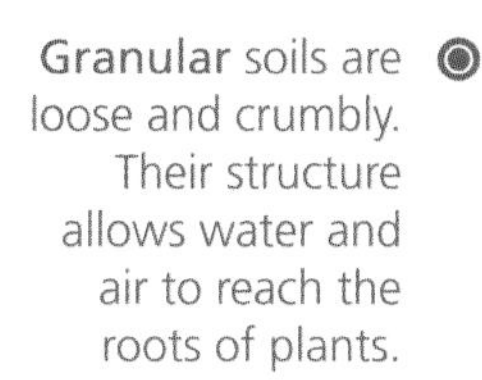

Granular soils are loose and crumbly. Their structure allows water and air to reach the roots of plants.

Platy soils look like stacked dinner plates. Platy soils do not allow air and water to flow freely.

Single-grained soils are loose. The particles do not stick together. Sandy soils are often single-grained.

Soil and Nutrients

Minerals and Rocks

Soil that is able to hold nutrients and make them available to plants is **fertile**. Soil with a lot of organic matter is rich. Rich soil has nutrients that plants need. Soil with very little organic matter is poor. Plants do not grow well in poor soil. Soil does not just need organic matter to be fertile. Plants also need minerals, water, air, and a good climate.

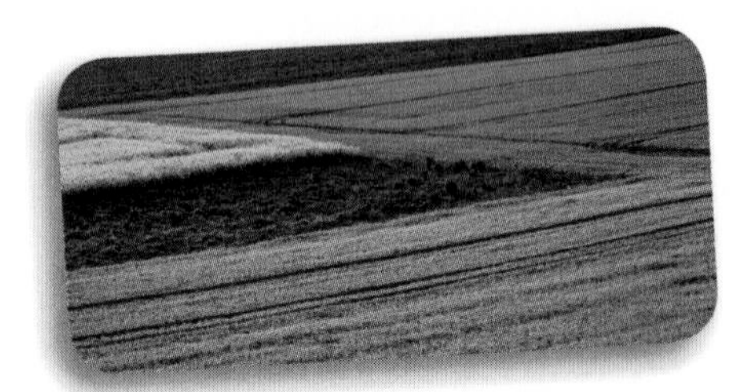

Grassland Grassland soil has a lot of humus. It holds water well. Rain does not wash minerals and other nutrients out of the soil. Grassland soil is good farmland.

Plants

Rainforest Rainforest soil is not very fertile, even though it is full of organic material. Heavy rains wash minerals and other nutrients from the topsoil into the subsoil.

Desert Few plants survive in the desert. Desert soils have very little decayed organic matter. Minerals are not washed out of the soil because deserts do not get much rain.

Arctic Arctic soil has very little humus. Organic matter breaks down very slowly in the cold climate. Arctic areas have very little rain or snow, but water does not drain easily. The soil can be soggy.

See also: page 228 Weather, Seasons, and Climate

Mountain Mountain soil is made up of rock, sand, and clay. Mountain soil often has no layer of topsoil because rains wash organic material and lighter soil downhill.

SOAKING IT UP

Compare how well different soils hold water. You will need a soil sample from near your home or school, a sample of potting soil from a garden center, a magnifying glass, a watering can, two 30 cm × 30 cm screens, and newspaper.

1. Use a magnifying glass to examine each sample of soil. What does each sample look like? Describe the color, texture, and contents in your science notebook.
2. Rub a small part of each soil sample between your fingers. What does it feel like? Squeeze each type of soil in your hand. What is the texture of the soil? Describe how each soil feels and the soil shape in your science notebook.
3. Cover each screen with a different type of soil. Make sure the soil is at least 5 cm deep. Place the screens on newspaper. Using a watering can, slowly pour two cups of water over each sample. Describe what happens.
4. Which sample held water best? How might this information be important?

People Use Soil

See also: page 136 Ecosystems

Farmers and gardeners depend on soil to raise crops. The kinds of crops that grow in an area depend on the climate and the soil.

Clay soil is not easy to farm, but it can be molded easily. Clay has been used to make pottery and ceramics for thousands of years.

Builders used bricks at least ten thousand years ago. The most common soil used for bricks is clay.

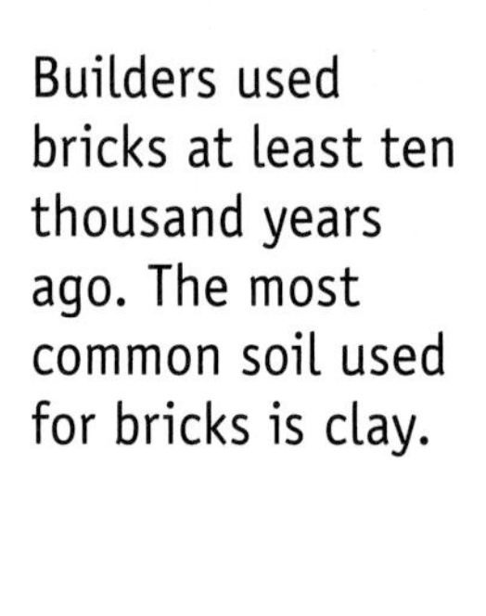

Is That a Fact?

Calling Dr. Soil Critter

Did you know that many **antibiotics** are developed from organisms that live in soil? Antibiotics slow the growth of bacteria. Many microscopic organisms produce antibiotics to fight the harmful bacteria that live in the soil with them. Penicillin was developed from a soil fungus. Tetracyclines were developed from a soil-dwelling bacterium.

Soils can come in many shades of yellow, red, and brown. Some of these soils are used as natural dyes.

Women in Mali use a traditional process to make bogolanfini (bo ho lahn FEE nee), or "mud cloth."

Plants Use Soil

Plants obtain moisture and minerals from the soil through their roots. Water seeps down through the soil horizons. Minerals in the soil are dissolved in water. Water is absorbed into the roots and then transported throughout the plant.

Have you ever seen a plant growing up through the crack in a sidewalk? If you could see down through that crack, you would find soil underneath. Plants depend on soil to grow. Soil provides the minerals plants need to survive. Minerals are important in building cell walls, cell growth, and reproduction.

If soil is damaged or lost, plants are not able to grow. Some poor farming practices can damage soil. Topsoil can wash away. Soil can lose nutrients. When plants no longer grow in soil, the soil loses a lot of moisture. With no plants and no moisture, fertile land can become a desert.

Plant roots protect soil from erosion. When plants and trees are removed, soil is exposed to wind and rain. Forest fires, hurricanes, and other natural disasters can remove a lot of vegetation and cause soil erosion. Drought can dry out topsoil, and good farmland can blow away.

Animals and Other Organisms Use Soil

A startled chipmunk can disappear into a hole in the ground. The hole is part of a network of tunnels that the chipmunk dug in the soil. Chipmunk tunnels can be 9–12 m (30–40 ft) deep. Chipmunks use chambers off the tunnels to store food, raise their young, and sleep. Soil provides underground homes to other mammals too. Rabbits, groundhogs, meerkats, and prairie dogs all make their homes in soil.

Most chipmunks dig their burrows in areas where there are a lot of trees. Much of the chipmunk diet includes nuts, seeds, mushrooms, berries, and insects found in leaf litter.

Snakes don't dig holes, but they often lay eggs in holes dug by other animals. Some snakes, frogs, and toads hibernate in the soil during the winter. Because they are cold-blooded animals, the soil helps keep their bodies at a safe temperature.

Some birds make burrows, too! Puffins nest underground. They dig out their homes with their bills and feet.

Some organisms live most of their lives in soil. These include worms, ants, and microscopic protists and bacteria. Soil-dwelling worms and insects provide food for birds and other larger creatures.

IT'S ALIVE!

Remember!

- Wash your hands after handling the soil sample and the organisms you find.
- Make sure the organisms in the funnel do not escape inside your home.
- Keep the lightbulb at least 15 cm (6 in) from the soil sample.
- Release the organisms into soil outdoors after you examine them.
- Dispose of the soil sample outdoors after you finish the activity.
- Safely dispose of the materials you used to make the Berlese funnel.
- Recycle both halves of the soda bottle.

You can separate soil organisms from a sample of soil with a Berlese (bur LAY zee) funnel.

1. Cut a 2-L soda bottle in half. Put a paper coffee filter or a folded paper towel into the bottom half. Turn the top half of the bottle upside down into the bottom half to form a funnel.
2. Place a piece of hardware screen over the hole in the funnel. Put about one cup of soil and litter carefully onto the screen. Try not to let the soil fall onto the filter or paper towel.
3. Place a gooseneck reading lamp above the funnel so that the light warms the soil. Be careful! The lamp will need only a 25-watt or 40-watt bulb. Place the bulb at least 15 cm from the soil sample. Soil organisms will crawl out of the soil and fall onto the coffee filter or paper towel.
4. Carefully separate the top half of the funnel from the bottom half. Remove the filter or paper towel. Use a magnifying glass to examine the creatures you find. Write your observations in your science notebook.

It may take three or four days for most of the organisms to crawl out. Check for organisms each day. After you have examined the soil dwellers, release them into soil outside.

The Solar System

What Is Our Place in Space?

Take a look up at the night sky. What do you see? If the sky is clear, you may see stars, the Moon, and hazy spots that are more difficult to identify. All you see and so much more make up what we call the universe.

The **universe** contains everything there is. This includes all the planets, stars, moons, and asteroids. It also includes space dust, comets, gas clouds, and vast areas of empty space.

All the matter in the universe is collected together into enormous systems called **galaxies**. Galaxies are held together by gravity. Scientists estimate there are about 125 billion galaxies in the universe.

We live in a spiral galaxy that we call the Milky Way. A spiral galaxy looks something like a pinwheel. This image of Spiral Galaxy M74 was taken by the Hubble telescope.

Our galaxy is about 100,000 light years across and 1,000 light years thick. There are about 400 billion stars in it. A light year is the distance light travels in one year. This distance is just about six trillion miles!

Is That a Fact?

Twinkle, Twinkle, Little Star…

What causes the stars to twinkle? Earth has a thick, turbulent atmosphere. Light rays that enter the atmosphere bend many times in different directions. This makes it look as though the stars are twinkling.

Parts of Our Solar System

A **solar system** is a group of planets and other objects orbiting a star. Our star is called the Sun. There are eight planets in our Solar System. There are also several dwarf planets and many asteroids, meteoroids, comets, and other objects.

The planets in our Solar System orbit (revolve around) the Sun in oval-shaped paths. These planets are, in order from their distance to the Sun, Mercury, Venus, Earth, Mars, Jupiter, Saturn, Uranus, and Neptune. Many of these planets have moons that also orbit around them.

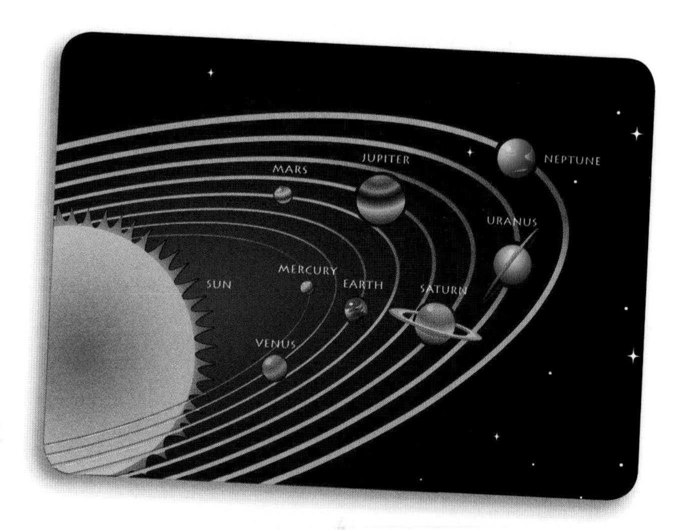

An **asteroid** is a chunk of rock that orbits the Sun. Asteroids range in diameter from 30 m (100 ft) to 966 km (600 mi)! Most asteroids are found in the asteroid belt. This is an area between Mars and Jupiter.

Objects too small to be asteroids are called **meteoroids**. These objects often enter Earth's atmosphere. The atmosphere is so hot that the meteoroids burn up in a bright flash of light. We call this flash a shooting star. Any part of the meteoroid that does not burn up falls to Earth as a **meteor.** If a meteor makes it to the Earth's surface, it is called a **meteorite.**

A **comet** is a chunk of dust and ice that travels into our Solar System from another part of the galaxy. When a comet gets close to the Sun, it starts to evaporate into water vapor. This water vapor forms the familiar long, glowing tail of the comet.

Inner Planets of Our Solar System

The four planets closest to our Sun are called the **inner planets.** They are grouped together because they have similar characteristics. These are rocky planets composed of minerals and metals. They have few moons, or none at all. Three of the four planets have a well-defined atmosphere.

Mercury

Mercury is the smallest planet in our Solar System. It is also the closest planet to the Sun. This makes Mercury hard to see in the night sky. The planet shares its name with the Roman messenger god.

Venus

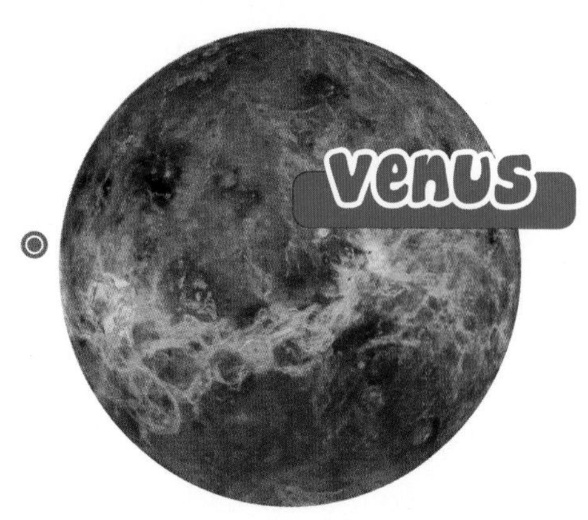

Venus is the second planet from the Sun. It is sometimes called Earth's twin because it is about the same size as Earth. Venus got its name from the Roman goddess of beauty.

Go Online

To learn more about the Solar System, check out these Web sites!

- **Solar System Exploration**
  http://solarsystem.nasa.gov/planets/index.cfm
- **The Solar System**
  http://www.bobthealien.co.uk/solar.htm

Earth is the third planet from the Sun. It is the only planet we know of that supports life. This is because oxygen and water are abundant on Earth. Our planet has one moon, and to keep things simple, we just call it "the Moon"!

See also: page 208 Space Exploration

Mars is the fourth planet from the Sun. We call it the "Red Planet" because it looks red from Earth. There have been several missions to Mars. Scientists want to learn whether life once existed there. Mars is named for the Roman god of war.

The Inner Planets: Fast Facts

Planet	Number of known moons	Average distance from the Sun (in km)	Average equatorial radius (in km)	Length of year (relative to a year on Earth)
Mercury	0	57.9 million	2,439.7	0.241
Venus	0	108.2 million	6,051.8	0.615
Earth	1	149.6 million	6,378.14	1
Mars	2	228 million	3,397	1.881

Outer Planets of Our Solar System

The four planets farthest from our Sun are called the **outer planets.** These planets are very large and are composed mainly of gases such as hydrogen and helium. Unlike the inner planets, each of these **gas giants** has many moons.

Jupiter is the fifth planet from the Sun and the largest planet in our Solar System. Jupiter is named for the Roman god of thunder.

SATURN

Saturn is the sixth planet from the Sun. It is best known for its many rings, which are composed of ice crystals. Saturn is named for the Roman god of agriculture.

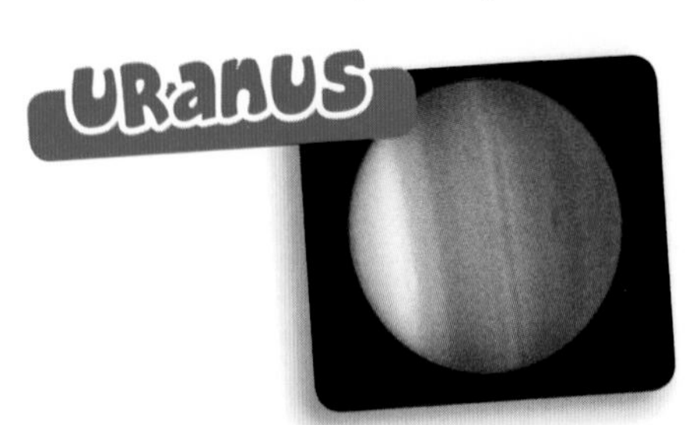

Uranus is the seventh planet from the Sun. Little is known about Uranus. One thing we do know is that it was named for Ouranos, the ancient Greek god of the sky.

Is That a Fact?

What Happened to Pluto?

Illinois astronomer Clyde Tombaugh discovered Pluto in 1930, and it became the ninth planet in the Solar System. In 2006, scientists reclassified Pluto as a **dwarf planet**.

Neptune

Neptune is the eighth planet in our Solar System and the farthest planet from the Sun. Neptune was named after the Roman god of the sea.

Space Exploration

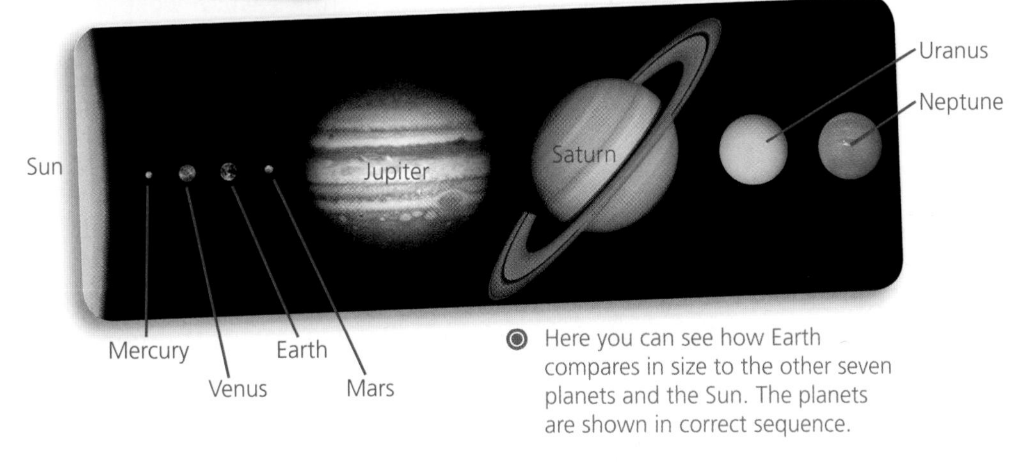

Here you can see how Earth compares in size to the other seven planets and the Sun. The planets are shown in correct sequence.

The Outer Planets: Fast Facts

Planet	Number of known moons	Average distance from the Sun (in km)	Average equatorial radius (in km)	Length of year (relative to a year on Earth)
Jupiter	62	778.4 million	71,492	11.857
Saturn	61	1.4 billion	60,268	29.4
Uranus	27	2.9 billion	25,559	84.02
Neptune	13	4.5 billion	24,764	164.79

Day and Night

Earth rotates (spins completely around) on its axis in one day. The **axis** is the imaginary line around which the planet spins. This line runs through both the North and South Poles.

During the day, the Sun appears to travel from east to west across the sky. This might make you think that the Sun orbits Earth. In fact, the opposite is true. Earth orbits the Sun, but our planet is so large that we can't feel or see its movement.

At night, the Moon and stars also seem to travel from east to west across the sky. The Moon doesn't seem to travel as fast as the stars, however. This is because the Moon orbits the Earth in the other direction. But because it orbits slower than the Earth rotates, the Moon still appears to go east to west. If it orbited at the same speed that the Earth rotated, it would seem to hold still in the sky!

This time-lapse photo shows how other stars seem to circle the star called Polaris, the "North Star." Earth's axis points directly toward Polaris. As a result, Earth "moves" with Polaris, and so the star appears not to move in the sky like other stars.

Seasonal Changes

Earth makes a complete orbit around the Sun in one year. Earth is tilted on its axis as it orbits the Sun. Because of this, some parts of Earth are closer to the Sun at different times throughout the year. This is what causes the change of **seasons.**

See also: page 228 Weather, Seasons, and Climate

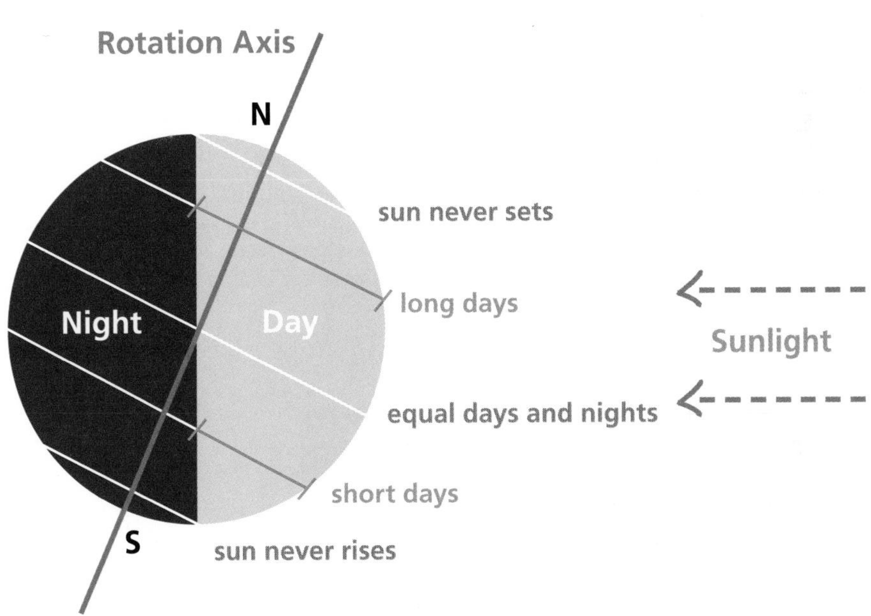

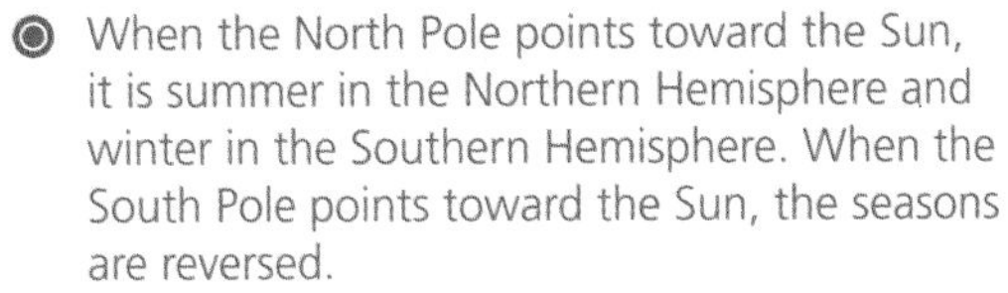

When the North Pole points toward the Sun, it is summer in the Northern Hemisphere and winter in the Southern Hemisphere. When the South Pole points toward the Sun, the seasons are reversed.

Earth's Moon

You can see the Moon when the night sky is clear, and sometimes during the day as well. It is about one-quarter the size of Earth and has a radius of 1,737 km (1079 mi). The Moon is roughly 384,000 km (238,607 mi) from Earth.

Unlike Earth, the Moon has no atmosphere. Because it is smaller than Earth, it has less gravity. This means that an object on the Moon would weigh about one-sixth as much as on Earth. For example, a person weighing 50 kg (110 lb) on Earth would only weigh about 8 kg (18 lb) on the Moon!

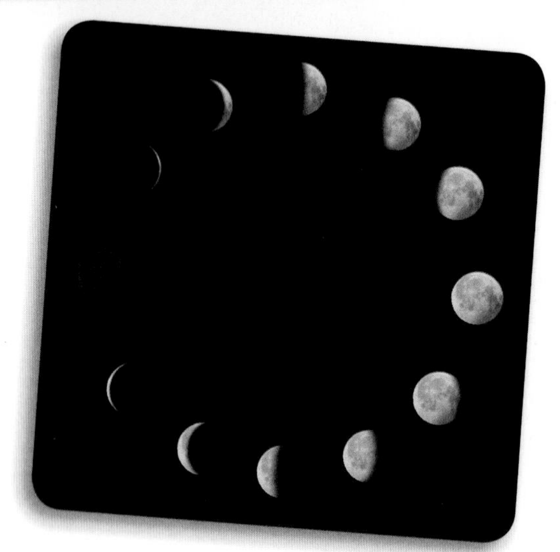

The phases of the Moon occur throughout the 27.3-day cycle of the Moon's orbit. As the Moon moves around Earth, we see different amounts of the Sun's light reflected off the Moon's surface.

Earth's Sun

The Sun is one of billions of stars in the Milky Way Galaxy. The Sun is the center of our Solar System. Without the Sun's energy, life would not exist on Earth. The Sun's radius is about 696,000 km (432,474 mi). More than one million Earths would fit inside the Sun. Would you believe that the Sun is actually a medium-sized star?

The temperature at the Sun's core, or center, is 15.7 million °C (28.3 million °F). The heat and pressure in the center is so intense that nuclear reactions take place there. The outermost layer of the Sun is called the corona. It can stretch millions of miles out into space.

See also: page 302 Light

CHECK YOUR ANSWERS ON PAGES 408–419!

MY ACTIVE SHADOW

See how the movement of Earth around the Sun affects your shadow.

1. At noon on a sunny day, go outside and look for your shadow. Stand still and put a rock on the ground at your feet.
2. Have a friend mark the top of your shadow with another rock. Measure the distance between the two rocks. This is the length of your shadow.
3. Predict the length of your shadow in 1, 2, and 3 hours. How will those lengths relate to the original length? How will they relate to each other?
4. Go outside in 1, 2, and 3 hours. Stand in the same spot as you did at noon. Have a friend mark the top of your shadow each time. Check whether your predictions were correct.
5. Would your shadow change in a different way if you do this activity in the morning? Why do you think so? Test your idea.

"STAR LIGHT, STAR BRIGHT" —HEY GUYS, WHAT MAKES A STAR BRIGHT?
STARS ARE MADE UP OF GASES THAT ARE SO HOT THAT THEY GLOW.
LOOK! THERE'S THE BIG DIPPER IN THE CONSTELLATION URSA MAJOR.
THE ANCIENT GREEKS GROUPED STARS INTO CONSTELLATIONS. MANY ARE NAMED FOR ANIMALS.
DID YOU KNOW THAT THE SUN IS A STAR, AND THAT IT'S THE ONLY ONE IN OUR SOLAR SYSTEM?

SURE! LIKE OTHER STARS, THE SUN IS EARTH'S SOURCE OF LIGHT ENERGY.

THERE ARE MANY MORE STARS IN OUR GALAXY, THE MILKY WAY. AND OUR GALAXY IS ONLY ONE OF MANY GALAXIES IN THE UNIVERSE!

MAKES YOU THINK, DOESN'T IT?

Space Exploration

Why Do We Study Space?

People have always wondered about things they couldn't see or didn't understand. Early humans imagined sea serpents lived in the ocean depths. Others thought that monsters lived in swamps that were too difficult to navigate. It is no surprise that the Sun and stars, so far away, captured the imagination of ancient peoples.

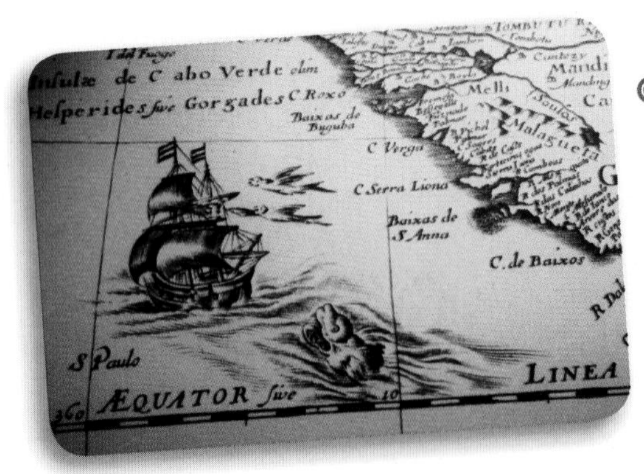

A sea monster approaches a sailing ship in this map, drawn around 1600. Brave explorers, curious about the unknown, overcame their fears and sailed to uncharted places. Their achievements helped to shape our world.

People today still explore mountains, oceans, and poles. Like early humans, we also want to know what lies in space. Some people hope to find new life forms. Others hope to discover new minerals. Some believe we might even find a planet similar to Earth. If we do, someday humans might move there!

Space has been called the "final frontier." It is vast. Most likely humans will never learn all there is to know about space. Yet we still want to explore. As we do, we learn more about our universe and how it is changing.

History Makers

Galileo Galilei (1564–1642)

Galileo Galilei was born in Pisa, Italy. As a young man he studied math and science. He improved the early telescope, and used it to examine the Moon and to discover sunspots. During Galileo's lifetime, most people believed the sun orbited the Earth. Galileo's discoveries helped prove that the planets orbit the Sun.

See also: page 196 The Solar System

In 1610, Galileo discovered the four largest satellites of Jupiter. In his honor, we call these the "Galilean moons." From top to bottom, this composite image shows Io, Europa, Ganymede and Callisto next to Jupiter.

The Andromeda Galaxy is the galaxy nearest our own, the Milky Way. It is slightly larger than the Milky Way. Like ours, it is a spiral galaxy. Powerful telescopes help us learn about our galaxy and others.

Early Space Exploration

The desire to explore space began after World War II. Both the United States and the Soviet Union wanted to make the most advancements in space exploration. The Soviet Union was the first to launch a manmade satellite into orbit. This satellite, known as ***Sputnik 1***, was launched in 1957. Not to be outdone, the United States launched its own satellite, named *Explorer 1*, in 1958. This was the beginning of the era of space flight, nicknamed the "Space Race."

This 1972 postage stamp from the Soviet Union commemorates the flight of *Sputnik 1*.

Manned space flight was soon to follow. In 1961, President John F. Kennedy set a landmark goal for the United States: An American would set foot on the Moon by the end of the decade.

In April 1961, Yuri Gagarin of the Soviet Union became the first person to orbit the Earth. In May of that year, Alan B. Shepard became the first American astronaut to travel into space. As technology progressed, astronauts were able to stay in orbit for a long time. In 1963, Gordon Cooper orbited the Earth a record 22 times.

Is That a Fact?

Ham in Space

Did you know that Ham went into space before people? It's true! Ham was a chimpanzee belonging to the United States Air Force, and he flew into space before any human astronauts. His 16-minute sub-orbital flight took place on January 31, 1961. His rocket reached an altitude of 253 km (157 mi)!

Reaching the Moon

NASA's Apollo 11 was the first mission to reach the Moon, on July 20, 1969. Astronaut Neil Armstrong became the first man in history to set foot on the Moon. Armstrong and fellow astronaut Buzz Aldrin explored the Moon's surface near their lunar module, the Eagle. They collected moon rocks and planted an American flag on the Moon's surface. An estimated 500 million people around the world watched the Moon landing live on television.

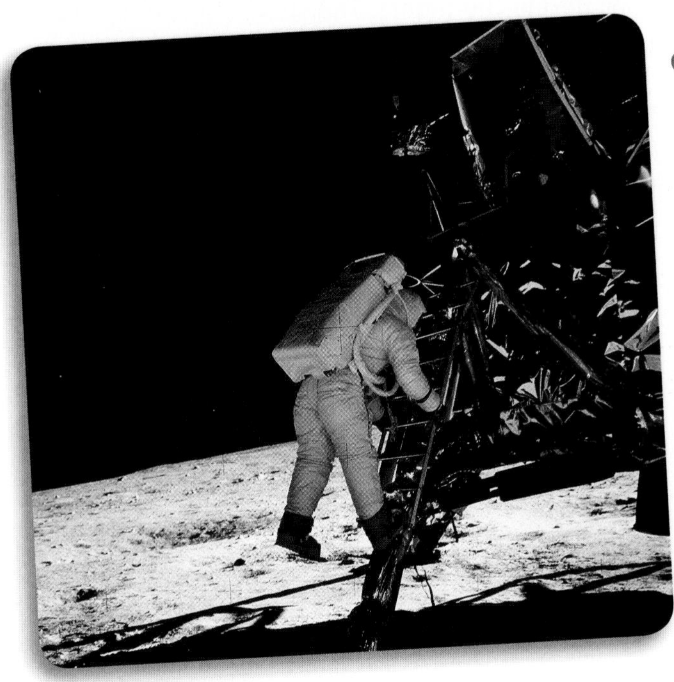

Buzz Aldrin climbs down from the *Eagle* to the Moon's surface during the Apollo 11 space flight.

GO ONLINE

To learn more about space exploration, check out these Web sites!

- **Space Week: Satellite of Solitude**
  http://www.cosmosmagazine.com/features/print/545/satellite-solitude

- **Space Week: Why the Moon Rocks**
  http://www.cosmosmagazine.com/features/online/2858/why-moon-rocks

Space Shuttle

In the early 1980s, engineers unveiled a new way to explore space. This was the space shuttle, the world's first reusable spacecraft. The space shuttle launches like a rocket, with powerful engines and boosters. However, space shuttles return to Earth in a much different way than earlier missions had returned. Earlier astronauts "splashed down" in the ocean in a small command module. It was the only remaining portion of their rocket (the rest of their rocket had detached in stages during the flight). The space shuttle returns through the atmosphere and lands on a runway like an airplane.

The Space Shuttle *Atlantis* landed in California on May 24, 2009.

On August 7, 1971, the Apollo 15 command module splashed down in the Pacific Ocean. Parachutes slowed its fall through the atmosphere.

The space shuttle typically carries a crew of seven astronauts. The flight deck and crew cabin are located in the forward part of the shuttle. Behind the shuttle's nose is the huge payload bay, where the shuttle carries its cargo.

International Space Station

Space exploration took a big step forward in 1998, when construction began on the International Space Station (ISS). Sixteen nations have collaborated on the project. The purpose of the ISS is to enable long-term exploration of space.

See also: page 196
The Solar System

The ISS typically houses a crew of six people. Crew members travel to the orbiting station in the space shuttle. The shuttle then docks with the ISS, and astronauts move from one craft to the other. There are 20 research units on the station, where the crew performs experiments in biology, physics, and astronomy. They also keep track of their own health. This information will help doctors learn about long-term effects of weightlessness on humans. Such information will be useful if we take future trips to Mars and beyond.

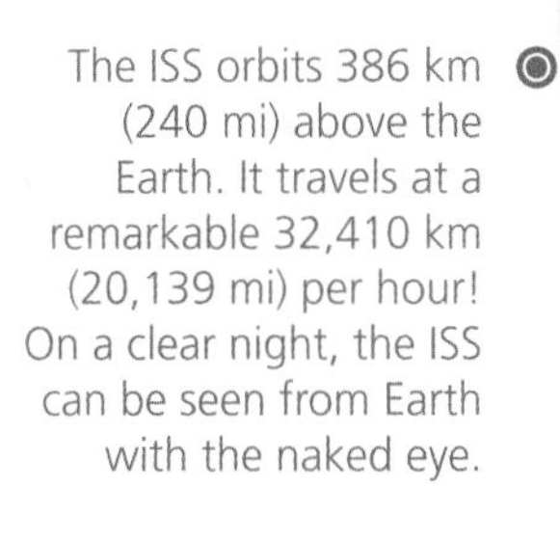

The ISS orbits 386 km (240 mi) above the Earth. It travels at a remarkable 32,410 km (20,139 mi) per hour! On a clear night, the ISS can be seen from Earth with the naked eye.

Astronauts work on the space shuttle while floating in a weightless environment.

How We Explore Space

We explore space from the Earth and from space itself. Huge observatory telescopes allow astronomers to see deep into outer space. One of the largest and most powerful telescopes is the Southern Astrophysical Research (SOAR) telescope in Chile. SOAR works by collecting and focusing light from space by using large mirrors.

The SOAR telescope in Chile.

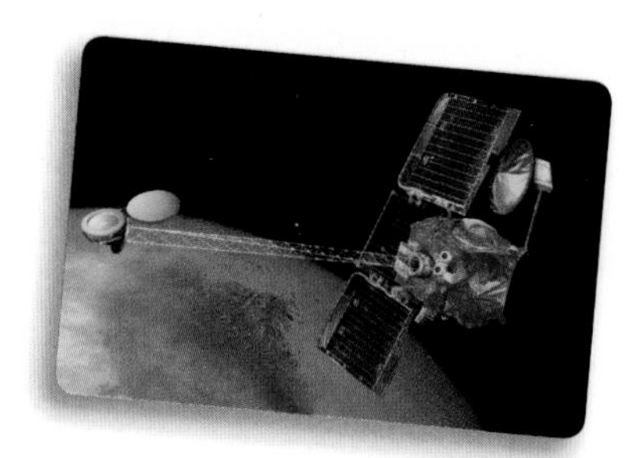

The *Mars Odyssey* in orbit.

We have learned about deeper space with unmanned spacecraft. One of these, the Mars Odyssey, left Earth in April 2001 and reached Mars six months later. The data it collects is sent back to NASA. From this mission, we have learned what materials make up the surface of Mars, including the planet's ice caps. We have also learned about the radiation surrounding Mars.

Voyager 1 and Voyager 2 are two of six unmanned spacecraft that were sent out to gather data about the outer planets. The two Voyager craft left Earth in 1977. They are now beyond our Solar System.

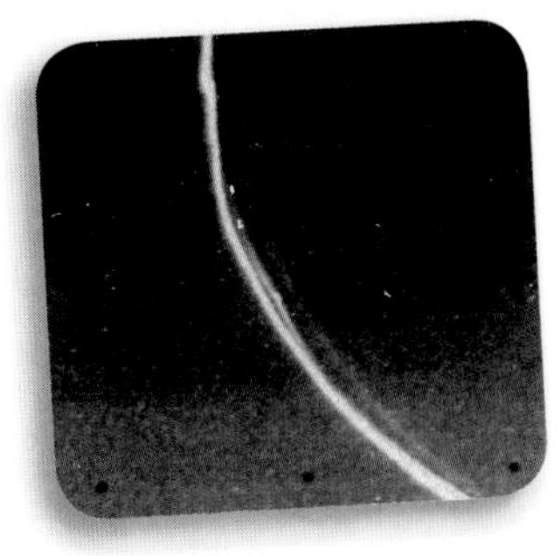

Saturn's braided F Ring, photographed by *Voyager 1* in November 1980.

Space is full of surprises. The more we explore, the more we learn! Voyager 1 and Voyager 2 have made interesting discoveries. One of Saturn's rings, called the "F Ring," is made of several strands that appear to be twisted or braided around each other.

Astronomers once believed all planets orbited their stars in the same direction the stars rotate. In 2009, a planet was found that rotates opposite to the rotation of its host star. It is thought that a near-collision may have caused this backward orbit. The star and planet, known as WASP-17, are about 1,000 light years from Earth.

CHECK YOUR ANSWERS, PAGES 408–419!

How Far Would You Travel?

Earth is a long way from other bodies in our Solar System. Imagine you are a space traveler. You are going to travel from Earth to Mars, and from there on to Jupiter.

- **Earth is 149,597,890 km (92,955,820 mi) from the Sun.**
- **Mars is 227,936,640 km (141,633,260 mi) from the Sun.**
- **Jupiter is 778,412,020 km (483,682,810 mi) from the Sun.**

1. How far will you travel from Earth to Mars, in kilometers? In miles?

2. How far will you travel from Earth to Jupiter, in kilometers? In miles?

Cool Space Stuff

Have you heard the saying "necessity is the mother of invention"? It means that when people need to do something, they will build things to help them do it.

Why do you think astronauts have to stick food items to the wall of the space shuttle?

The desire to explore space was the "mother" of some technologies we use today. Dehydrated foods were improved for astronauts to eat while in space. These foods are light and take up little room on a crowded spacecraft. Flame-retardant fabrics were originally invented for astronauts. Fire is especially dangerous in space, and clothing made of these fabrics do not easily catch fire.

Cool Space Stuff

Usage	Object
Around the House	cordless power tools
	smoke detectors
	water purifiers
Personal Items	athletic shoe insoles
	ear thermometers
	invisible braces
	scratch-resistant coatings for eyeglasses
Technology	lasers
	medical imaging machines

Space exploration has brought about an amazing variety of objects we use every day. Web sites such as Google Earth and Bing Maps use satellite imagery. Photographs taken of the Earth from orbiting satellites are made available on the Internet. This allows people to have a "space-eye view" of sites anywhere on the globe.

Do you know what a GPS receiver is, or have you used one? The Global Positioning System (GPS) is a system of orbiting satellites and land-based monitoring stations. With a GPS receiver you can locate where you are at any given time. The receiver displays a map with your location. GPS is used in navigation, tracking systems, map-making, surveying, and in many other applications.

GPS receivers are also used for fun activities such as geocaching. Geocaching (GEE oh CASH ing) is a worldwide hide-and-seek game in which people hunt for hidden treasures.

GO ONLINE

To learn more about space exploration, check out these Web sites!

- **NASA – Spinoffs That Rock!**
  http://www.nasa.gov/audience/forstudents/k-4/home/spinoffs_feature_k_4.html
- **Space Shuttle Spinoffs**
  http://www.sti.nasa.gov/tto/shuttle.htm

Air

What Is Air?

Take a deep breath. Hold it for a moment. Let it out. What did you inhale when you took that breath? You might answer air.

Air is a mixture of gases that surrounds Earth. Air is made up mainly of nitrogen and oxygen. It also has smaller amounts of argon, carbon dioxide, hydrogen, neon, helium, and other gases. Air, moisture, and small particles of matter such as dust surround Earth in a thick layer called the **atmosphere**.

Life on Earth depends on air. Animals use oxygen to break down food in the body so it can be used as fuel. The fuel gives the body energy to grow, repair cells, keep warm, and perform other tasks. After the body uses energy, it gets rid of wastes in the form of carbon dioxide and water.

Active sports such as swimming or running require much more energy than resting does. When you exercise, you take in more oxygen as the need for energy increases.

When plants make sugar, oxygen is a waste product. Plants release the oxygen into the atmosphere. Animals breathe the oxygen. This is part of the oxygen cycle.

Plants use oxygen and carbon dioxide from the atmosphere. Carbon dioxide and energy from sunlight are used to produce a sugar called **glucose** as food. Oxygen is used to burn the sugar for fuel so the plant can grow, repair cells, and perform other life functions.

Other planets in our solar system have atmospheres, but they are quite different from ours. Human beings cannot breathe the air on those planets.

- The atmosphere of Mars is made up mainly of carbon dioxide. It also has small amounts of nitrogen, argon, and oxygen.

- Mercury lost most of its atmosphere soon after it was formed. Solar winds bring hydrogen and helium, which eventually escape back into space. Other gases in its atmosphere include oxygen, sodium, calcium, and potassium. There is also a very small amount of water.

- Jupiter is the largest planet in our solar system. Its atmosphere is made up of ammonia, methane, helium, hydrogen, sulfur, and water.

The Solar System

What's in the Air?

Earth and Jupiter have very different air. Which gas is in Earth's atmosphere but not in the atmosphere of Jupiter? Write your answer in your science notebook.

HELIUM AMMONIA

OXYGEN HYDROGEN

Layers of Our Atmosphere

Air has mass and weight. **Mass** is the amount of matter in an object. We cannot see air, but air is made up of molecules. Molecules have mass. **Weight** measures the pull of **gravity** on an object. Air seems weightless, but gravity pulls air toward the surface of Earth. About one kilogram of air per square centimeter (almost 15 pounds per square inch) pushes against everything on Earth, all the time. Air pushes in all directions. The force of air pushing on the surface of objects is called **air pressure.** There is less air pressure the higher up you go because there is less air.

Machu Picchu, in the Andes Mountains of Peru, is 2,450 m (7,710 ft) above sea level. Air has less oxygen at high altitudes. People from the Andes have adapted to high elevations. They carry more oxygen in each blood cell than people at sea level do.

Earth's atmosphere is divided into layers.

The **troposphere** (TRO puh sfeer) is close to the surface of Earth. It rises to about 17 km (11 mi) above sea level. Most life on Earth lives in the troposphere. As altitude increases, the air gets colder.

The **stratosphere** (STRA tuh sfeer) is above the troposphere. It goes to 50 km (31 mi) above sea level. Few clouds form there. At the top of the stratosphere the air is about 1,000 times thinner than it is at sea level. In the stratosphere, air gets warmer as the altitude increases.

This plane is flying in the lower stratosphere. You can see clouds in the troposphere.

The **mesosphere** (MEH zoh sfeer) starts at the top of the stratosphere. It rises to about 85 km (53 mi). The air is colder as you go higher through this level. The mesosphere is the coldest layer of the atmosphere.

The **thermosphere** (THER muh sfeer) starts at the top of the mesosphere. It rises to about 640 km (about 400 mi) above sea level. The temperature here is very high.

The **exosphere** (EX oh sfeer) starts in the top of the thermosphere. It goes to about 10,000 km (6,200 mi) above Earth. This layer is farthest from Earth's surface. The thin air merges into outer space.

Layers of the Atmosphere

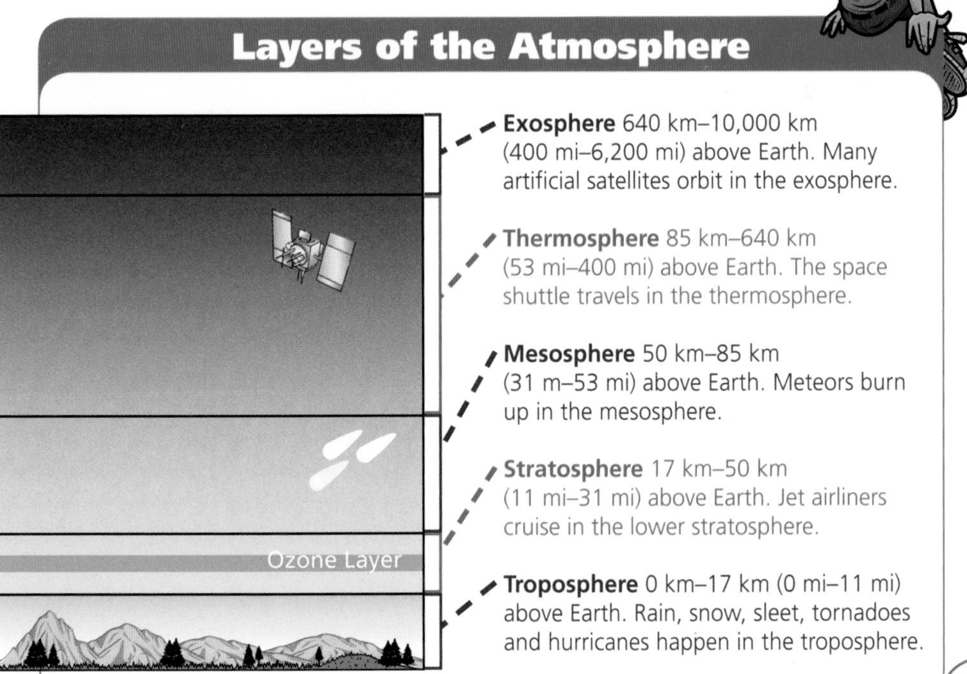

Exosphere 640 km–10,000 km (400 mi–6,200 mi) above Earth. Many artificial satellites orbit in the exosphere.

Thermosphere 85 km–640 km (53 mi–400 mi) above Earth. The space shuttle travels in the thermosphere.

Mesosphere 50 km–85 km (31 m–53 mi) above Earth. Meteors burn up in the mesosphere.

Stratosphere 17 km–50 km (11 mi–31 mi) above Earth. Jet airliners cruise in the lower stratosphere.

Troposphere 0 km–17 km (0 mi–11 mi) above Earth. Rain, snow, sleet, tornadoes and hurricanes happen in the troposphere.

Air Resistance and Falling Objects

Objects that fall through the air are pulled down by gravity. **Air resistance** pushes objects up as gravity pulls them down. As the objects meet air resistance, they slow down.

Air is made up of molecules. As objects fall, they push through all the air molecules that are in the way.

The faster things fall, the more molecules they push, so there is more air resistance.

The part of an object that meets the air molecules is the surface area. The more surface area objects have, the more molecules they push, so there is more air resistance.

Leaves don't usually fall straight down. They drift. A leaf is small and light. A leaf cannot push through air molecules as easily as a coin. The curved shape of the leaf can work like a parachute. The curve makes it hard for air molecules to get out of the way.

If you drop coins, they will usually fall straight down. Coins are small but heavy. They are able to push through air molecules easily.

See also: page 50
Atoms, Molecules, and Elements

Have you ever watched someone skydive from a plane? Sometimes a skydiver opens the parachute within a few seconds of jumping. Air fills up the parachute. There is more air inside the parachute, so there is more air resistance. The air pushes against the parachute and slows the skydiver down.

A meteor is a space rock that falls through Earth's atmosphere. Meteors fall very fast and push through a lot of air molecules. There is a lot of air resistance. Air resistance causes so much friction that meteors burn up.

EXPLORE MORE

CHECK YOUR ANSWERS, PAGES 408–419!

THE INCREDIBLE PARACHUTING EGGS!

Observe how a parachute affects the speed at which an egg falls to the ground.

Materials: 3 large lightweight plastic trash bags, a meter stick or ruler, 12 50 cm (20 in) pieces of string, 3 plastic sandwich bags, 3 raw eggs

1. Cut 3 squares out of the trash bags. Make one square 30 cm × 30 cm (12 in × 12 in), one square 61 cm × 61 cm (24 in × 24 in), and one square 91 cm × 91 cm (36 in × 36 in).
2. Make three parachutes out of the three squares, one at a time. Tie a string to each corner of each square. Then attach the loose ends of the string to the top corners of a plastic sandwich bag.
3. Place one egg in each bag. Seal each bag completely!
4. Predict which of the eggs has the best chance to survive a 3 m (10 ft) drop. Why do you think so? Write your answer in your science notebook.
5. Drop each egg parachute from a height of 3 m. Did all of the eggs survive? Did you make a correct prediction?
6. What forces acted on the parachutes? How was the fall of the large parachute different from the fall of the small parachute? Write your answers in your science notebook.

Air at Work

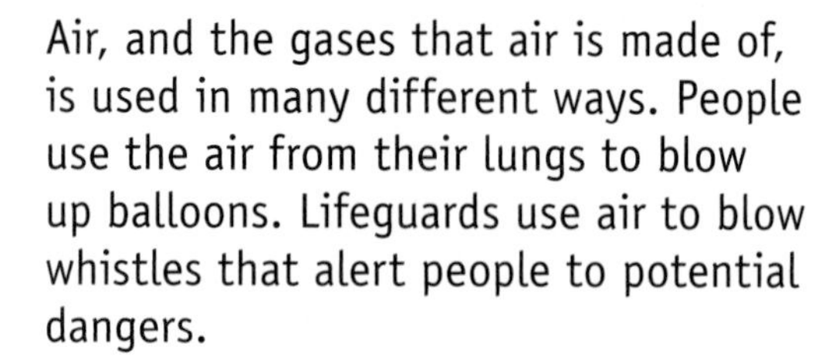

Air, and the gases that air is made of, is used in many different ways. People use the air from their lungs to blow up balloons. Lifeguards use air to blow whistles that alert people to potential dangers.

Hot air rises because it is lighter than cool air. Helium rises because it is lighter than the other gases that make up air.

Air that has been squeezed into a small space is called **compressed air**. When you blow up a balloon, you fill it up with air molecules. Air pressure builds up inside the balloon. If you fill the balloon with too much air, it will explode. The air will be released with a lot of force.

Hot air balloons get their lift from hot air forced into the balloon.

A gas called helium lifts blimps.

The air inside a balloon is compressed air.

A bicycle pump forces comressed air into the tire. When the correct pressure is reached, it make the bike easier to ride.

Air is put to work in hospitals. Some patients with heart or lung problems need more oxygen than they are able to get from the air around them. Oxygen tents are used to give patients oxygen-rich air. A pump takes out the carbon dioxide that the patient exhales.

Gases called **anesthetics** (an uhs THEH tiks) are used to put patients to sleep. The patient inhales the compressed gases through a mask during surgery.

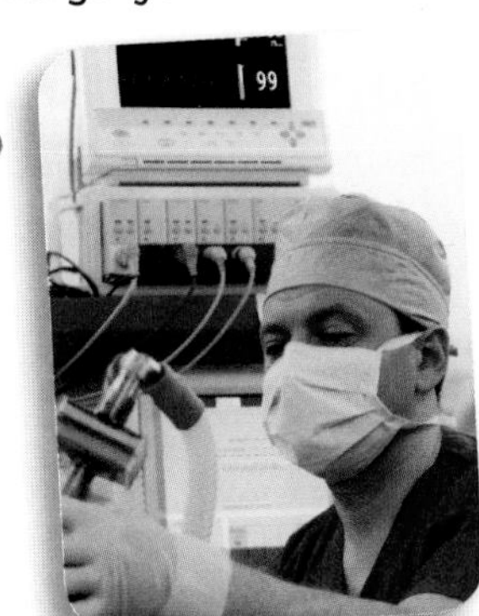

A doctor who administers anesthetics must constantly watch the amount of gases the patient receives.

Is That a Fact?

Early Risers

A French scientist, Pilatre de Rozier, was the first human passenger on a hot air balloon. In November 1783, he and the Marquis d'Arlandes made a 25-minute trip. The first passengers were a sheep, a rooster, and a duck. They stayed airborne for 15 minutes.

Make the Connection

Check your answers on pages 408–419!

Hey Air! Get Back to Work!

Use what you've learned about air at work to identify the many uses of air in daily life at home or at school.

1. Take your science notebook on a mini field trip around your home or school.
2. Write down the ways you see air being used. You may see it being used by plants, animals, machines, or in other ways.
3. You can't see air. How did you know it was at work?

Protecting the Air

Harmful substances in the air are **air pollution**. Some air pollution is caused by natural events. Volcanoes and forest fires cause air pollution. Dust, pollen, and mold in the air cause air pollution. Most air pollution is caused by human activities.

Tiny pieces of dust, ash, and soot, and drops of water in the air, cause **particulate** (par TIK yoo lit) **pollution**. Large particles irritate your nose and throat. Very small particles get deep inside your lungs.

Burning wood and fossil fuels can put **noxious** (NOK shuhs) **gases** and **noxious chemicals** into the air. Noxious chemicals and gases make it hard to breathe and might also get into your blood.

Car exhaust mixed with air and sunlight makes ozone. Ozone is helpful when it is high in the atmosphere. It protects us from dangerous radiation from the Sun. Ozone at ground level causes breathing problems. It can also hurt crops and trees.

Chemicals in car exhaust + air + sunlight = ozone

Ozone + more car exhaust + no wind = smog

Ozone mixed with car exhaust forms smog. Air filled with smog is hard to breathe. The mountains around Los Angeles trap smog. The air is filled with smog many days each year.

Water mixed with gases from burning fuels and chemicals forms **acid rain**. Acid rain can wear away or dissolve hard materials like metal and stone. Acids can kill fish and other living things in lakes and ponds. Trees all over the world have been damaged or killed by acid rain.

Many factories now remove pollutants before they get into the air. Cars, trucks, and buses use **catalytic converters** to turn harmful gasoline exhaust into carbon dioxide and water.

Acid rain killed these spruce trees.

Which Kind of Air Pollution?

Find answers on 406–419

Here are some causes of air pollution:

Wind blows over a field and picks up topsoil.

Electricity is made by burning fuel in a power plant.

Volcanic ash covers cars and houses.

Cars and trucks burn gasoline and emit carbon monoxide.

Which of these is particulate pollution? Which is noxious gas or chemical pollution? Write your answers in your science notebook.

Weather, Seasons, and Climate

What Is Weather?

Earth's **atmosphere** is a mixture of gases, water vapor, and tiny bits of matter. The atmosphere is kept in place by the force of gravity, and it protects Earth from dangerous solar radiation.

Weather is the condition of the atmosphere at a given time and in a given place. Weather includes daily changes in precipitation (mist, rain, sleet, or snow), air pressure, temperature, and wind conditions. These changes can happen rapidly.

Climate describes the average weather conditions of a place over a long period of time. Earth has three main climate zones: polar, temperate, and tropical. The polar climate zones tend to be covered by snow and ice. The temperate climate zone can experience wide variations in temperature. The tropical climate zone is usually warm and wet.

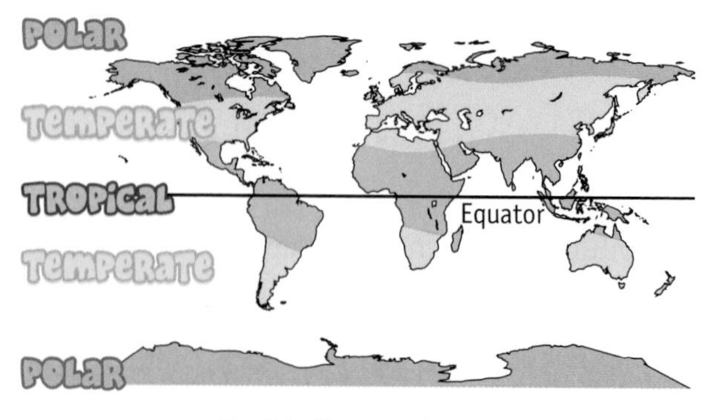

Earth's three major climate zones

Several factors affect climate. One factor is the distance the area is from the Equator. A second factor is how high the area is above sea level. A third is how far the area is from a large body of water. Still another factor is whether the land is mountainous or flat.

Weather and Seasons Affect Living Things

Some climates experience little change in weather from one season to another. Other climates vary quite a bit over the course of the year. We divide the calendar year into four three-month **seasons** that we call spring, summer, fall, and winter.

See also: page 330 Forces and Motion

Weather affects people and other living things. Humans tend to wear light clothing during hot summers. Some animals, such as frogs, turtles, and skunks, hibernate in winter. Sleet may cause people to drive more slowly. During thunderstorms, people go inside to avoid lightning strikes. Plants do not grow well during a drought.

The weather can affect your activities. A heavy snowfall may close schools or block roads. It can also mean time for sledding!

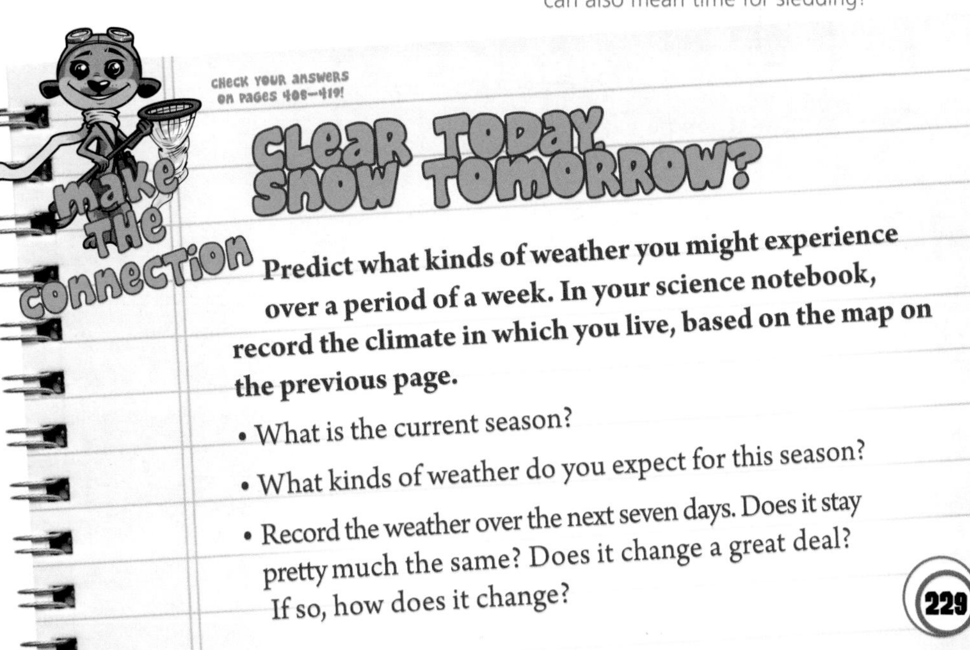

Clear Today, Snow Tomorrow?

Predict what kinds of weather you might experience over a period of a week. In your science notebook, record the climate in which you live, based on the map on the previous page.

- What is the current season?
- What kinds of weather do you expect for this season?
- Record the weather over the next seven days. Does it stay pretty much the same? Does it change a great deal? If so, how does it change?

Water Cycle and Precipitation

All the water on the Earth and in our atmosphere makes up what we call the **hydrosphere** (HI dro sfeer). Water is recycled over and over again through the hydrosphere by a process called the **water cycle**. Most of the Earth's surface is covered with water. Most of this water is in the oceans. Oceans are an important part of the water cycle. They are connected to all the other reservoirs of water by evaporation and precipitation.

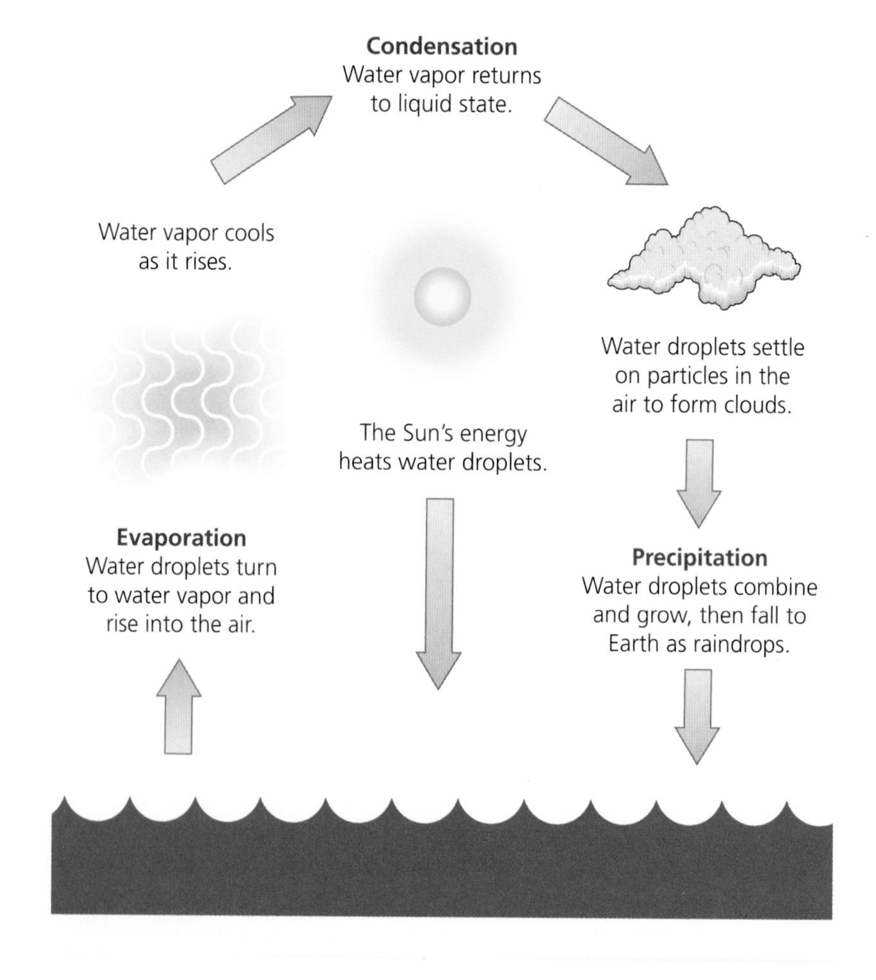

Precipitation (pree sip i TA shun) is the weather word for any form of water that falls from the sky. The form that water takes will change depending on the weather, season, and climate. Sleet and snow tend to occur during the cold seasons. Hail occurs in hot weather. Rain can occur anytime throughout the year.

Earth and Its Landforms

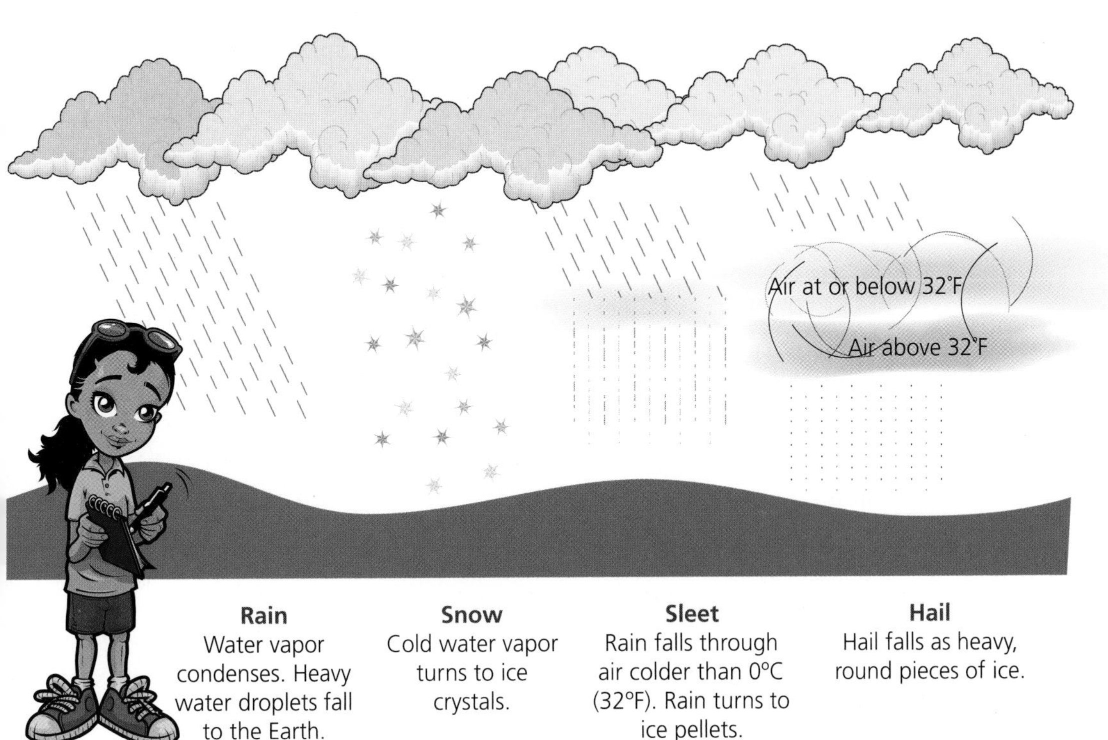

Rain
Water vapor condenses. Heavy water droplets fall to the Earth.

Snow
Cold water vapor turns to ice crystals.

Sleet
Rain falls through air colder than 0°C (32°F). Rain turns to ice pellets.

Hail
Hail falls as heavy, round pieces of ice.

GO ONLINE

To learn more, check out these Web sites!

- **NOAA Education: Specially for Students**
  http://www.education.noaa.gov/students.html
- **Web Weather for Kids**
  http://eo.car.edu/webweather

The Sun's Role in Our Weather

Energy from the Sun heats Earth's surface. However, the surface heats unevenly. There are several reasons for this. First, because Earth is a sphere, different areas of the Earth receive more steady heat energy than others. The Equator gets more heat than the poles. Another reason is that water surfaces and ground surfaces heat at different rates. This uneven heating causes air movements that result in changing weather.

As you know, cool air sinks and warm air rises. Cool air sinks toward the Earth's surface. The ground, which has been heated by the Sun, warms the air. The heated air then rises. As it rises, it begins to cool off again. Then it sinks. This cycling of air is called a **convection current**. Wind occurs when nearby cool air rushes in horizontally to the space where the rising warm air had been.

Air pressure is the weight of the atmosphere in a particular place on the Earth's surface. Convection currents cause air pressure to change. Small convection currents produce winds and rain. Large currents can produce hurricanes and severe thunderstorms.

During a thunderstorm, winds that are flowing in different directions at different levels in the atmosphere may line up and begin to spin around. This air circulation, along with a strong updraft, can form a tornado.

Clouds

When water vapor mixes with air, the air rises and expands at altitudes with lower air pressure. When this air cools, the water vapor changes to tiny droplets. These droplets usually form around floating dust particles. Together, millions of droplets form a **cloud.** Here are some common types of clouds:

- Cumulus clouds are fluffy white clouds. They indicate fair weather. They form and vanish quickly.

- Cumulonimbus clouds are dark and ready to rain. Adding the suffix "nimbus" to a cloud name means it is a rain cloud.

- Cirrus clouds are found high in the atmosphere. Instead of water, they are formed of ice crystals. Cirrus clouds usually predict fair weather.

- Stratus clouds usually form low in the atmosphere. They have a long, horizontal appearance. An overcast day is caused by stratus clouds.

Is That a Fact?

You *Can* Touch the Clouds!

Did you know that fog is a cloud that is suspended close to the Earth's surface? When you walk through fog, you get damp because clouds are made up of water droplets.

Air Masses and Fronts

An **air mass** is a large volume of air that has a certain temperature and amount of moisture. Air masses can be described as warm or cold, wet or dry. Warm air masses form in tropical climates. Cold air masses form in polar or temperate climates. Dry air masses form over land, and wet air masses form over water.

The movement of air masses causes weather to change. When air masses move, they bump into each other along an imaginary line called a **front**. This is where most storms occur. A cold front forms where a cold air mass moves under a warm air mass. Cold fronts bring storms that are followed by cool, fair weather. Warm fronts form when a warm air mass moves over a cold air mass. Warm fronts bring rain that is followed by warm, humid weather.

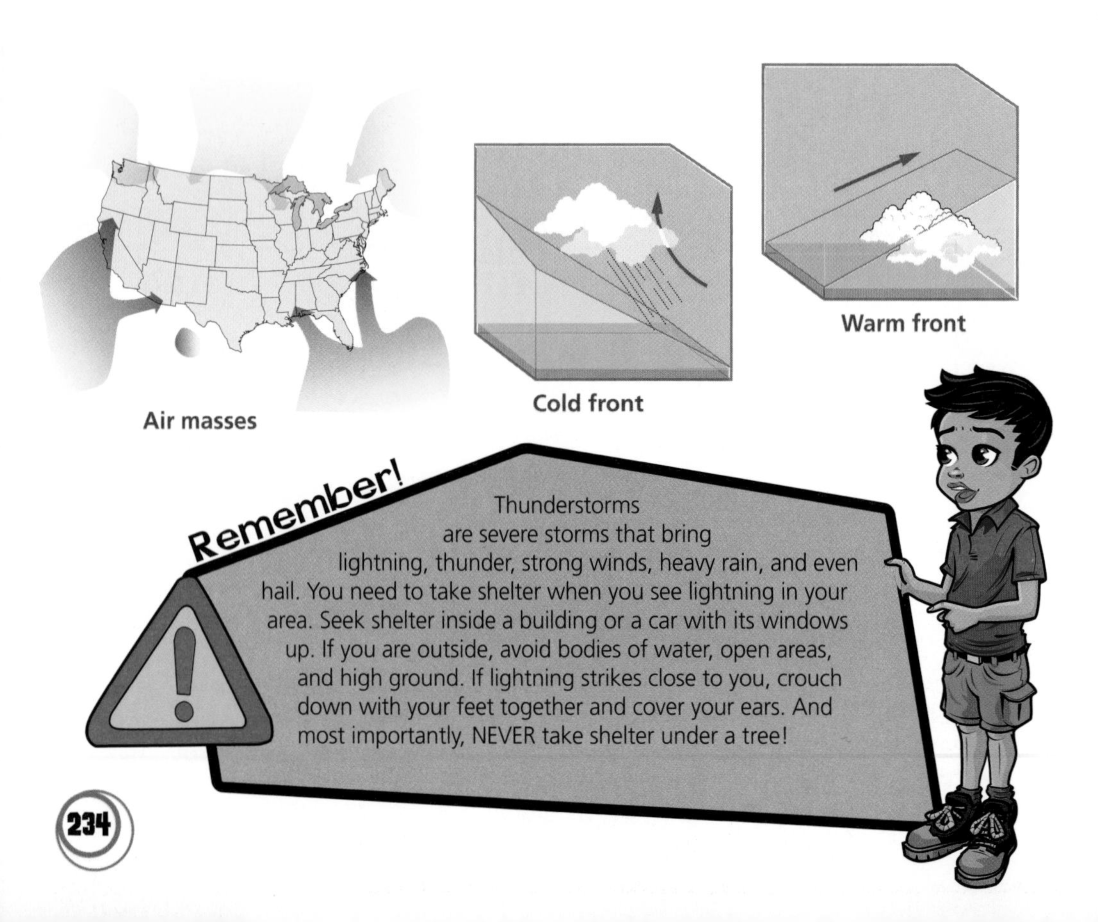

Thunderstorms are severe storms that bring lightning, thunder, strong winds, heavy rain, and even hail. You need to take shelter when you see lightning in your area. Seek shelter inside a building or a car with its windows up. If you are outside, avoid bodies of water, open areas, and high ground. If lightning strikes close to you, crouch down with your feet together and cover your ears. And most importantly, NEVER take shelter under a tree!

WHY ARE YOU ALL WEARING RAIN GEAR?
WE JUST HEARD A SEVERE WEATHER ALERT ON THE RADIO AND WE WANT TO BE PREPARED.
I FILLED MY BACKPACK WITH THINGS WE MIGHT NEED IF THE POWER GOES OUT.
IF THERE IS A BLACKOUT IN YOUR TOWN:
· USE FLASHLIGHTS INSTEAD OF CANDLES.
· USE A BATTERY-POWERED RADIO TO LISTEN TO WEATHER REPORTS.
· KEEP YOUR REFRIGERATOR AND FREEZER DOORS CLOSED, TO HELP KEEP FOOD FROM SPOILING.
· IF YOU SMELL GAS IN YOUR HOUSE, LEAVE IMMEDIATELY! GO TO A NEIGHBOR'S HOUSE AND CALL 911 TO REPORT A POSSIBLE GAS LEAK.
· NEVER GO NEAR DOWNED POWER LINES! CALL 911 TO REPORT THEM.

Weather Maps and Instruments

How do people measure weather? It's easy to go outside and stick a ruler into the snow to see how deep it is. But how can we measure rain? What about wind speed? What about air pressure? And how do these measurements help us?

Scientists who study and report on the weather are called **meteorologists** (mee-tee-or-AHL-oh-gists). They collect data from the Earth's surface and atmosphere, and create weather maps using the data. Weather maps help them forecast the weather.

Thermometers, such as the ones invented by Gabriel Fahrenheit, record temperature. Most thermometers record both the Fahrenheit scale (often used in the United States) and the Celsius scale (used in other places around the world and in science).

Barometers measure air pressure. Falling air pressure usually means stormy weather. Rising air pressure usually means fair weather on the way. What types of weather do you see labeled on this barometer?

An **anemometer** (an-uh-MAHM-uht-uhr) measures wind speed. As you've learned, air moves from areas of high pressure to areas of low pressure. This creates winds. If there is a great difference in air pressures, the wind will blow faster. Global winds blow around the Earth, covering large areas. Local winds blow across smaller regions.

Weather maps use symbols to show the weather. They usually include temperature, air pressure, fronts, and cloud cover. This data comes from weather stations across the country. The National Weather Service (NWS) collects the data from these stations. Then the NWS makes the weather maps.

Doing the Work of Scientists

Predictable patterns of weather affect temperature, wind direction, wind speed, precipitation, air pressure, and cloud cover. These are things people want to know before they plan a picnic or cross-country trip to visit relatives. Why might you want to know what weather is on the way?

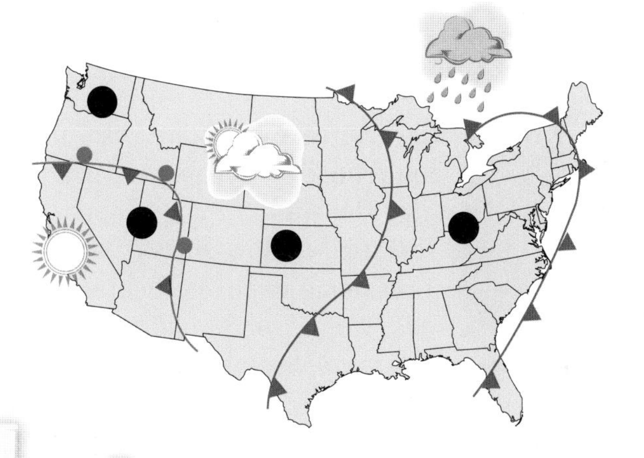

Keeping a Science Notebook

CHECK YOUR ANSWERS, PAGES 408–419!

CREATE YOUR OWN AIR MASS

You'll need one cup filled with ice cubes and one cup containing chilled water from the fridge.

1. Fill a cup halfway with ice cubes. Wait five minutes.
2. With one hand, pour the chilled water into the cup with the ice. As you pour, keep your free hand low over the cup with the ice.
3. Record what you feel.
   If the air you felt were an air mass, how would you describe it?

Natural Resources

What Are Natural Resources?

You're helping your family make a pizza for dinner. What ingredients do you need? You'll need flour for the crust. Cheese and mushrooms and green peppers for the topping. And don't forget the tomato sauce!

Natural gas is an energy resource. It is used mainly to heat buildings and for cooking.

Natural resources are the "ingredients" for almost everything we use in daily life. The wood in your floors comes from forests. Window glass is made with sand. Your home might be heated with natural gas. Wood, sand, and natural gas are all natural resources.

There are two types of natural resources. Energy resources produce energy. Fossil fuels, wood, water, and wind are examples of energy resources. So is the Sun.

See also: page 272 Alternative Energy

Material resources are used to make other products. They are like the ingredients in a pizza. In fact, the ingredients in pizza come from material resources! The cheese comes from milk, which comes from cows. The cows eat grass, which needs carbon dioxide from air and water and minerals from soil. Animals, plants, air, water, soil, and minerals are all material resources that affect our lives every day.

Material resources include both living and nonliving things.

Renewable and Nonrenewable Resources

Not all natural resources are alike. Some can be replaced. Others can be used only once.

Renewable Resources

Plants, animals, water, air, and soil are renewable resources. Because they can replace themselves, they can be used many times.

Nonrenewable Resources

Rocks, minerals, and fossil fuels such as coal and petroleum products are nonrenewable resources. They can't be replaced or would take too long to replace naturally.

Earth is overflowing with natural resources. It's our responsibility to care for them.

Is That a Fact?

Extreme Resources

Some places don't seem to have any natural resources. Like the North Pole—it looks like it's just a big chunk of snow and ice. But it's brimming with life! Polar bears hunt seals on the thick ice covering the Arctic Ocean. Millions of fish and tiny organisms live in the ocean's icy waters. The water and ice at the Pole are resources, too. But a lot of this ice is melting because the Earth is getting warmer. These changes are disturbing the Arctic ecosystem. If this continues, we might lose many important resources.

Fossil Fuels

Fossil fuels come from the remains of dead plants and animals that have been buried for more than 300 million years! Fossil fuels are very useful, but they cause problems. Mining them pollutes soil and water supplies. Burning them pollutes the air. It also forms a lot of carbon dioxide, which can cause serious climate changes.

Oil is a liquid fossil fuel. It is found deep below the ground between layers of rock. Special pumps, like this one, pull it out of the ground.

Natural gas is a fossil fuel formed as a gas from oil deep underground. Like oil, natural gas is collected with pumps. It is a common fuel for heating homes. Some stoves burn natural gas to cook food.

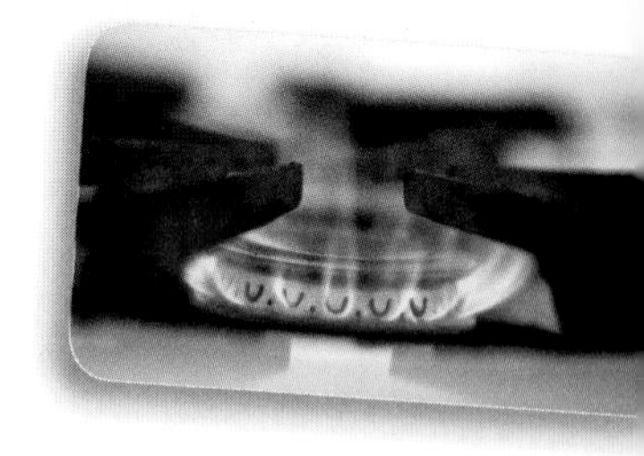

Gasoline is a liquid fuel that is made from oil. It is the main fuel used to power cars and trucks in the United States.

See also: page 176
Fossils

Coal is a solid fossil fuel. It is found near the ground's surface. We get coal by removing it from the surface or through underground mines. Many power plants burn coal to generate electrical energy.

Water

Water is one of our most precious natural resources. Every living thing needs water to survive. Most organisms need liquid fresh water. But very little of Earth's water comes in that form.

Most of the Earth's water is salt water. Salt water is an important resource. It is held in oceans, bays, and seas. These bodies of water are important resources themselves. They form ecosystems that contain plants, animals, minerals, and other resources.

Almost 97% of the water on Earth is salt water.

Only three percent of the Earth's water is fresh water. Most of this is frozen in ice and glaciers near the poles. Most fresh water collects underground as **groundwater**. Humans can dig wells to get groundwater. But this isn't always possible.

The main source of water for humans and other organisms is **surface water**. This is liquid fresh water that collects on Earth's surface. It is in lakes, rivers, streams, ponds, swamps, and marshes.

Most fresh water on Earth is frozen in glaciers and ice caps. Humans and other organisms need fresh water in liquid form.

Did You Know?

Don't Waste Water!
Water is a renewable resource because it cycles between Earth and the atmosphere through the water cycle. But that doesn't mean that some areas can't run out of water! Water supplies can dry up when there isn't enough rain. As the Earth gets warmer, water may evaporate faster than usual. And many human activities pollute water. So we must all try to keep water clean and to not use more than we need.

Air

Air is a mixture of gases that includes **carbon dioxide** and **oxygen**. These gases are very important for living things. The good news is that air is a renewable resource. And you help renew it every time you breathe!

When you inhale, or breathe in, you take in oxygen (O_2) from the air. When you breathe out, or exhale, you release carbon dioxide (CO_2) to the air.

What happens to that carbon dioxide you exhaled? It's used by plants! Remember that plants need carbon dioxide to make their own food through **photosynthesis.** They get the carbon dioxide they need from the air. Plants form oxygen at the end of photosynthesis. Then they release this oxygen into the air.

Plants form oxygen at the end of photosynthesis. They release that oxygen into the air. And that's where you come in again—because you take in that same oxygen when you inhale. And the cycle starts again!

See also: page 116 Plants

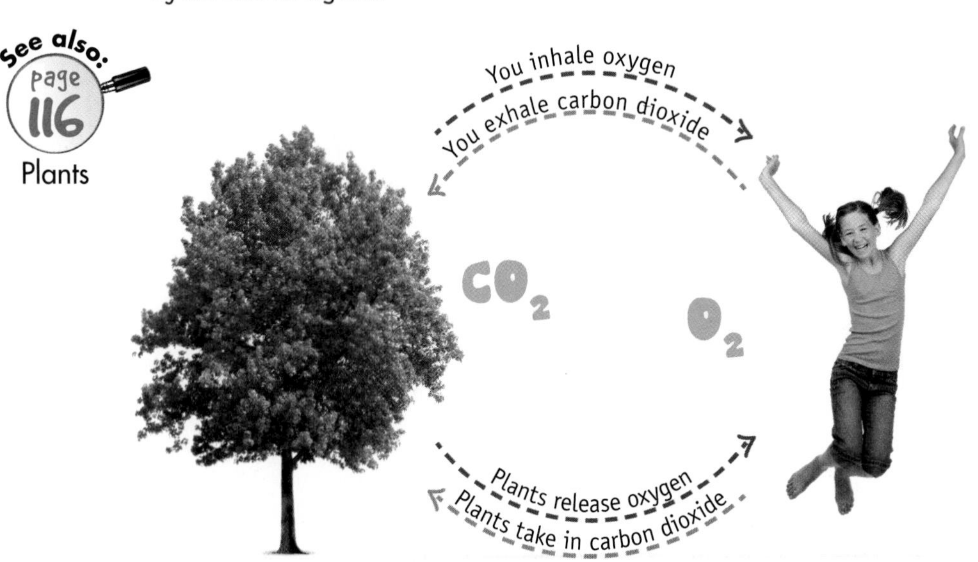

Plants and Animals

Plants and **animals** are very important resources. And they're renewable! Plants provide our food. In fact, they provide food for everything on the planet! That's because plants are **producers**. They form the base of almost every food chain in most ecosystems.

Animals give us meat and dairy products. That includes the cheese on your pizza, the milk on your cereal, and the turkey in your sandwich. These are just a few of the foods that come from animal resources.

Plants and animals provide materials used to make many other items, too.

PLANT AND ANIMAL RESOURCES

Plant Resource:

- **Wood** is used for fuel, furniture, paper, and building supplies.
- **Cotton** is used for clothes, towels, and sheets.
- **Rubber** (rubber tree sap) is used for tires, shoes, and toys.
- **Medicines** are made from plant sources to treat heart disease and cancer.

Animal Resource:

- **Leather** (animal skins) is used for shoes and handbags.
- **Wool** (sheep hair) is used for clothes and blankets.
- **Bone** and horns are used to make carvings and buttons.
- **Medicines** are made from animal sources to treat diabetes and high blood pressure.

Tiny Resources

Bacteria and fungi are living resources just like plants and animals are. Some bacteria are used to make yogurt and cheese. Mushrooms and yeast are fungi. Yeast is used for baking bread. Mushrooms are used for food. Many bacteria and fungi are used to make medicine. Some bacteria and fungi are **decomposers**. So are some insects. Decomposers break down dead plants and animals. This helps recycle matter through ecosystems.

Soil, Rocks, and Minerals

Soil may just seem like dirt to you but it's a very important resource. Soil holds plant roots in place so plants can grow on land. The nutrients in soil are very important for living things. Plants take up nutrients from soil through their roots. Animals get these nutrients when they eat plants.

These crops are growing in soil that is loaded with minerals and other nutrients.

Rock takes on different forms as it moves through the rock cycle. These forms are important material resources. Limestone is used for buildings and statues. So are granite and marble. Slate is used for floors and roofs.

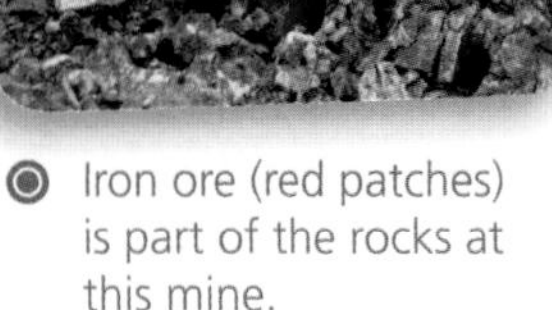

Iron ore (red patches) is part of the rocks at this mine.

Minerals are removed from the Earth for many purposes. Silica is a mineral found in sand that is used to make glass. Metals such as copper and iron are very important resources. Iron is used to make steel, which is used to make bridges and buildings. Copper is used to make electrical wires and pipes that carry water. These are just a few of the ways that minerals are used.

Soil is renewable. It is formed when **decomposers** break down decayed plants and animals. These mix with bits of rocks and minerals to form soil. Rocks and minerals are not renewable. Once they are removed from the ground, they are gone forever.

GO ONLINE

To learn more, visit these Web sites!

- **Natural Resources Conservation Service**
  http://www.nrcs.usda.gov/feature
- **Mine Safety and Health Administration Kid's Page**
  http://www.msha.gov/kids

Pollution

Factory smoke is a major cause of air pollution.

Pollution happens when something is released into the environment that can harm living things. Burning fossil fuels is a major cause of air pollution. Factories release smoke filled with harmful chemicals and dust. Car fumes contain poisonous gases like carbon monoxide.

For hundreds of years, people dumped wastes in rivers and other bodies of water. It is illegal to do this today. But problems and accidents still occur. Fertilizers from farms drain into streams and disturb ecosystems. Ships that transport oil have accidents at sea that spill oil into the water. Oil spills cause terrible problems for animals and plants that live in the area.

Many people still throw trash in bodies of water, although this is against the law.

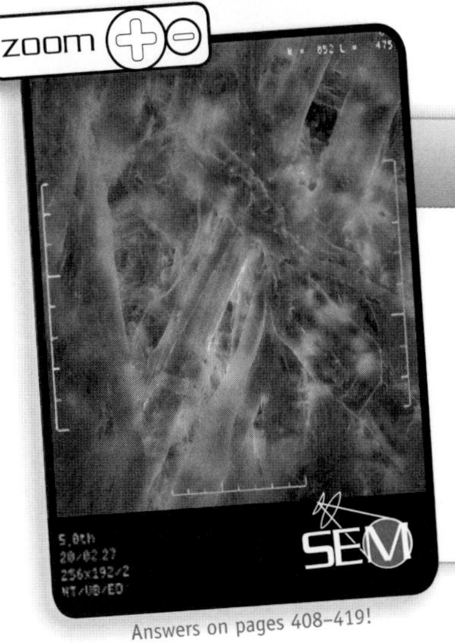

Answers on pages 408–419!

Thinking BIG™

www.carolinacurriculum.com/ThinkingBig

Record your ideas in your science notebook.

- What state of matter? Why do you think so?
- Write 3 physical properties you see.
- Look for shapes and patterns. What do you see?
- Might water or air go through this? Why or why not?

1) I am a sheet, flat and thin.
2) I'm used today and recycled tomorrow.
3) I'm made of recycled matter and wood pulp.
4) Early morning or late night delivery trucks bring me to newsstands, gas stations, and home doorsteps.
5) Today you can also find me on the internet.

Conservation

You may be wondering why we must protect resources that are renewable. Remember that renewable means something can replace itself by nature. And nature takes time! If we use a resource faster than it can replace itself, we will destroy it. Just because something is renewable doesn't mean it can't be used up.

See also: page 218
Air

Take fish for example. Fish populations only produce a certain number of offspring each year. Imagine that a lot of people catch a lot of fish every year. The fish population will get smaller each year. Soon there won't be enough fish left to reproduce. The population will go extinct. And a resource will be gone forever.

But there are ways to protect resources and use them without using them up. This is called **conservation.**

Fishermen can only catch a certain amount of some fish species. This helps prevent these species from going extinct.

Did You Know?

Power Down!
To conserve our resources, we all must make good choices. And these aren't that hard! Turning off the lights when we leave a room helps conserve energy. Unplugging your computer after you log off will save energy too. That leaves more energy for someone else to use!

Thinking BIG™

GO ONLINE

Look online for micrographs of a brown paper towel and glossy paper.
http://www.carolinacurriclum.com/ThinkingBig

Reduce / Reuse / Recycle

A good way to protect our resources for tomorrow is to cut back on what we use today. You can do this in three ways: Reduce. Reuse. Recycle.

Reduce what you use by using less. Help your family shop for items that don't use extra packaging. Talk to your family about unplugging your TV and DVD player at night. Turn off the water while you brush your teeth. Then turn it on again when you need to rinse your mouth.

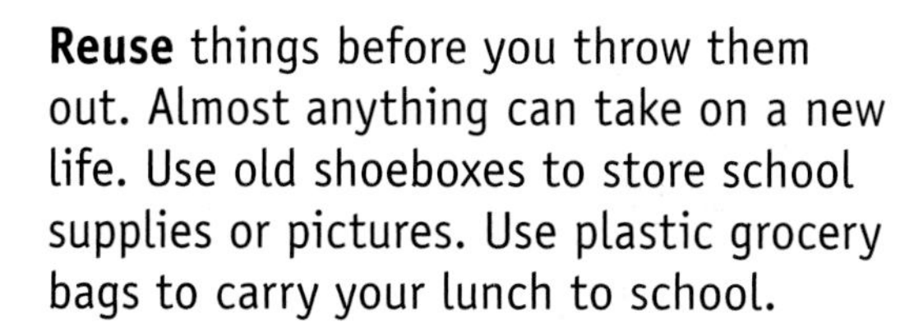

Reuse things before you throw them out. Almost anything can take on a new life. Use old shoeboxes to store school supplies or pictures. Use plastic grocery bags to carry your lunch to school.

Recycle whatever you can. You can recycle paper, glass, and aluminum. Most types of plastic can be recycled, too.

Go Online

For more information, visit these Web sites!

- **The Green Squad**
  http://www.nrdc.org/greensquad/
- **Kids Recycle**
  http://www.kidsrecycle.org
- **NIEHS Kids' Pages**
  http://www.kids.nieh.nih.gov/recycle.htm

THE "3 R'S"-
REDUCE/REUSE/RECYCLE-
ARE ONE WAY TO REDUCE YOUR
CARBON FOOTPRINT!
I CAN MAKE
GREAT
FOOTPRINTS
WITH THESE
SHOES!

YOUR CARBON
FOOTPRINT
MEASURES HOW
MUCH YOU "POLLUTE"
THE EARTH
EVERY DAY.
POOR CHOICES
MAKE YOUR FOOTPRINT
BIGGER. GOOD CHOICES
MAKE YOUR FOOTPRINT
SMALLER.

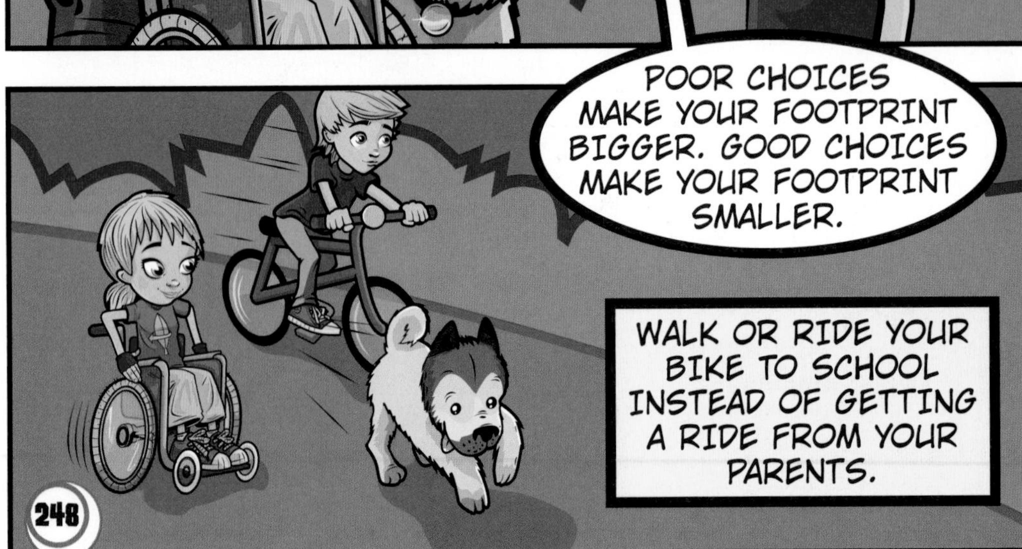
WALK OR RIDE YOUR
BIKE TO SCHOOL
INSTEAD OF GETTING
A RIDE FROM YOUR
PARENTS.

RECYCLING BOTTLES, CANS, AND PAPER SENDS LESS TRASH TO LANDFILLS.

TURN OFF THE LIGHTS WHEN YOU LEAVE A ROOM. AND DON'T FORGET TO UNPLUG YOUR LAPTOP AND YOUR MP3 PLAYERS AT NIGHT, TOO!

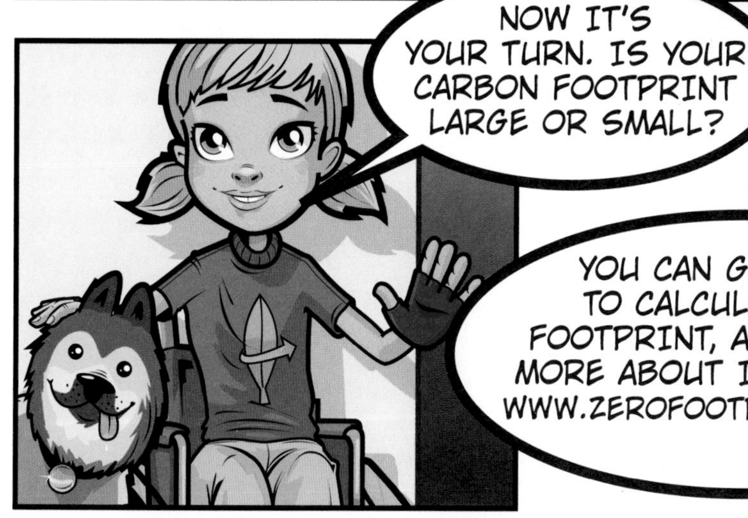
NOW IT'S YOUR TURN. IS YOUR CARBON FOOTPRINT LARGE OR SMALL?
YOU CAN GO ONLINE TO CALCULATE YOUR FOOTPRINT, AND TO LEARN MORE ABOUT IT. CHECK OUT WWW.ZEROFOOTPRINTKIDS.COM

HIYAAH! PHYSICAL SCIENCE COMES IN HANDY IF YOU'RE BREAKING BOARDS OR RUNNING A HALF-PIPE!

Physical Science

What is physical science? **Physical science** is the study of what things are made of and how things work. Physical science explores magnets, friction, and electricity. It explores how a seesaw works and why a boat floats. Why can't we touch energy? How does a bike stop? Everything we do involves physical science.

Matter

What Is Matter?

Did you ever sit at a wobbly desk? One way to fix the wobble is with a piece of paper. If you fold the paper several times, it becomes thick. You can place the folded paper under the desk leg that is causing the wobble. It holds the leg up. The desk leg and the paper cannot be in the same place at the same time. That is because they are both made up of matter.

Matter is anything that has mass and takes up space. A piece of paper is made up of matter. So is a rock. You may not be able to see it, but the air around you is made up of matter, too. Even animals, plants, and people are made up of matter.

See also: page 238 Natural Resources

All matter is made up of building blocks called **atoms**. Atoms are very, very small. They are so small that you cannot even see them. One piece of paper has the same thickness as a stack of about 100,000 atoms!

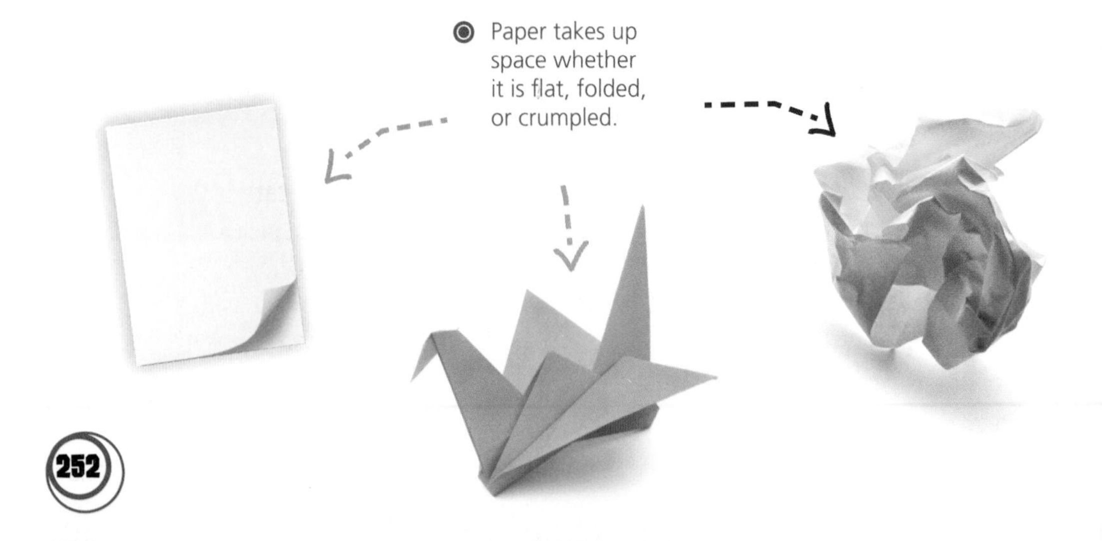

Paper takes up space whether it is flat, folded, or crumpled.

Physical Properties

All matter is made up of atoms. However, not all matter is the same. Think about a marble, an orange, a basketball, and magnetic marbles. How are they the same? All of these things have the same shape. They are balls. But most other things about them are different. All these characteristics are physical properties.

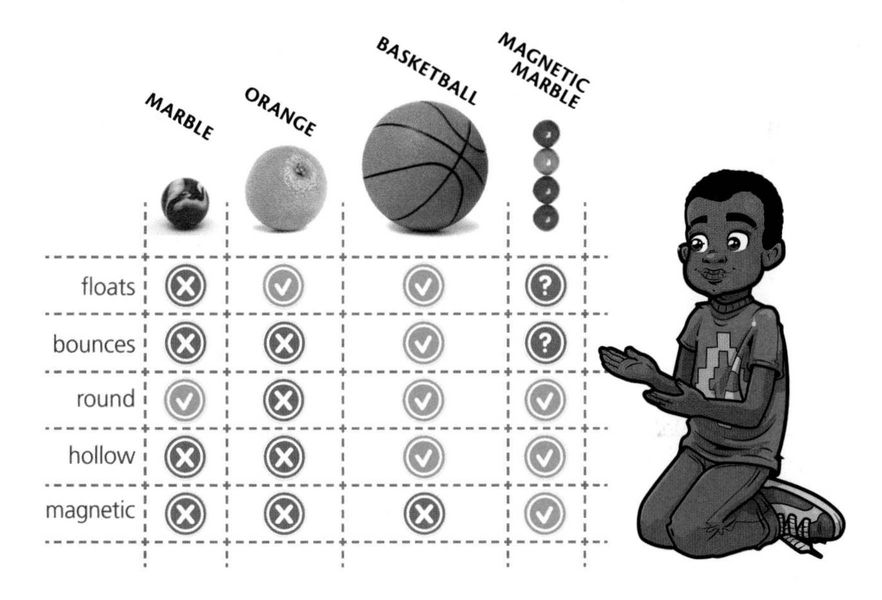

Different kinds of matter can have the same shape but have other physical properties that differ.

A **physical property** is a characteristic of matter that you can observe without changing the type of matter that it is. Physical properties describe matter. You can use them to compare different kinds of matter. Color, shape, and hardness are physical properties. Some things have the physical property of being drawn by a magnet. Whether something can float in water is also a physical property.

Mass, Volume, and Temperature

You can measure some physical properties. Three physical properties that you can measure are mass, volume, and temperature.

Two objects balance if they have the same mass.

Mass is a measure of how much matter there is. Things that have a lot of mass are heavier than things that have little mass.

Volume is a measure of how much space something takes up. You can measure the volume of juice using a measuring cup. You can measure the volume of a cube by multiplying. Volume is length times width times height.

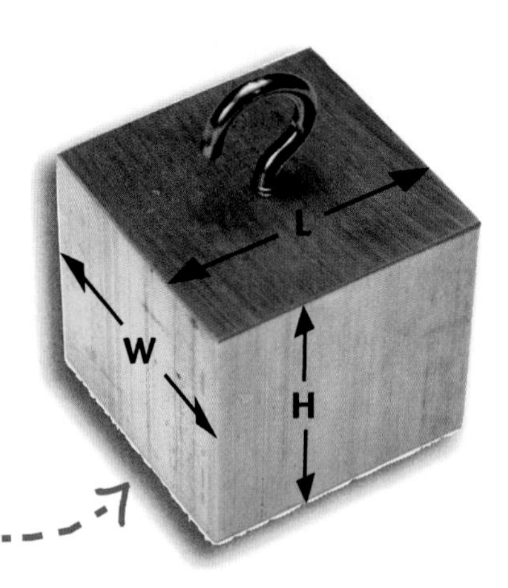

Volume = length x width x height

Most household fish need to live in 23°–25°C (74°–78°F) water.

Temperature is a measure of how hot or cold something is. Fish must live at the right temperature to stay healthy. You can use a thermometer to find out if the water in an aquarium is the right temperature.

Density

A wooden boat will float in water. A small rock will sink. Why does one thing float and the other sink? The reason has to do with density. **Density** is a measure of how much mass there is in a given volume. You can find density by dividing mass by volume. Things that have a lower density than water will float. Things that have a higher density than water will sink.

Density = mass ÷ volume

This type of rock is called pumice. Pumice has a very low density for a rock. It can actually float in water!

PAGES 408–419!

Will It Float?

Find the density of a block of wax. Use the density to predict if the wax will float or sink in water.

1. Use a ruler to find the length, width, and height of the block of wax in centimeters (cm).
2. Multiply the length, the width, and the height. This is the volume of the block of wax.
3. Use a balance to find the mass of the block of wax in grams (g).
4. Divide the mass by the volume. This is the density of the block of wax.
5. Water has a density of 1. Does wax have a higher or lower density than water? Will it float or sink in water?
6. Place the block of wax in a tub filled with water to check your answer.

Physical Changes

Imagine a cube of sugar. If you crush it, it will still be sugar. The crushed sugar has many of the same physical properties as the sugar cube. It is white. It is made up of crystals. The only difference is that it has a different shape. Crushing is a **physical change**. A physical change is a change in matter that does not change the type of matter that it is. Crushing sugar does not change the sugar into something else. The crushed sugar is still just as sweet as a sugar cube.

Crushing is only one type of physical change. Bending, mixing, and melting are others. No matter what physical change is made to an object, the material is the same before and after the change.

Did You Know?

Jewelry makers use physical changes to change a natural gold nugget into a gold ring. The gold nugget can be melted. The molten gold can be molded into a ring shape. After each physical change, it is still gold.

States of Matter

You know that water doesn't always look the same. Sometimes you can pour it. Sometimes it is cold and hard. When you heat it in a kettle, it will become a cloud of hot steam. In each of these cases, the water is still water. It is just water in a different state. State is a physical property. The **state of matter** is whether it is a solid, a liquid, or a gas. Water flowing in a stream is liquid. Water frozen as an ice cube is solid. Water in the air on a humid day is a gas. Water in each state has different physical properties.

Matter Has Three States

SOLID

A **snowflake** is solid water. It keeps its shape, even when it moves.

LIQUID

Hot chocolate is a liquid. It can be poured from one container to another.

GAS

Air is a gas. The air inside bubbles is pushing on the soap film.

Which Is Which?

WOOD

ROCK

OXYGEN

Name the state of matter that describes each material listed above. Write your answers in your science notebook.

FIND ANSWERS ON 408–419

Changes of State

Matter can change from one state to another. Such a change is called a **change of state**. A change of state is a physical change. Matter that goes through a change of state is still the same type of matter as it was before the change. It just has a few different physical properties. Freezing, melting, boiling, and condensing are changes of state.

Freezing is the change of state from a liquid to a solid. When you cool a liquid, it can freeze. When you make popsicles in the freezer, you are changing the state of the juice. You cool the liquid juice until it freezes. The frozen juice is a solid.

When you make popsicles, the juice freezes to form a solid.

Rock that melts underground is called magma. When the melted rock spills out of a volcano, it is called lava.

Melting is the change of state from a solid to a liquid. When you heat some solids, they will melt. An ice cube taken out of the freezer will melt in a warm room. Even rock can melt if it is heated enough. Lava is rock that has been melted at very high temperatures deep underground.

Boiling is the change of state from a liquid to a gas. If you want to make tea, you put water in a kettle on the stove. After being heated, the water starts to boil. Bubbles of gas form in the boiling water. The gas escapes from the liquid. If you boil the water long enough, all of it will change to a gas. There will be no more liquid left in the kettle.

When you boil water, the liquid water turns into a gas.

Condensing is the change of state from a gas to a liquid. Have you ever noticed drops of water forming on a cold glass? The water didn't come from the drink inside the glass. It came from water in the air around you. When water in the air is cooled by a cold glass, it condenses. It changes from a gas and forms drops of liquid.

Water in the air condenses on a glass. The gas turns into a liquid.

EXPLORE MORE

CHECK YOUR ANSWERS, PAGES 408–419!

HOW COOL? HOW COOL!

Observe a change of state. Use a thermometer to find the temperature at which the change takes place.

1. Place an ice cube in a cup. Wait 10 minutes.
2. What is happening to the ice cube? What change of state is taking place?
3. Use the thermometer to find the temperature of the air inside the cup. Use a piece of masking tape to hold the thermometer to the side of the cup. The bottom of the thermometer should not touch the water or the ice.
4. Wait 20 minutes. What is the temperature now?

Mixtures

Mixing is a physical change. When you mix two kinds of matter, you get a mixture. A **mixture** is a combination of two or more different types of matter. Mixtures can be solid, liquid, or gas. Air is a mixture of different gases. A salad is a mixture of different kinds of solids. Salad dressing, on the other hand, is a mixture of oil and vinegar—two different liquids. Mixtures can even be made up of two different states of matter. Smoke is a mixture of a solid and a gas. When you stir solid salt into liquid water, you make salt water.

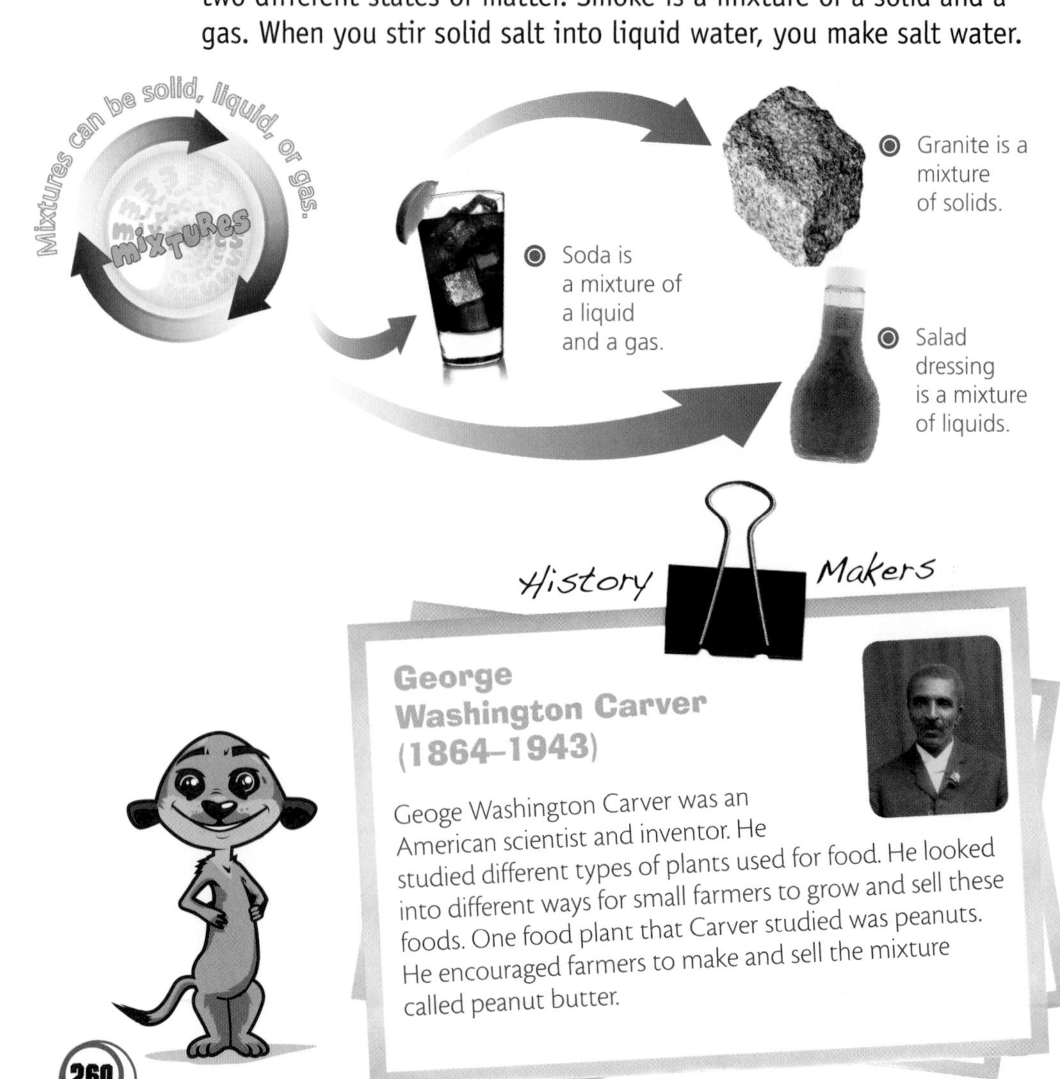

Granite is a mixture of solids.

Soda is a mixture of a liquid and a gas.

Salad dressing is a mixture of liquids.

George Washington Carver (1864–1943)

Geoge Washington Carver was an American scientist and inventor. He studied different types of plants used for food. He looked into different ways for small farmers to grow and sell these foods. One food plant that Carver studied was peanuts. He encouraged farmers to make and sell the mixture called peanut butter.

Solutions

Have you ever stirred a powder into water to make pink lemonade? Pink lemonade is a special type of mixture called a solution. A **solution** is a mixture in which the different kinds of matter are spread out evenly. They are mixed so evenly that they look like one kind of matter. Yet you can taste both parts of the mixture in pink lemonade!

A solution has two parts: the solvent and the solute. The solvent makes up the biggest part of the solution. Many solutions are liquids that have a liquid solvent. In pink lemonade, water is the solvent. The pink lemonade flavoring and coloring are solutes. Salt water is also a solution. The solvent in salt water is also water.

Pink lemonade is the solution. The powder mix is the solute. Water is the solvent.

Solutions are not always made by mixing a solid and a liquid. Carbon dioxide gas dissolved in water makes soda water for fizzy drinks. Carbon dioxide is the solute. Water is the solvent.

In the air, oxygen and other gases are dissolved in nitrogen. Oxygen is the solute. Nitrogen is the solvent.

Carbon + Iron

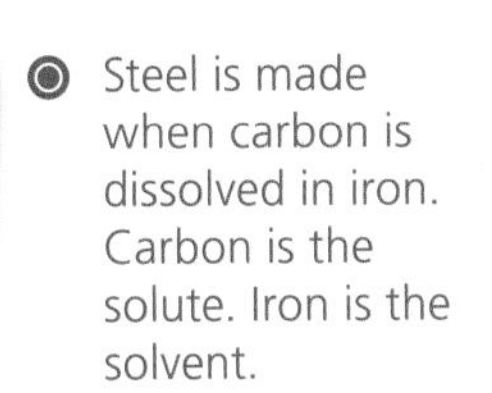

Steel is made when carbon is dissolved in iron. Carbon is the solute. Iron is the solvent.

Separating Mixtures

When two things are mixed together, they keep their identity. Imagine making a fruit salad. You might mix together slices of apples, bananas, and grapes. The apples are still apples even when they are in the mixture. You can pick them out from the fruit salad. They will be exactly the same as before you made the mixture.

You can use physical properties to separate mixtures. When you pick apples out of a fruit salad, you are separating the mixture by color, size, or shape. To separate a mixture of oil and water, you can use density. Over time, the mixture will form two layers. Oil is less dense than water, so it will float on top. You can pour off the top layer of oil to separate the mixture. Magnetism is another physical property that can be used to separate mixtures. A magnet can be used to separate a mixture of steel cans and aluminum cans at a recycling center. The magnet will draw out the steel cans. The aluminum cans will not stick to the magnet.

See also: page 238
Natural Resources

Physical changes can also help separate mixtures. People who collect salt from seawater make use of changes of state. They let the water in the seawater mixture turn to a gas and mix with air. The solid that is left behind is crystals of sea salt. The salt can be used in cooking.

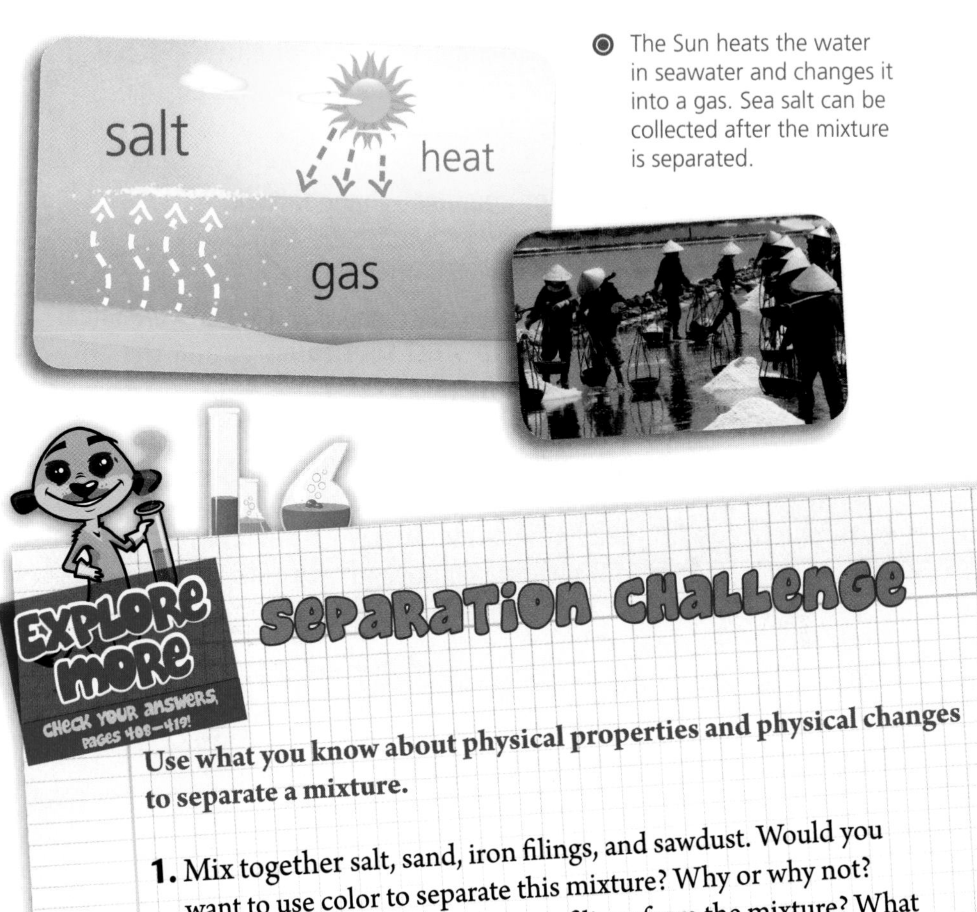

The Sun heats the water in seawater and changes it into a gas. Sea salt can be collected after the mixture is separated.

EXPLORE MORE

CHECK YOUR ANSWERS, PAGES 408–419!

SEPARATION CHALLENGE

Use what you know about physical properties and physical changes to separate a mixture.

1. Mix together salt, sand, iron filings, and sawdust. Would you want to use color to separate this mixture? Why or why not?
2. How could you separate the iron filings from the mixture? What physical property does iron have that salt, sand, and sawdust do not have?
3. How could you use water to separate the rest of the mixture? What will happen to each part of the mixture when you add water?
4. Separate the mixture. Write down what you did to separate each part. Describe the physical properties that you used to separate them.

Chemical Changes

So far you have learned about physical changes. Remember that during a physical change, matter does not change its identity. If you grind a block of wood into a pile of sawdust, it is still wood. However, if you burn the wood, it becomes something else. It becomes ashes and smoke. This type of change is called a **chemical change**. During a chemical change, matter changes from one type to another. It has a new identity.

Burning is just one type of chemical change. Another type of chemical change is rusting. When you leave a bicycle outside for too long, the metal in the frame will rust. It changes from a silver color to a red-brown color. The metal before the chemical change is steel. The red-brown stuff that forms during the chemical change is called rust.

Chemical changes can happen all around you. Chemical changes even take place in the garbage. Rotting is a chemical change in which old food or dead things break down. They eventually can form soil. Some people use a compost pile to turn dead leaves and food scraps into rich soil for their gardens.

Gross! Rusting and rotting are two types of chemical changes.

Chemical Properties

Most soup cans are made out of steel. If you leave an empty soup can outside for a year, it will rust. Soda cans, on the other hand, are often made out of aluminum. If you leave an empty soda can outside, it will not rust. Why are these metals different? The reason has to do with their chemical properties. A **chemical property** is the ability of matter to go through a chemical change. Steel has the chemical property of being able to rust. Aluminum has the chemical property of not rusting.

A chemical property describes whether something can go through a chemical change. It also describes whether something cannot. For example, many things are flammable. They can burn. Some things are nonflammable, which means that they cannot burn. Chemical properties explain why an oven is made of metal and not wood. The wood is flammable. A wooden oven would likely catch on fire!

Is That a Fact?

Busting Rust

Steel has the chemical property of being able to form rust. Stainless steel does not have this chemical property. What makes one metal get rusty and the other rust-free? Regular steel is a metal mixture of iron and other chemicals. The iron forms rust when it gets near oxygen and water in the air. Stainless steel has the metal chromium mixed in with the iron. Chromium forms a layer on the outside of the stainless steel. It keeps the oxygen and water from reaching the iron. This means that things made of stainless steel do not rust easily.

This lighted sign is made of glass tubes filled with neon gas. Neon is a gas that does not easily go through chemical changes.

Energy

What Is Energy?

Energy is a property of matter. Unlike matter, energy cannot be touched. But it can affect matter. A crumpled paper ball is a piece of matter. If you throw it across a room, you are using energy to move it. Energy has an effect on the paper. **Energy** is the ability to make things change or move.

Look at it this way: if matter is the what, then energy is the happened. Put them together, and what do you have?

Matter + energy = what happened

The powerful winds of a major storm have the energy to flatten buildings and uproot trees. You can't touch the energy, but it's there.

Check out these other forms of energy!

- **Sound energy** is carried by sound waves.
- **Heat energy** depends on the temperature of matter.
- **Light energy** is transferred by electromagnetic waves (including light waves).
- **Electrical energy** is delivered by electric charges.
- **Chemical energy** is stored in chemical bonds.
- **Nuclear energy** is produced when nuclear bonds form or break.

Basic Forms of Energy

When you talk about energy, you are talking about a very large topic. Sports is a broad topic. You don't talk about "sports," right? You talk about basketball. Or baseball. When we talk about energy, it helps to get specific.

Matter

Energy takes many forms. **Mechanical energy** is the energy in matter. It has two forms: energy that is stored and energy that is moving.

Light

Mechanical energy

Potential energy

Kinetic energy

Potential energy is energy that is waiting to happen. The swimmer on the starting block has potential energy in his muscles. This energy is ready to move him when he dives.

Kinetic energy is energy that moves. Once these swimmers pushed off their starting blocks, the potential energy in their muscles changed into kinetic energy to move them forward. They are in motion and in the race!

Potential Energy

Potential energy is stored energy. It is energy that is waiting to happen. The potential energy in an object depends on what it's made of and where it's located. A book on a desk has potential energy because it has the potential to fall off the desk. Place the same book on a high shelf, and presto! The book now has more potential energy! Why? Because it has farther to fall.

There's a lot of potential energy in the water behind this dam! Notice that the water is stored above a wall. When the floodgates open, potential energy in the water will help move the water down that wall.

The higher an object, the greater its potential energy.

A bathroom sink full of water has potential energy, too. When you pull the sink stopper, potential energy will propel the water down the drain.

Quick Question

Look at the pictures of the dam and the sink on this page. Which has more potential energy? How do you know?

Find answers on 408–419

Kinetic Energy

Kinetic energy is potential energy in motion. It is in all of the types of energy you use every day. You know these as light, heat, and sound, to name a few.

Potential energy always affects kinetic energy. The more potential energy an object has, the more kinetic energy it produces. The potential energy in the water held by a dam changes into kinetic energy as the water is released.

These marbles pick up speed and release more kinetic energy as they fall from level to level.

Remember how size affects potential energy? The same is true for kinetic energy. A large dam releases more kinetic energy than a small dam does.

Heavy objects have more potential energy than lighter objects have. So heavy objects release more kinetic energy when they fall than lighter objects release. Look at these marbles. The farther they fall, the faster they move. The faster they move, the more kinetic energy they release!

See also: page 330
Forces and Motion

Conversion of Energy

See also: page 312 Simple Machines

Energy can change from one form to another. This is called an **energy conversion.** The energy system shown below is a good example of energy conversion at work.

You know that lamps use electricity. But where does that electricity come from? And how does it get to your lamp?

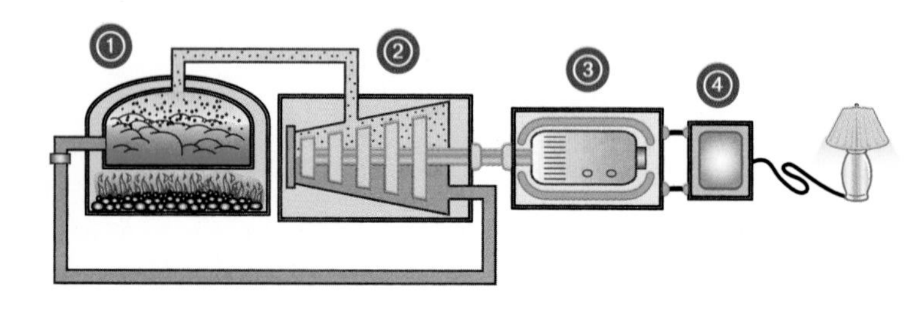

See also: page 272 Alternative Energy

1. Coal in the boiler has chemical energy. Burning the coal converts its chemical energy to heat energy which boils water.
2. The boiler releases heat energy as steam. When steam turns the turbine blades, heat energy is converted to mechanical energy.
3. The turbine drives the generator. The turbine and the generator both use mechanical energy to move. Although energy is transferred, it does not change form.
4. Mechanical energy from the generator is converted to electrical energy in the transformer. This electrical energy is transferred to your lamp when you turn it on.

Quick Question

Every time energy undergoes a conversion, it changes from one type of energy to another. How many conversions take place in the process shown above?

Find answers on 408–419

Energy Conversion

There are many types of energy conversion.
Here are some examples:

PLANTS

Solar energy
(light)
↓
Photosynthesis
↓
Chemical energy
(stored as starch)

ANIMALS

Chemical energy
(from food)
↓
Metabolism
↓
Mechanical energy
(to move and grow)

MACHINES

Kinetic energy
(from water)
↓
Mechanical energy
(to move)

To learn more, check out these Web sites!

- **Downhill Race**
  http://www.exploratorium.edu/snacks/downhill_race/index.html
- **Energy of Motion**
  http://www.physics4kids.com/files/motion_energy.html
- **The Energy Story**
  http://energyquest.ca.gov/story/index.html
- **Power Up With Tortillas**
  http://spaceplace.nasa.gov/en/kids/st5/st5_tortillas1.shtml
- **Energy Kids**
  http://www.eia.doe.gov/kids/

Alternative Energy

What Is Alternative Energy?

Most of the energy you use comes from burning fuels. A fuel is a material that gives off heat energy when it is burned. Power plants burn fuel to produce the electricity that runs your computer. Burning fuel produces heat energy that cooks food on a stove, dries clothes in a dryer, and heats water for your bath.

The fire produced by burning wood gives off heat energy that can cook food and keep you warm.

The most common fuels used today are fossil fuels. These include oil, coal, and natural gas. Fossil fuels provide a lot of energy, but they come with problems. One big problem is that they are not renewable. That means that someday they'll be used up.

The good news is that scientists are developing alternative forms of energy. An alternative means "another choice." An **alternative energy** is an energy source that does not use fossil fuels.

Renewable, Nonrenewable, and Inexhaustible Energy

Not all energy sources are created equal, but they all have their good points and bad points. Some are abundant. Some create lots of pollution. Others are "clean." Some types of energy can be "recycled." Some types will last forever. But others may soon be used up.

Natural Resources

A **renewable energy** is an energy source that can be replaced. Wood is renewable because it comes from trees. When these are cut down to provide fuel, they can be replaced by planting new trees.

	RENEWABLE ENERGY	NONRENEWABLE ENERGY	INEXHAUSTIBLE ENERGY
Energy Source	Biofuels: trees, plant materials	Fossil fuels (oil, coal, natural gas) and nuclear energy	Wind, solar, water (hydro), and geothermal
Good Points	If used up, can be replaced by replanting	Fossil fuels: dependable Nuclear: doesn't pollute air Both: pack a lot of energy into small units	Clean! Not expensive! Can't use it up!
Bad Points	Expensive Burning biofuels adds a lot of pollution to the atmosphere USUALLY A GOOD CHOICE!	Fossil fuels: • Millions of years to form. Once used—gone forever! • Very polluting! Nuclear: • Needs special handling • Hard to store HIGH ENERGY, BUT MESSY!	Not available everywhere THE GOOD GUYS!

continued

Nonrenewable energy comes from an energy source that cannot be replaced once it is used up. Fossil fuels are nonrenewable: it takes so long for them to form that they can't be replaced in our lifetime.

Some forms of energy can never be used up. They are replaced as soon as they are used. Or there is so much that we could never run out. Energy that can't be used up is called **inexhaustible energy**. For example, wood is renewable but it's not inexhaustible. You could use up all the wood on Earth if you don't plant new trees. Solar energy is inexhaustible because you can't use up the Sun!

Energy

Number Crunch

Energy use in the United States

Nuclear Energy 8%
Renewable and Inexhaustible Energy 7%
Coal 23%
Oil 39%
Natural Gas 23%

Check your answers, pages 408–419!

Use the data in the pie chart to understand how energy is used in the United States.

1. Each "slice" of the pie chart stands for a different source of energy. What percentage of energy is from natural gas?

2. What color on the pie chart stands for renewable and inexhaustible energies? What percentage is this?

3. If you add up all the percentages, what number do you get? What does this number stand for?

4. Which source of energy is used the most?

5. Which types of energy are fossil fuels? What total percentage of energy used in the United States comes from fossil fuels?

Hydroelectric Energy

Hydroelectric energy uses the energy in moving water to make electricity. One way to produce hydroelectric energy is with a **dam.** A dam holds up the flow of water in a river. It makes a lake above the dam. The water flows downhill through pipes in the dam. These send the water to turbines below the dam. The moving turbines then power the generators that make electricity.

The Hoover Dam between Arizona and Nevada powers turbines in a hydroelectric power plant.

Hydroelectric energy can cause environmental problems. Dams block river flow, which changes habitats and disturbs the river ecosystem. Designing hydropower plants that cause less harm to ecosystems is an important area of research.

Hydroelectric energy is one of the best alternative energy sources available. Moving water is an inexhaustible resource. And hydroelectric power plants produce very little carbon dioxide, so they don't pollute the air!

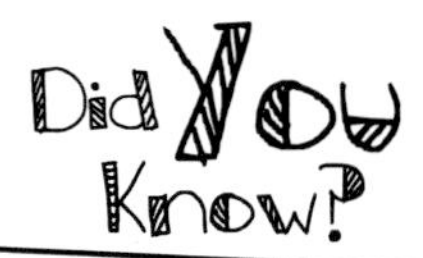

Power in the Oceans Ocean water naturally rises and falls in patterns called tides. There are also "rivers" of faster moving water in the ocean called currents. Both of these types of moving water can power turbines to make hydroelectric energy.

Solar Energy

Solar energy is captured directly from the Sun's rays. Large solar panels can be set up at a central solar station to gather energy. Then the energy is transmitted to businesses and homes.

Did You Know?

Solar panels can also be attached directly to homes! We have them at our house. Every time I recharge my laptop computer, I load it up with Sun Power!

It's hard to get energy from the Sun on a cloudy day. And you can't use solar energy at night unless you store energy you captured when the Sun was out. Scientists are working to solve these problems. Solar energy is one of our best alternative energy sources. It doesn't pollute the air. And it's an inexhaustible resource, so it can't be used up.

CHECK YOUR ANSWERS ON PAGES 408–419!

SUNNY SIDE UP

Use what you know about solar energy to cook food.

1. Fill four jars with equal amount of water. Label the jars A, B, C, and D. Record the water temperature of each jar. Then cover each jar with a plastic sandwich bag.
2. Place Jar A on a sunny sidewalk. Next to this, place Jar B on black paper, and Jar C on a sheet of foil. Place Jar D on the sidewalk in the shade.
4. Wait one hour. Check the water temperature in each jar. Did the temperature change? Was it different for each jar?
5. Use your results to design a solar cooker. Use your cooker to make a melted marshmallow and chocolate graham cracker treat.

Geothermal Energy

Geothermal energy comes from heat stored in the Earth. Processes deep inside the Earth release energy that heats underground rocks and water. This energy pushes clouds of steam and boiling water to Earth's surface. The hot water and steam erupt through small cracks in the ground called geysers.

To harvest geothermal energy, scientists drill holes deep into the ground. These holes fill with underground water. The water and steam erupt with enough energy to move a turbine. The turbine then makes electricity.

Geyser

Geothermal energy is found mainly near earthquake zones and places where heat rises to the Earth's surface. So it's not available to everyone. But for those who can use it, it's inexhaustible and clean. And that benefits all of us!

Go Online

To learn more, check out these Web sites!

- **Energy Kids Page**
  http://tonto.eia.doe.gov/kids/
- **Energy Star Kids**
  http://www.energystar.gov/index.cfm?c=kids.kids_index
- **Renewable Energy Basics**
  http://www.nrel.gov/learning/re_basics.html
- **Kids Saving Energy**
  http://www.eere.energy.gov/kids/renergy.html
- **Energy Quest**
  http://www.energyquest.ca.gov/

Wind Energy

Wind energy is captured from the wind. The Sun heats Earth's atmosphere unevenly. Hot air rises and cool air falls and this movement causes wind. Large turbines are set in open fields to harvest this energy. The turbines look like giant fans with very long blades. Wind moves air across a turbine, turning its blades. The blades transfer energy from the moving air to a large generator. The generator then makes electricity.

Wind energy isn't always the best energy source for a community. Wind depends on weather, and some weather brings no winds. But when it can be used, wind energy is a great energy solution. It's clean and it's inexhaustible. And it can be captured wherever the wind blows!

Humans have used wind energy for thousands of years! In the United States, windmills were used on farms and ranches to grind grains and pump water. Some Western ranches still use windmills to pump water for livestock.

THAR SHE BLOWS

Make your own wind turbines to demonstrate how the blades capture air movement.

If you have trouble getting your pin to hold the paper to the eraser, stick a small piece of modeling clay to the point of the pin. You now have a pinwheel that looks something like the picture!

1. On a square piece of paper, draw two diagonal lines that cross in the center. They should make a giant X. Each arm of the X will touch a corner of the paper.
2. From each corner of the paper, cut along the line about ¾ of the way to the center.

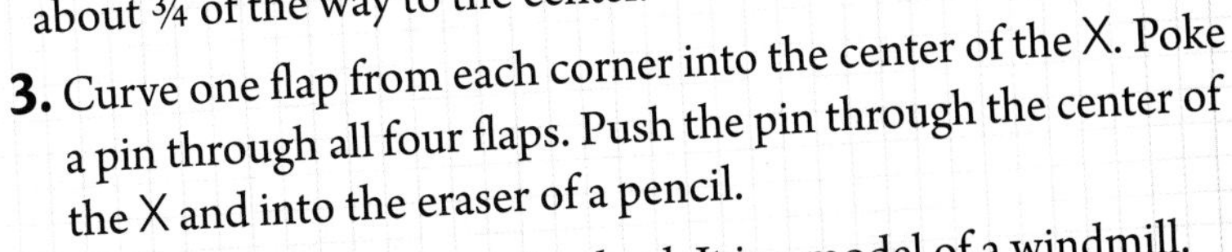

3. Curve one flap from each corner into the center of the X. Poke a pin through all four flaps. Push the pin through the center of the X and into the eraser of a pencil.
4. Experiment with your pinwheel. It is a model of a windmill. What happens when you blow on it? Does it matter which direction you blow? How hard you blow? Think of a way to make it useful.

Biofuels

You know that fossil fuel comes from plants and animals that have been dead for a very long time. **Biofuel** is a fuel made from organisms that lived recently. Biofuel energy comes from burning organic matter.

Crops like corn, beets, and sugar cane are used to make liquid fuels that can be burned. Biofuels also include oils that are removed from soybeans and palm plants. Some cars can run on liquid biofuels like these.

Corn is one of the main crops used to make biofuel.

Solid biofuels come from wastes. Dead trees, yard waste, and wood chips are examples of solid biofuels.

Biofuel energy is renewable. Crops raised for biofuel can be grown again in a short amount of time. Biofuels have one big problem, though: they produce a lot of air pollution when they are burned.

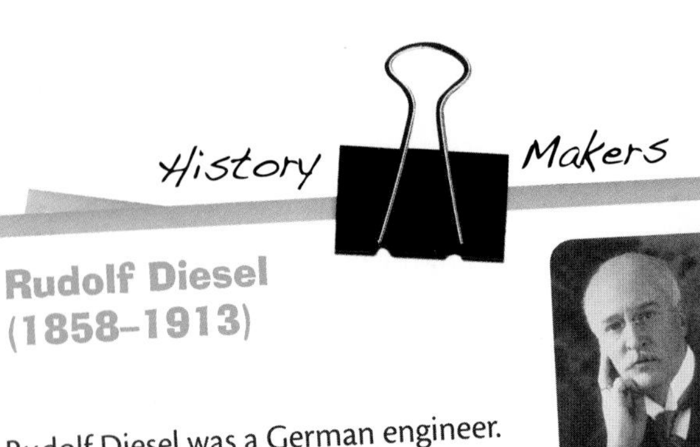

History Makers

Rudolf Diesel (1858–1913)

Rudolf Diesel was a German engineer. He experimented with engines that could run on peanut oil. He invented the diesel engine in the 1890s. Many cars and trucks today have diesel engines. They run on diesel fuel, which is a type of fossil fuel. But some of them can also run on plant oils.

Nuclear Energy

Nuclear energy is produced by splitting the atoms of some elements. One of these is uranium. Atoms release a large amount of energy when they are split. This energy can turn turbines and make electricity.

Atoms, Molecules, and Elements

Nuclear energy is nonrenewable, but it is also inexhaustible. You can't replace used uranium. But very little uranium is used to produce energy. Uranium makes so much energy that it can't be used up.

Nuclear power doesn't pollute the air, but it forms dangerous wastes. These wastes are very harmful to living things. Nuclear waste also can permanently harm ecosystems. An accident in a nuclear power plant can cause a lot of harm. For these reasons, most people do not consider nuclear energy an alternative energy, even though it is not a fossil fuel.

Steam comes out of cooling towers at a nuclear power plant.

Which-Is-Which?

Types of Energy Name whether each type of energy is renewable, nonrenewable, or inexhaustible. Write your answers in your science notebook.

Sound

What Is Sound?

When a drummer plays a snare drum, the drumhead vibrates and creates sound waves.

Sound is a form of energy that is created when an object vibrates. Vibration creates pressure waves, or sound waves, that travel away from an object in all directions. Your ears detect these waves and send information about them to your brain. Your brain interprets this information as a sound.

Energy waves are longitudinal waves. They are like the lines of longitude that you see on a map or a globe. Lines of longitude travel north and south across the globe. Energy waves travel lengthwise, up and down.

What's Happening?

A tuning fork emits sound when it vibrates. Was this tuning fork making sound when it was photographed? How do you know?

Find answers on 408–419

Sound Patterns and Waves

Sound waves cannot exist in a vacuum. They must pass through a medium. A **medium** is a gas, a liquid, or a solid. The type of medium determines the speed at which sound passes through that medium. Sound usually travels faster through a solid than it does through a liquid. It travels faster through a liquid than through a gas. Temperature also affects the speed of a sound wave. Sound waves slow down in a cold medium, and they speed up in a hot medium.

The chart below shows that the speed of sound varies depending on the medium that sound waves pass through. The speed of sound is measured in meters per second, or m/s.

Speed of Sound in Different Media

MEDIUM	SPEED OF SOUND (m/s)
rubber	60
air (40°C)	355
lead	1,210
lake water (25°C)	1,497
ocean water (25°C)	1,550
copper	3,100
glass	4,540
stone	5,971
aluminum	6,420

See also: page 42
Patterns

Amplitude and Loudness

The **amplitude** of a sound is the distance in height that a sound wave travels from its starting point. Amplitude is also known as **volume**, or loudness. As the amplitude of the sound wave increases, so does the volume of the sound. A sound wave with a high amplitude is louder than one with a low amplitude.

Is That a Fact?

Protect Your Ears!

A lot of people enjoy listening to music with headphones. However, you should pay attention to the volume of the music. Listening to loud sounds for a long time can permanently damage your hearing.

Loudness is measured in units called **decibels**. Humans can detect sounds as low as zero decibels. Sounds over 120 decibels can damage your hearing. Have you seen airport workers directing planes near the gates? Did you notice that they wear headphones? It is usually very loud next to a plane, so workers wear headphones to reduce the decibel level of the sounds around them.

Testing... Testing...

Quiz other members of your group to determine whether they can detect changes in amplitude. Make the changes reasonable. The amplitudes should be neither too similar nor too different.

1. Choose five sounds that have different amplitudes.
2. Record each sound.
3. Ask others in your group to rank the sounds in order of amplitude, from loudest to softest. Did everyone agree? If not, the amplitudes of your recordings may have been too similar.

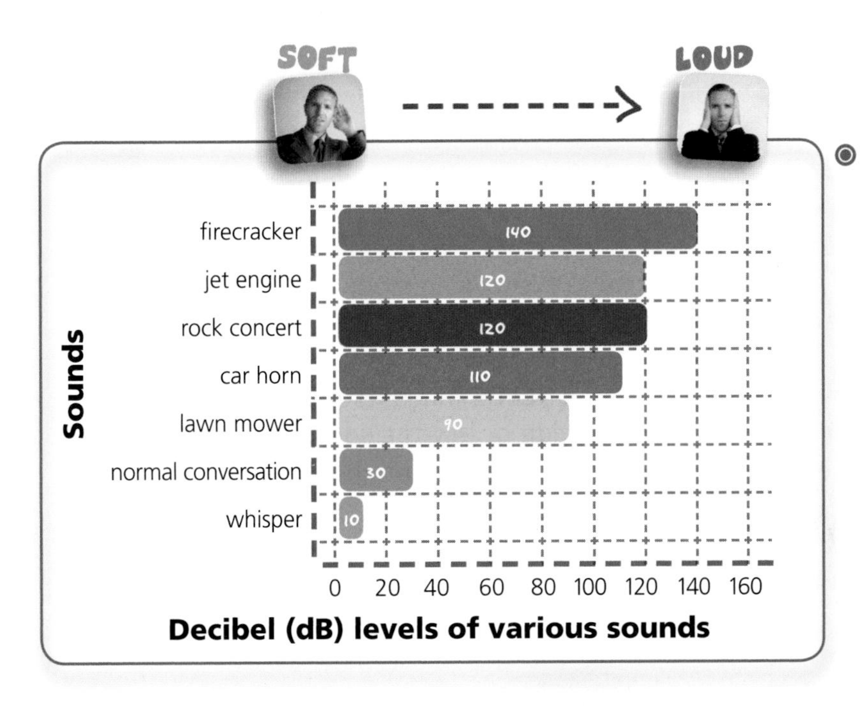

This graph compares the decibel levels of various sounds.

See also: page 266 Energy

Fireworks do not harm your hearing because they explode high in the sky. However, the workers who stage the fireworks must wear ear protection because they stand so close to the launching area.

People who work near airplanes wear headphones to protect their hearing.

Some musicians have damaged their hearing after continued exposure to loud music at rock concerts.

Frequency and Pitch

You have already learned that sound is produced when an object vibrates. The number of times that an object vibrates per second is called its **frequency**. Objects that vibrate quickly have a high frequency. Those that vibrate slowly have a low frequency.

The frequency of an object determines its pitch. The **pitch** of a sound is how high or low the sound is. The higher the frequency, the higher the pitch. Generally, a small object vibrates at a higher frequency than a large object. For example, the pitch of a flute is higher than the pitch of a bassoon.

Flute

Bassoon

Let's Make Music!

You may have had fun doing a version of this activity before. You likely had no idea you were experimenting with sound waves and pitch!

1. First, clean 6 glass juice bottles, all the same size. Any size between 355 ml (12 oz) and 710 ml (24 oz) should work fine.
2. Fill each one with a different amount of water, in graduating amounts. The least full should contain just about an inch of water and the most full should be filled almost to the top.
3. Arrange your bottles in a row, from lowest to highest water level.
4. Use a wooden spoon handle to lightly tap the empty part of each bottle.

You should hear a different tone, or pitch, in each bottle. The pitch depends on the amount of empty space in the bottle.

Dog whistle

Do you know what a dog whistle is? When you blow a dog whistle, you will not likely hear anything, but your dog will respond to the sound. This is because dogs can hear higher frequencies than humans. Now you know that you can call your dog without anyone else hearing you!

See also: page 78 Animals

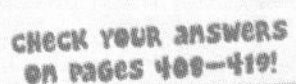

Make the Connection

Can You Hear Me?

Have you ever wondered whether your goldfish could hear? What about an insect? This graph shows the range of frequencies that various organisms can hear.

Where do humans fall in this list? Discuss with your friends which animal has the best hearing.

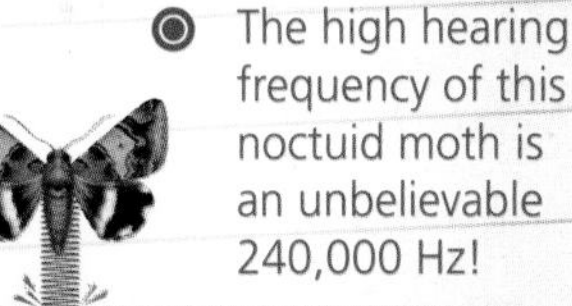

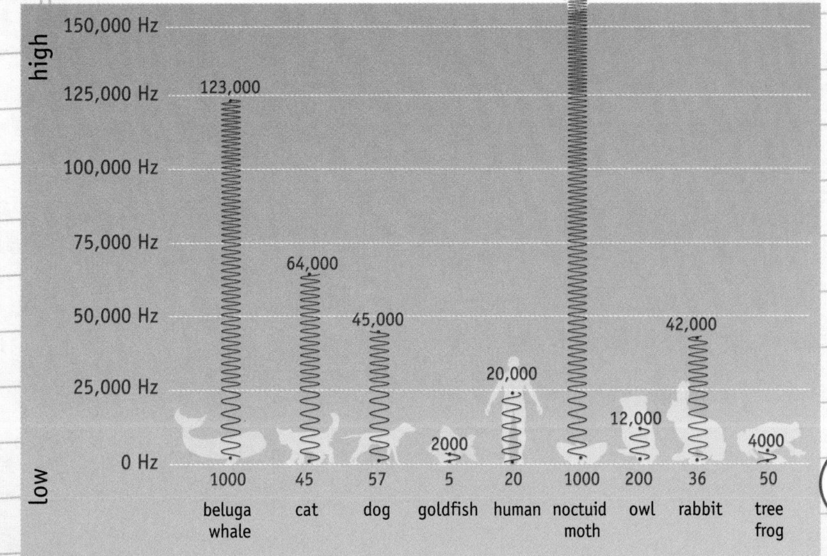

Echo

An **echo** is created when sound waves hit a surface and bounce back. Large open areas with high, straight walls make ideal settings for echoes. Hallways, gyms, canyons, caves, and tunnels are some places where you can hear echoes.

Is That a Fact?

Did You Hear That?

Bats are nocturnal animals, which means that they are most active at night. They have poor vision, so they rely on their excellent hearing to help them fly. A bat makes high-frequency clicks that humans cannot hear. These clicks bounce off of nearby objects, and then echo back toward the bat. The bat judges how far it is from the object and immediately adjusts its flight pattern to avoid hitting the object.

Quick Question

Find answers on 408–419

Bryce Canyon in Utah is a great place to experience echoes. Why do you think this is so?

Science Notebook

Does it Bounce?

Go outside for this exercise and get one or two partners. Take all of the items listed in the chart with you. One by one, hold each item up at arm's length. Cup your other hand around your mouth and shout at the item. Your partner should stand at an angle about three feet away from you with his or her hand cupping the ear facing you to "catch" any bouncing sounds. Your partner will make observations on which objects "bounced" or echoed sounds better than others.

1. Which items echoed the sounds better?
2. Do you know why, based on what you have learned about echoes?
3. Change places with your partner and repeat this exercise. Do you get the same results? Discuss your observations.

ITEM	OBSERVATION
Cookie Sheet	
Cafeteria Tray	
Spiral Notebook	
Pillow	
Poster board	
Chalk board	

Uses of Sound Energy

Sound energy helps us to communicate with our friends, listen to music, and watch television. Tools that use sound energy can also keep us healthy. Your doctor uses a stethoscope to listen to your heart and lungs. The disc that rests on your chest or back captures sound waves. The waves travel up the tube and into the earpieces.

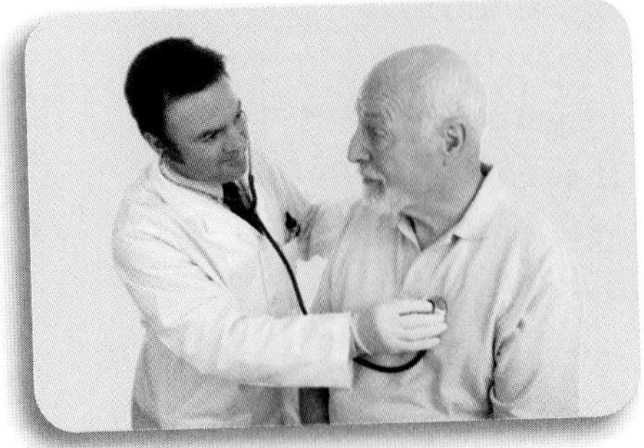

The stethoscope makes it easier for a doctor to hear sounds that are otherwise too quiet to hear well.

Ultrasound is sound energy that is far above the limit of human hearing. It can be used to create images of the inside of the body. Internal images are especially helpful to doctors because they enable doctors to diagnose some illnesses without surgery.

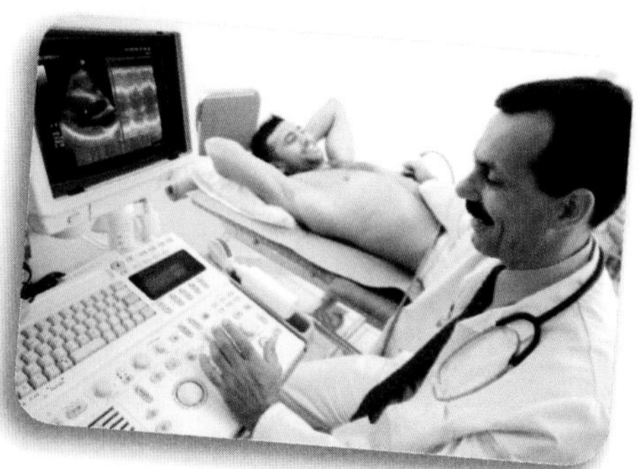

This technician is using an ultrasound machine to create an internal image of the patient's abdominal organs.

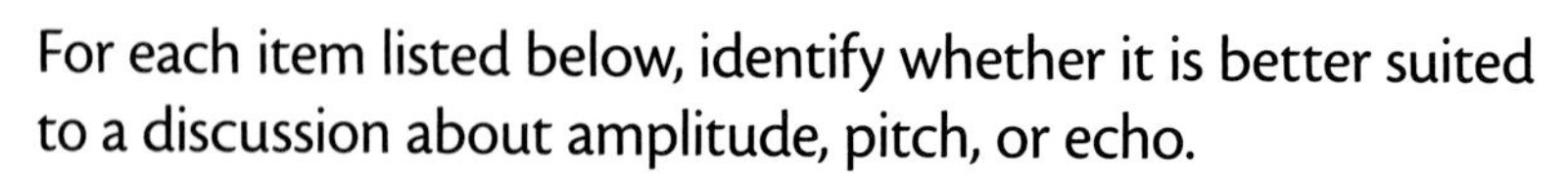

For each item listed below, identify whether it is better suited to a discussion about amplitude, pitch, or echo.

Bouncing Sound Waves

Soft Sounds

Find answers on 408–419

Uses of Sound Energy...

How is sound energy used in these activities?

Wireless Computing
A wireless router sends out radio waves that the computer picks up and interprets.

See also: page 266 Energy

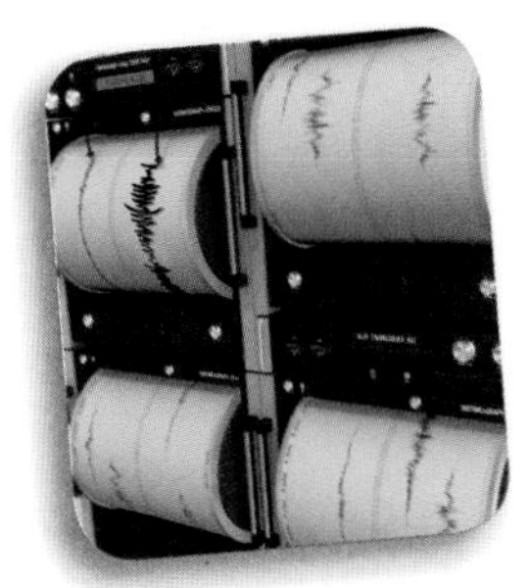

Earthquake Research
Seismic waves emitted by earthquakes can help scientists determine if and when aftershocks will occur.

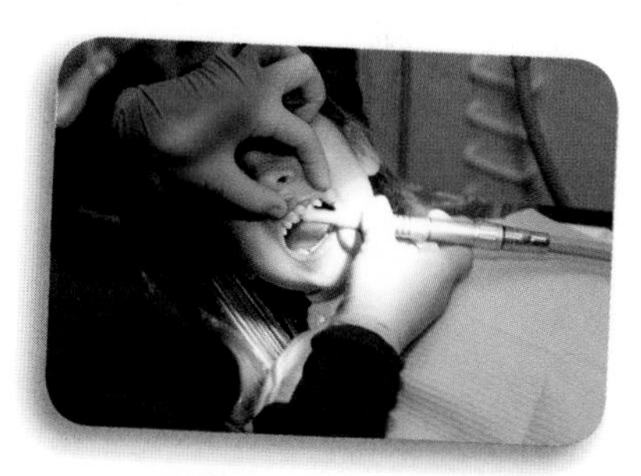

Dental Hygiene
Dental tools use sound energy to remove tartar from teeth.

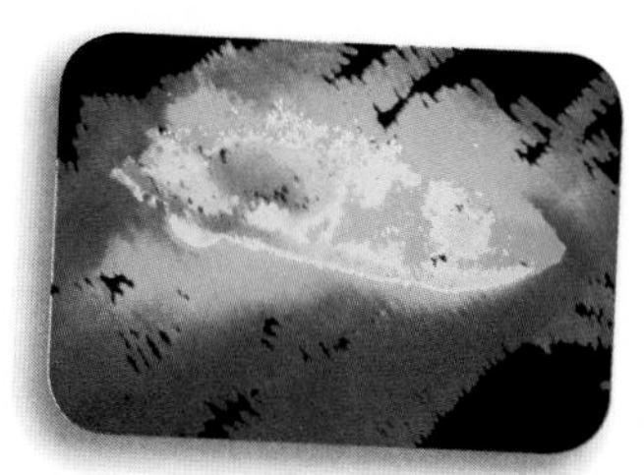

Seafloor Mapping
A ship sends sound waves downward, which bounce off the sea floor. Depth is calculated by the time it takes for the echo to reach the ship.

Go Online

To learn more about sound, check out these Web sites!

- **Interactive Sound Ruler**
  http://www.nidcd.nih.gov/health/education/decibel
- **Sound Wave Applet**
  http://www.grc.nasa.gov/WWW/K-12/airplane/sndwave.html

Heat

What Is Heat?

You want to go outside to play soccer. Should you wear a heavy jacket? Should you wear a hat, gloves, and long pants? Or should you wear shorts and a T-shirt? The answer depends on heat.

When atoms move very quickly, the temperature is high. It is warm outside. When atoms move more slowly, the temperature is low. It is cold outside.

Heat is energy. Heat energy is also called thermal energy. **Thermal energy** is the *total* amount of energy in a substance. **Heat** is the movement of energy from one thing to another due to a difference in temperature. Heat and temperature are not the same. **Temperature** is a measure of the *average* heat or thermal energy.

Matter is made up of particles called atoms and molecules. They bump into each other. They slide past each other. They vibrate back and forth. The more heat energy an object has, the faster the atoms and molecules move.

All of the particles in matter are moving. They are not all moving at the same speed. The more mass there is in a substance, the more heat energy it has. Temperature does not depend on the mass of a substance.

A pot of hot water has a lot of heat energy. The atoms and molecules in the hot water move faster than the atoms and molecules in cooler water move. A pot of hot water has more thermal energy than the same pot of water has at a cooler temperature.

The temperature of boiling water is 100°C. The temperature of ice is 0°C. The atoms and molecules in ice move very slowly. The iceberg has a lot more mass than the water boiling in the pot has. The iceberg has more thermal energy.

Heat and Energy Transformations

Heat is energy. Energy can change from one form to another.

A hair dryer turns electrical energy into heat. The heat dries your hair. Some ovens use electrical energy. The oven changes electrical energy into heat to cook food. Chemical energy stored in wood becomes heat energy when the wood burns. And heat energy can be given off during an energy change. That is why some light bulbs get hot when they have been on for a while.

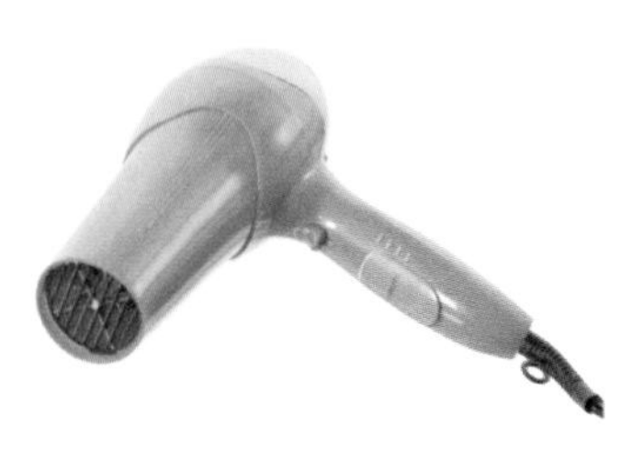

Chemical energy from food changes into heat energy to keep your body warm.

A light bulb uses electrical energy to make light. Energy that is not used for light is changed into heat.

Light energy from the Sun becomes heat energy when it reaches the atmosphere and the surface of Earth.

Heat and Friction

Friction is a force between two surfaces that touch. When you use brakes on a bicycle, the brake pads press against the wheel. Friction between the brake pads and the wheel slows the wheel down. The wheel and brake can get very hot. Some of the energy of motion changes to heat energy.

See also: page 330 Forces and Motion

A car needs oil because of friction. When oil is used, there is less friction between moving parts that touch. It keeps the engine cooler. Without oil, a car will overheat.

When you rub sandpaper on wood, the friction between the two surfaces can make heat.

Transfer of Heat Energy

Heat moves all the time. How hot or cold something feels has to do with the direction that heat energy flows. Heat flows from places where the temperature is higher to places where the temperature is lower. When the temperatures in both places are the same, heat stops moving.

Heat energy flows from your warm tongue to a frozen ice pop. The ice pop feels cold because your tongue is hotter than the ice pop.

If soup is hotter than your body temperature, heat energy flows from the hot soup to your tongue. The soup feels hot because your tongue is colder than the soup.

HOT OR COLD?

Feel how body temperature affects how hot or cold something feels.

1. Put both hands in a bowl filled with room temperature water. Does the water feel hot or cold? Do your right and left hands feel the same temperature?
2. Put your right hand into a bowl of water that has been cooled with ice. At the same time, hold your left hand under warm running water from the sink. (Be careful. The water should not be too hot.) Let both hands soak in this way for one or two minutes.

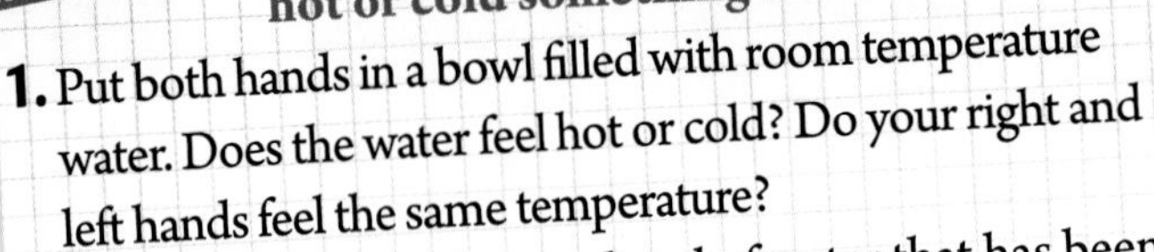

3. Now, dip both hands in the first bowl of room-temperature water. How does the water feel to each hand? Is it hot or cold?
4. Heat energy flows from warm things to cold things. Explain why each hand felt the way it did.

Conduction

Heat energy can move from one thing to another in three different ways: by conduction, by convection, and by radiation.

Conduction is the flow of heat energy between things that are touching. Everything is made of atoms. When things touch, their atoms touch one another. The fast vibrations of the atoms in the hot object will speed up the atoms of the cold object. The faster moving atoms make the cold object warmer.

You can feel conduction of heat when you have a hot drink. When you place a spoon in a cup holding a hot drink, the spoon gets warm because of conduction. The cooler spoon touches the warmer hot drink. Heat energy moves from the drink to the spoon. If you touch the handle of the spoon, it will feel warm. That is because conduction causes the heat energy to flow from the spoon to your hand.

Ice cream can melt in your mouth when you eat it. Heat energy is conducted from your warm mouth to the cooler ice cream.

When your bare feet touch the sidewalk near the pool, it feels hot. Heat energy is conducted from the sidewalk to your bare feet.

Heat Insulators and Conductors

Heat flows from hotter areas to cooler areas. **Insulators** are materials that do not allow heat energy to flow through them easily. Wood, plastic foam, and many kinds of cloth are insulators. A hot substance protected by an insulator does not get cold very quickly. Heat transfer is slowed down.

Insulators keep heat from flowing from one object to another. A picnic cooler will keep ice and drinks cold longer than a metal container.

The plastic foam inside the cooler is an insulator. It slows the flow of heat energy. The drinks and ice stay cold. Hot air cannot get inside.

Some materials conduct heat better than others do. A metal spoon in a hot drink feels hot. A plastic spoon stays cool. A **conductor** is a material that allows heat energy to flow through it easily. Metals are often good heat conductors. Plastics are usually not.

Water is a good conductor of heat energy. Your body is made up mostly of water. That is why heat energy flows easily from a hot object to your hand when you touch it.

The pancake griddle is made of metal. Metal allows the heat from the stove to move easily into the pancake batter.

Convection

Convection is the flow of heat energy through the movement of atoms in **fluids**. Liquids and gases are fluids. Water is a fluid. So is air.

Heat rises. Warmer fluid rises into cooler fluid. Cooler fluid sinks into warmer fluid. Over time, this movement causes all parts of the fluid to mix and be the same temperature. A small heater can warm the air in an entire room through convection. A swimming pool heater can heat all of the water in the pool in this way.

Convection causes some of the weather on Earth. Hurricanes, or tropical cyclones, are caused by convection.

Hurricanes form over very warm ocean water. Water must be about 27°C (80°F). That is why hurricanes often form between June and November. The heat energy that is stored in the ocean water flows to the cooler air in the atmosphere. The water in the air turns into clouds. Heat from the air moves higher into the atmosphere. As the hot air rises, cooler air rushes in to take its place. This causes very strong winds.

Radiation

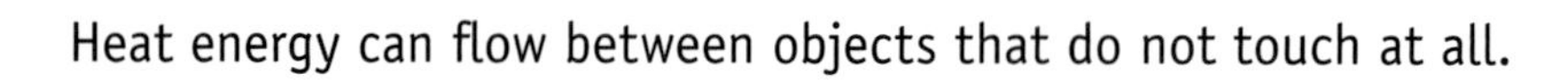

Heat energy can flow between objects that do not touch at all.

Heat energy comes off a hot object in all directions. This type of heat energy flow is called **radiation.** Heat energy can travel through empty space in this way. The Sun warms Earth through radiation.

Energy can move between particles of matter through radiation.

On a sunny day, the air inside a greenhouse is hotter than the air outside. The sunlight coming through the windows heats the air inside through radiation.

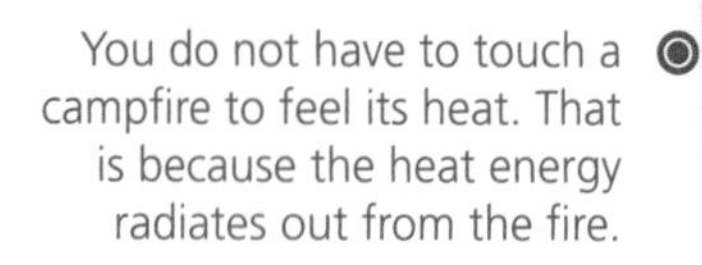

You do not have to touch a campfire to feel its heat. That is because the heat energy radiates out from the fire.

GO ONLINE

To learn more about heat, check out these Web sites!

- **Trapping heat**
  http://pbskids.org/dragonflytv/
- **Hurricanes**
  http://www.spaceplace.jpl.nasa.gov/en/kids
- **Heat Energy**
  http://portal.acs.org/

Light and Heat

Light and heat often go together. The Sun gives off light that heats up Earth. That is why you usually feel warmer in the sunlight than in the shade. A fire gives off both light and heat. A candle can light up a room, but it can also burn your fingers.

Energy

A special kind of lamp called a heat lamp gives off light that is good at heating things. The light is not very bright, but it is very hot. Heat lamps are used to keep food warm at some restaurants. They are also used to keep baby farm animals warm when their mother cannot be there to care for them.

A heat lamp gives off heat energy that warms the piglets.

Is That a Fact?

Glowing Hot

Some very hot things give off so much heat that they glow. This happens because their atoms vibrate very quickly. The vibrations can give off energy in the form of light. They also give off heat energy. When a blacksmith heats iron, it can get so hot that it glows orange. When glass is heated to 900°C, it can be molded into shapes. The hot glass glows bright orange.

Light

Which-Is-Which?

SUNBEAM WARMING A SLEEPING CAT

HOT OVEN WARMING A KITCHEN

Types of Energy Name the type of heat energy transfer in each situation. Write your answers in your science notebook.

Light

What Is Light?

Light affects every part of life. It lets you see your home and your friends. But there's much more to light than what meets the eye. So ... what is light?

Light is a form of energy that travels in waves. It is produced when one form of energy is converted to light energy. Because light is a form of energy, you cannot touch it. But light is very real.

The light coming from a flashlight looks like a single beam of white light. But it actually contains all the colors of the rainbow. You can see these by passing light through a **prism** (PRIZ um).

Did You Know?

People once thought prisms actually changed white light into colored lights. Sir Isaac Newton was the first to show that colors are part of white light. In 1666 he used a prism to break white light into its colors. Then he used a second prism to collapse the colors back into white light!

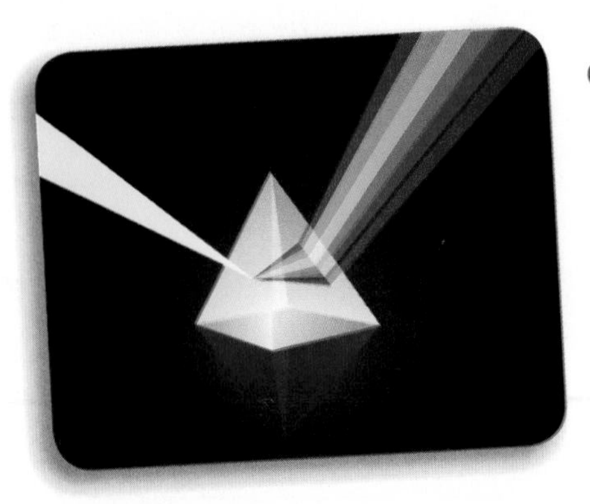

The colors in white light always appear in the same order: red, orange, yellow, green, blue, indigo, violet. If you write down the first letter of each color, they spell out a name:

Roy G. Biv!

Remember that name, and you'll remember the colors.

The Electromagnetic Spectrum

The colors that make up white light are called **visible light**. Most forms of light are not visible. Visible and nonvisible light make up the **electromagnetic spectrum** (ih lek troh mag NET ik SPEK trum).

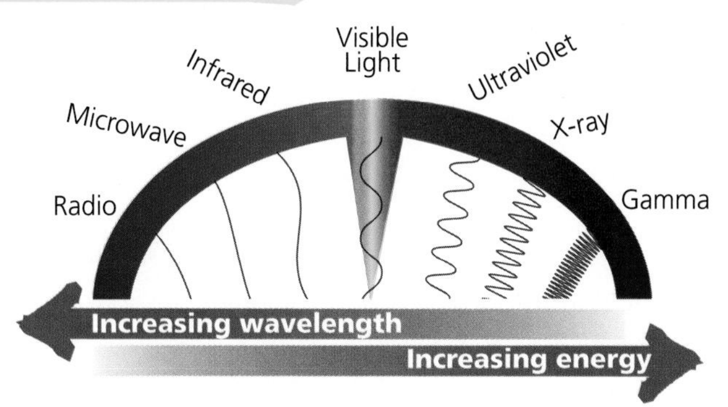

See also: page 266 Energy

Different forms of light have different wavelengths. Radio waves and microwaves have very long wavelengths. Gamma rays and cosmic rays have very short wavelengths. Light with longer wavelengths has less energy than light with shorter wavelengths.

Visible light is in the middle of the electromagnetic spectrum. Red light has the longest wavelength of visible light. Violet has the shortest. These colors always appear in the same order when you separate white light with a prism.

A glass paperweight, the bottom of a glass, or a watch crystal all work well as prisms.

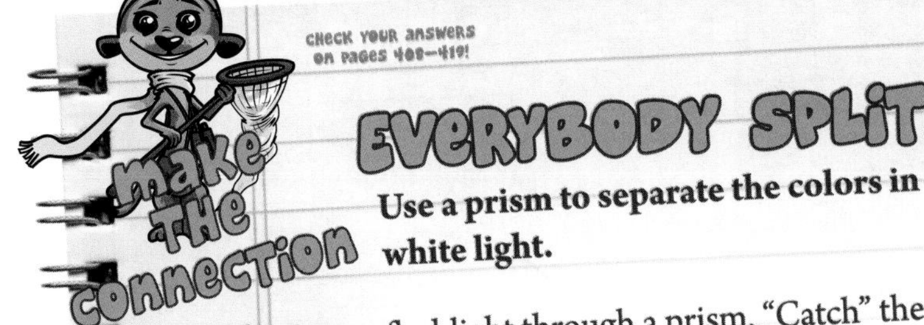

Check your answers on pages 408–419!

Everybody Split!

Use a prism to separate the colors in white light.

1. Shine a flashlight through a prism. "Catch" the light coming out the other side with a piece of white paper.
2. Write the colors you see in the order they appear from top to bottom.
3. Repeat Steps 1 and 2 with different prisms.
4. What conclusions can you draw from your observations?

Reflection of Light

Reflection occurs when light rays bounce off the surface of an object.

Light always travels in a straight line … until it meets matter. When light rays strike an object, some rays may bounce off the object's surface. The bouncing of light off a surface is called **reflection.**

The way light is reflected depends on the surface it strikes. If light hits a smooth, shiny surface such as a mirror, all of the rays are reflected together at the same angle. This is called **specular reflection.** It produces sharp, mirror-like images.

Light waves bounce off the smooth surface of this water and reflect the buildings nearby.

Moving water can't act like a mirror. Light rays bounce off this rough water and scatter in different directions. This produces a blurry image of sunlight on the water.

When light hits a dull or rough surface, the rays bounce off in different directions. Instead of a clear image, the scattered rays form a soft or hazy pattern. This is called **diffuse reflection.**

A **laser** is a special type of light that breaks many rules about light behavior! White light contains different wavelengths of light that spread in different directions. Laser light contains light rays that are all the same wavelength. And they travel parallel to each other. That's why laser light forms a focused point. You won't find laser light in nature. It is made in labs by humans. The word **laser** stands for **l**ightwave **a**mplification by **s**timulated **e**mission of **r**adiation.

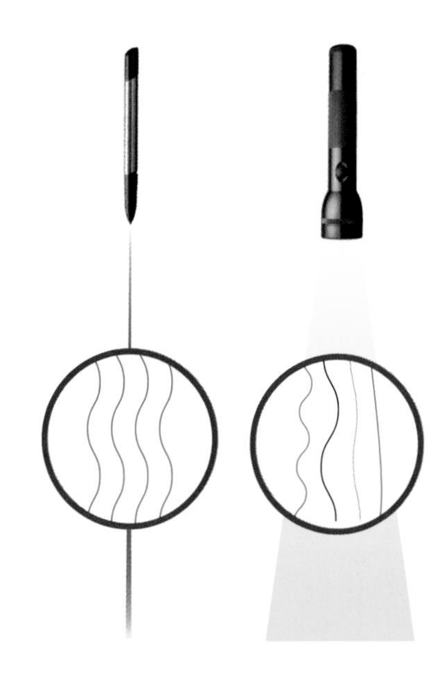

Make the Connection

Check your answers on pages 408–419!

Shiny, Shinier, Shiniest

Compare the way light reflects from different surfaces.

1. Use a small mirror, a shiny metal object such as a shiny pen or metal badge, and a blank CD. Plan to do this inquiry on a sunny day.
2. Stand near a window in your classroom. Hold the mirror so it catches sunlight from outside and reflects it onto a wall. Notice where the reflected light hits the wall.
3. Repeat Step 2 using your shiny metal object, and again with the blank CD. Record your observations.
4. Did light reflect the same way with each object? If not, how did it act?

Remember!

Do not shine the reflected sunlight into anyone's eyes; it can cause permanent damage.

Absorption of Light

When light strikes an object, some of its waves are reflected. But what happens to waves that are not reflected? Some waves that are not reflected are **absorbed**, or taken in by the object.

Absorption and Reflection

BLACK

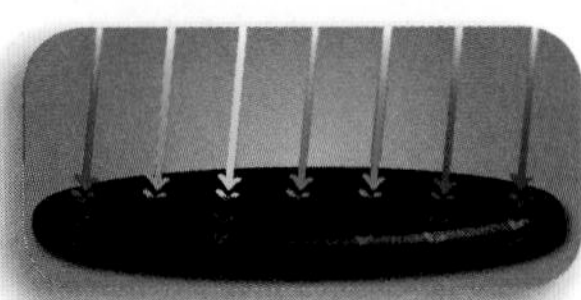

The way light is reflected or absorbed by an object determines the **colors** we see. This object looks blue because it reflects the blue wavelengths of light.

If an object reflects all wavelengths of incoming light, then no light is absorbed. In that case, the object appears to be white. Remember: white light contains all the colors of the visible spectrum.

If an object absorbs all wavelengths of incoming light, then no light is reflected. In that case, the object appears to be **black**.

This lemon looks yellow because it reflects yellow light and absorbs the other colors of visible light. The leaf looks green because it reflects green light.

GO ONLINE

To learn more, check out these Web sites!

- **Star Light, Star Bright**
  http://nasascience.nasa.gov/kids/the-universe
- **Amazing Facts/Why Is the Sky Blue**
  http://spaceplace.nasa.gov/en/kids/amazing_facts.shtml
- **Roofus' Solar & Efficient Home**
  http://www1.eere.energy.gov/kids/roofus/

Energy can change into different forms when it is absorbed by matter. Light energy that is absorbed by an object is converted to thermal energy. **Thermal energy** is a form of energy that produces heat.

Light energy can be put to work! Solar panels on the roof of this house convert light energy into thermal energy and electrical energy.

Alternative Energy

You may already know how to use color to reflect or absorb light. Do you wear a white shirt or a dark shirt on a hot sunny day? Most likely you said white. Dark clothes absorb light rays from the Sun and convert these to thermal energy. But light colors reflect the Sun's rays and keep you cool.

Did You Know?

Solar energy is a clean and "green" energy source. Some people put solar panels on their homes. These panels can produce almost all of the heat and electricity the family uses. That means they don't have to burn fossil fuels!

EXPLORE MORE

CHECK YOUR ANSWERS, PAGES 408–419!

CONVERT WITH COLOR

You will use color to change light energy to thermal energy.

1. Fill two glasses with cold tap water. Record the temperature of the water in each glass.
2. Completely cover one glass with a black cloth. Cover the other glass with a white cloth.
3. Set the glasses in a sunny window. Check the temperature in each glass every half hour for three hours. Write these in your science notebook.
4. Which glass warmed up faster? Why do you think this happened?

Transmission of Light

Some light rays are not reflected or absorbed when they strike an object. Instead, these light rays pass right through matter. This is called **transmission**. The amount of light transmitted through an object depends on how clear that object is.

Transparent surfaces allow most light to pass through. You can see through transparent matter. Glass, clear plastic, and clean water are transparent.

Translucent surfaces allow some light to pass through. You can see through translucent matter, but things may look cloudy. Sheer fabric and tracing paper are translucent.

Opaque surfaces block all light from passing through. You cannot see through opaque matter. A brick wall is opaque. So is a rock.

You can see through transparent materials, such as the glass and the water in it.

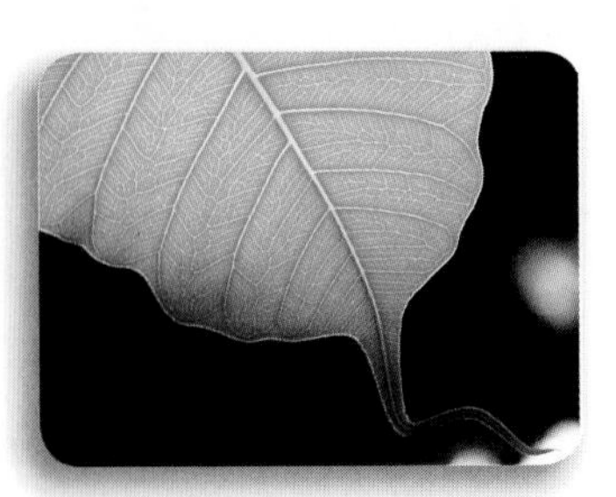

This leaf is translucent—it is neither clear nor opaque. You can see some light coming through the leaf.

A blackboard is opaque because no light passes through it. Chalk is opaque, too.

Which-Is-Which?

Identify each item as transparent, translucent, or opaque. Write your answers in your science notebook.

Find answers on 408–419

You know that light travels in a straight line until it strikes something. As light rays pass from one medium to another, the rays bend or change direction slightly. This is called **refraction**.

Light bends as it moves from air to liquid. That's why your legs seem bent when you're standing in a swimming pool.

Did You Know?

When we say *medium* here, we don't mean popcorn size! In science, a *medium* is a substance that can transmit something. Water is a different *medium* than air. *Medium* has a lot of other meanings, too. Check it out!

Make the Connection

CHECK YOUR ANSWERS ON PAGES 408–419!

Is That a Bendable Straw?

You will observe the law of refraction in this exercise.

1. Fill a clear glass or clear plastic glass with water.
2. Slowly lower a straight straw into the water. Write down what you see.
3. What did you see when you submerged the straw? Use what you know about refraction to explain what you observed.

Light and Eyes

Sight begins with the eyes and ends with the brain. When you look at something, an image enters your eye as light. But it's your brain that tells you what it is.

See also: page 58

Cells, Tissues, and Organs

Let's look step-by-step at how we see:

1. Light from the image enters the cornea and moves through the pupil to the lens.
2. The lens focuses the light and passes it to the retina.
3. The retina focuses the light and sends information about it to the optic nerve.
4. The optic nerve carries this information to the brain.
5. The brain interprets the information and "tells" you what the image is.

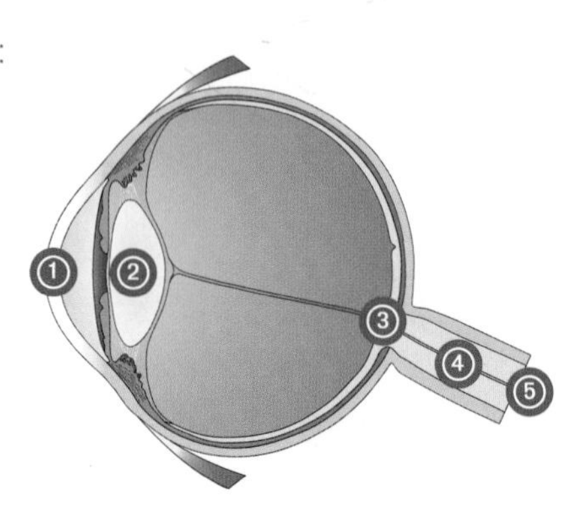

To see clearly, light must focus on the retina. People with poor vision wear glasses with lenses that bend light so that it focuses correctly.

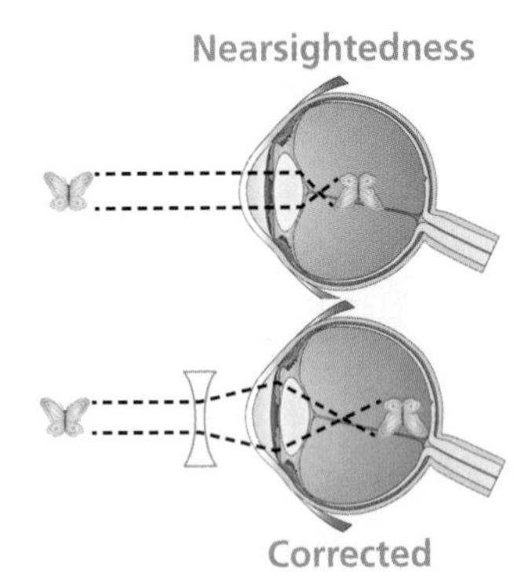

In nearsightedness, light focuses in front of the retina. Close objects are clear but distant images are blurred. Nearsighted people wear glasses with concave lenses. A **concave lens** is "caved in" or curved inward.

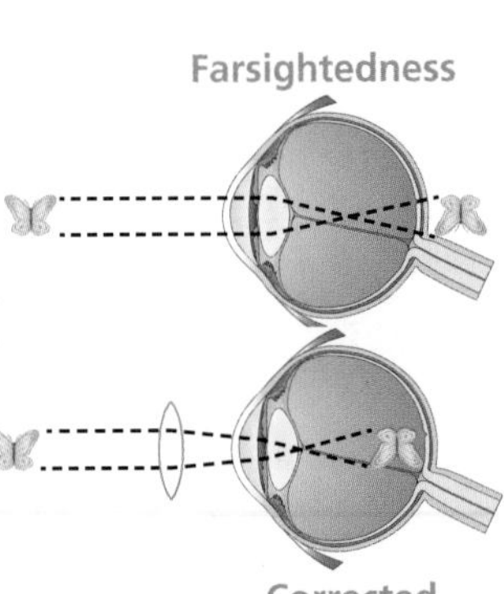

In farsightedness, light focuses behind the retina. Distant images are clear but close images are blurred. Farsighted people wear glasses with convex lenses. A **convex lens** is rounded and curves outward.

RODS AND CONES ARE SPECIALIZED EYE CELLS. RODS DETECT SHAPES AND MOTION IN THE DARK. CONES DETECT COLORS IN DAYLIGHT. ALL ANIMALS HAVE BOTH RODS AND CONES. BUT OWLS HAVE MORE RODS THAN ANY OTHER ANIMAL - INCLUDING US!

Simple Machines

What Is a Simple Machine?

A **machine** is a tool that uses energy to do work. When someone asks you to describe a machine, what do you say? You might mention something that runs on gasoline or electricity. But do you know that there are six machines that don't use a key, a motor, or fossil fuels, and have been in use for thousands of years? Read on to learn more about these simple machines.

Many simple machines work by letting you use less force to move an object. The trade-off is that the object must be moved over a greater distance. Sometimes, however, a greater force must be used, but over a shorter distance. The six simple machines are:

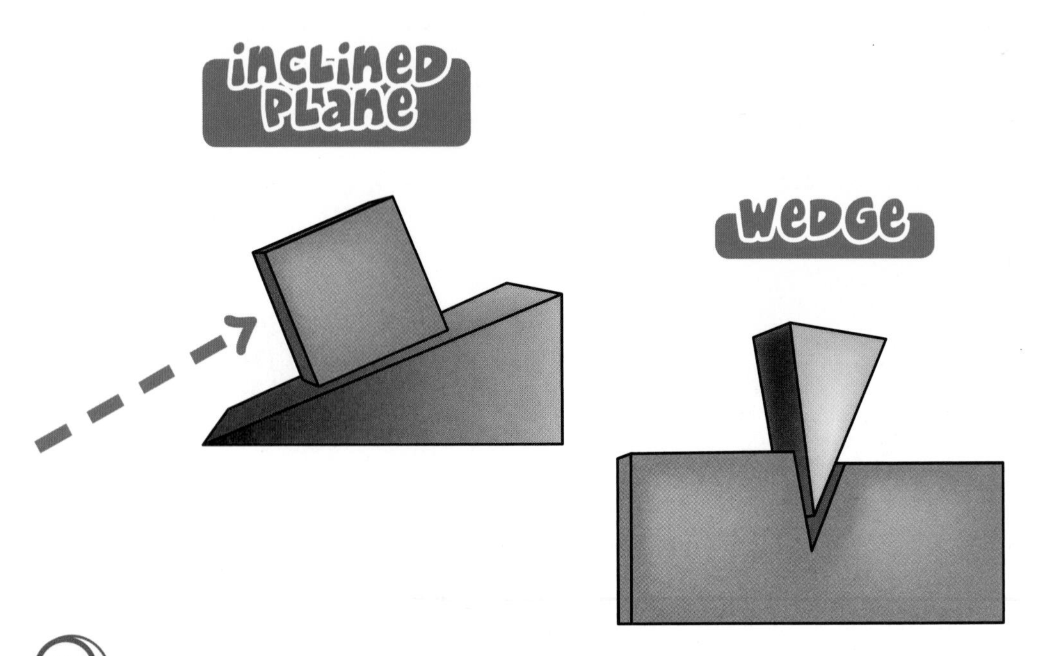

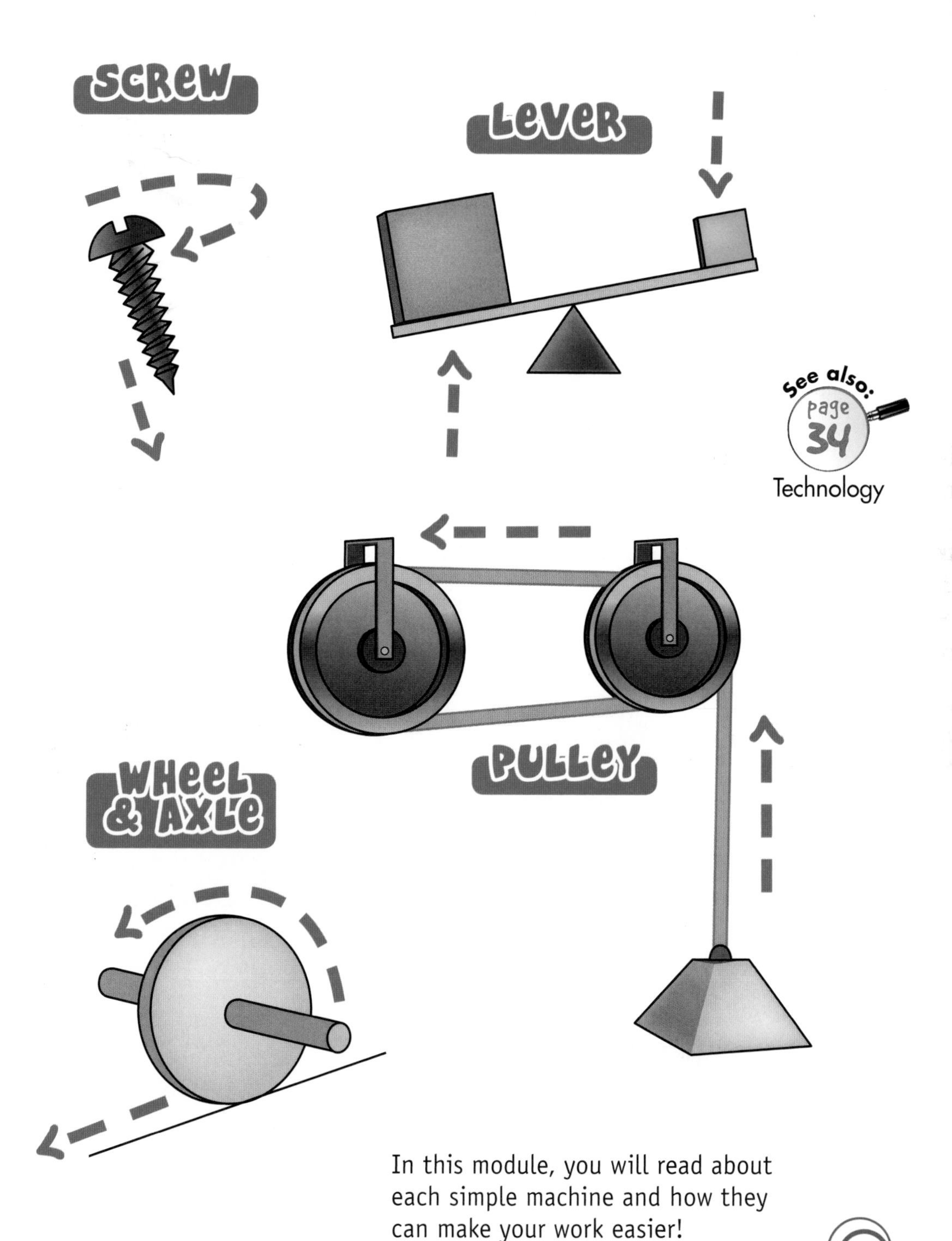

See also: page 34 Technology

In this module, you will read about each simple machine and how they can make your work easier!

Inclined Plane

An **inclined plane** is a flat surface that slants. You see these every day, and perhaps you use one yourself. A good example of an inclined plane is a wheelchair ramp. This type of inclined plane enables a person in a wheelchair to move from one level to another without relying on an elevator.

See also: page 350
Science and Math Together

A mover uses a ramp to roll heavy furniture on and off a van. It might take three people to lift a crate that weighs 150 kg (331 lb) several meters (feet) off the ground and push it onto the truck. By using a ramp, one mover can handle the same load.

NUMBER CRUNCH

CHECK YOUR ANSWERS, PAGES 408–419!

How Long? How High?

You are helping to build two wheelchair ramps at your school. Each ramp cannot rise more than 1 inch for every 12 inches it travels forward. The first ramp needs to travel up 2 steps. Each step is 4 inches high. What is the minimum length of the ramp?

1. Calculate the total height the ramp must travel:
(2 steps) × (4 inches per step) = 8 inches

2. Calculate the length of the ramp in inches:
(8 inches high) × (12 inches forward) = 96 inches

The ramp must be 96 inches long.

A second ramp needs to travel up 3 steps. Each step is 3 inches high. What is the minimum length of the ramp?

Wedge

A **wedge** is a triangular object that is essentially an inclined plane. Wedges are used to move or split objects, or to keep objects in place. A good example of a wedge is a doorstop. If you have a door in your house that won't stay open, you can place a wedge-shaped doorstop underneath the door, and it will stay put!

The head of this axe is a wedge. When force is applied, it can split a huge log in half! The energy comes from muscle power magnified by the motion of swinging the axe.

GO ONLINE

To learn more about simple machines, check out these Web sites!

- **Simple Machines Activities**
  http://www.edheads.org/activities/simple-machines
- **Understanding Simple Machines**
  http://fi.edu/pieces/knox/automaton/simple.htm

Screw

You know what a screw is ... or do you? In its most basic form, a **screw** is an inclined plane wrapped around a cylinder. The inclined plane is called the thread of the screw. The more times that the plane wraps around the cylinder, the easier the screw is to turn.

See also: page 330 Forces and Motion

A simple wood screw can hold two pieces of wood together.

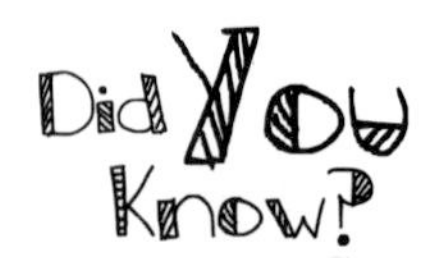

Did you know that a nail is actually a wedge? The angled edges at the end of the nail form a point, or wedge. When you hammer a nail into a block of wood, you drive the pointed end into the wood.

Check your answers on pages 408–419!

Hammer or Screwdriver?

Imagine that you are on a construction site. Which tools will you use? Hammer and nails or screwdriver and screws? Nails are often used when there are many pieces to install over a large area. Nails are fast; a good carpenter can sink a nail with two hits. Floorboards are put down with nails. Screws are used in areas where a tight fit and the ability to stand up to heavy use are important. Door hinges need screws. Now, what do you think roofers use in their work? Why?

Lever

When was the last time you rode a seesaw? How easy was it to send a friend sailing up high just by pulling yourself downward? Then you could bounce up with just a little push of your feet.

Energy

A seesaw is a lever. A **lever** uses a board and a support to lift heavy objects. The board that you sit on is the lever's long bar. The base of the seesaw is the fulcrum, or pivot, that the bar rests on.

What would happen if you moved the fulcrum one way or the other? The position of the **fulcrum** determines the amount of force needed to lift the load.

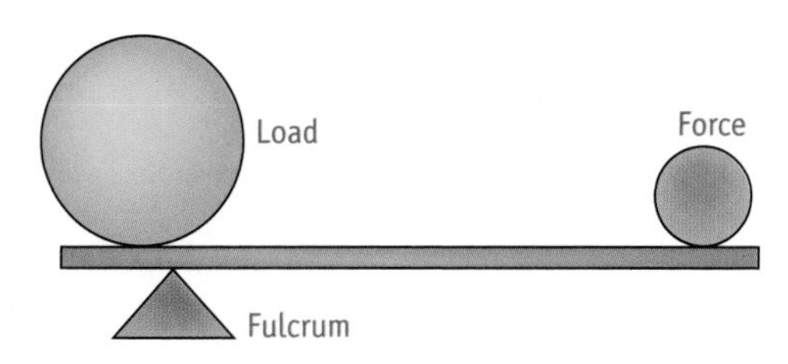

If the fulcrum is moved closer to the load, less force is needed to lift the load.

History Makers

Archimedes (287–212 b.c.e.)

Archimedes was a Greek mathematician, engineer, and inventor. He was the first person to explain the principle of how a lever works. He discovered that he could use a lever to move something that would be too heavy for him to lift or carry by himself.

Wheel and Axle

A **wheel and axle** consists of a wheel with a rod (an "axle") inserted through the center. The wheel and axle work together as a unit. When a force is applied to the axle, the wheel turns. Force that is applied to the wheel is transferred to the axle. Depending on the direction of the force, the wheel will move forward or backward. A good example of a wheel and axle is found on a bicycle.

The axle is attached to the center of the bike wheel. When you apply force to the axle in the direction you want to go, the wheel moves and sends you on your way.

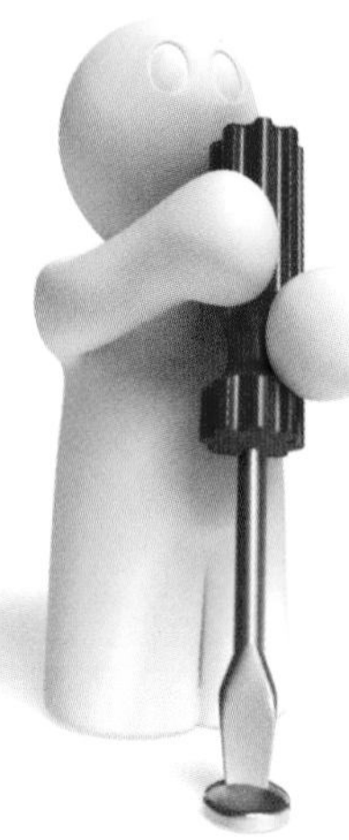

When a small force is applied to a wheel, the force is magnified in the axle. A not-so-obvious example is the screwdriver. The handle actually serves as the wheel. The shaft is the axle. As you turn the handle, the shaft has the greater power to turn the screw into the wood.

The handle of the screwdriver acts as the wheel. When it is turned, the shaft (or axle) moves with it, causing the screw to loosen or tighten.

Another example of a wheel and axle is a doorknob. The doorknob works as the wheel, and the shaft inside the door is the axle. When you turn the doorknob, it moves the shaft, which unlatches the door.

Pulley

A **pulley** is a simple machine that uses a small effort to lift a heavy load. A pulley is a grooved wheel with a rope or chain wrapped around it. The wheel is mounted on an axle. The axle is turned by a crank. On some large pulley systems, the crank is powered by a motor.

There are three types of pulley systems. Look at the systems on this page, and note the differences as you read about them.

A **fixed pulley** does not move. This type of pulley does not lessen the amount of force needed to lift the load. It does change the direction of the force. The fixed pulley on a sailboat hoists the sail when force is applied.

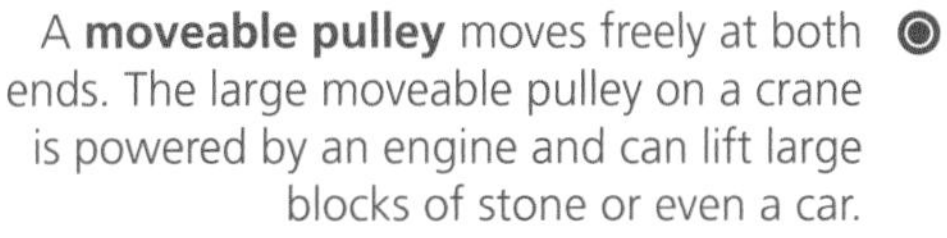

A **moveable pulley** moves freely at both ends. The large moveable pulley on a crane is powered by an engine and can lift large blocks of stone or even a car.

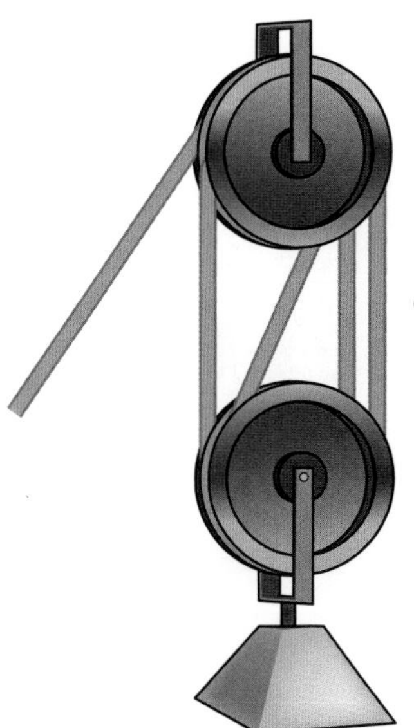

The **block and tackle** is a system of pulleys. A rope can be wrapped over one pulley, then up and over a second, and even a third pulley. Each pulley lessens the force needed to lift the load.

Electricity and Magnetism

What Is Electricity?

Electricity results from the movement interaction of positive and negatively charged atoms. Positive charges push away from each other. So do negative charges. But a positive charge and a negative charge attract each other. This is known as the **law of electric charges.**

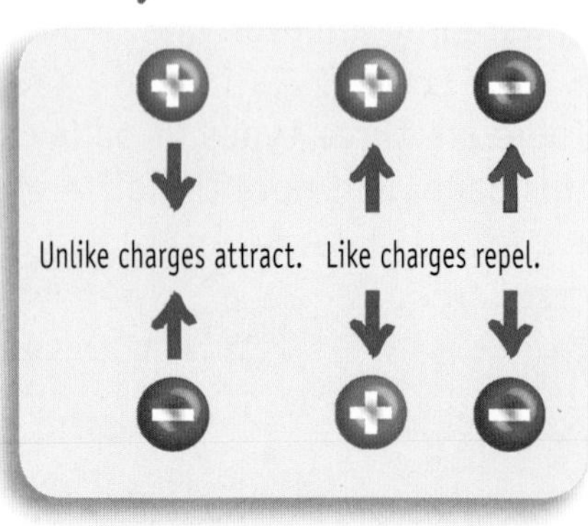

This illustration demonstrates the law of electric charges.

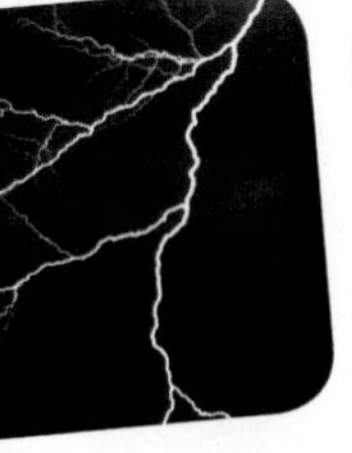

Static electricity is the buildup of electric charges on a surface. It can be harmless, like the shock you get from a doorknob after walking across a carpeted floor. It can also be dangerous, like a lightning strike.

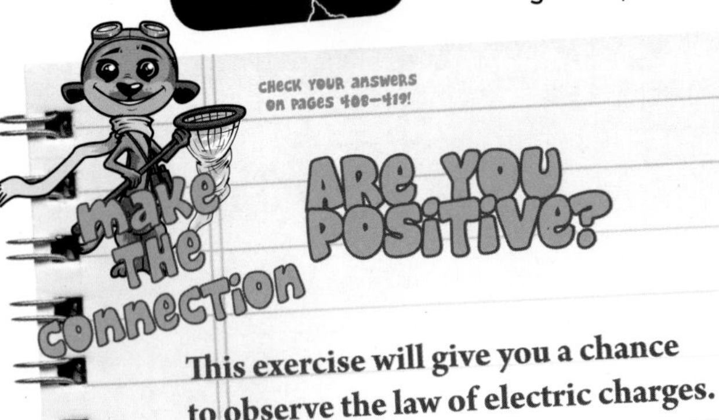

Check your answers on pages 408–419!

This exercise will give you a chance to observe the law of electric charges.

1. Blow up a balloon and tie it.
2. Rub the balloon several times against your hair.
3. Now hold the balloon near your head, but not touching it. What happens?
4. Why do you think this happens?

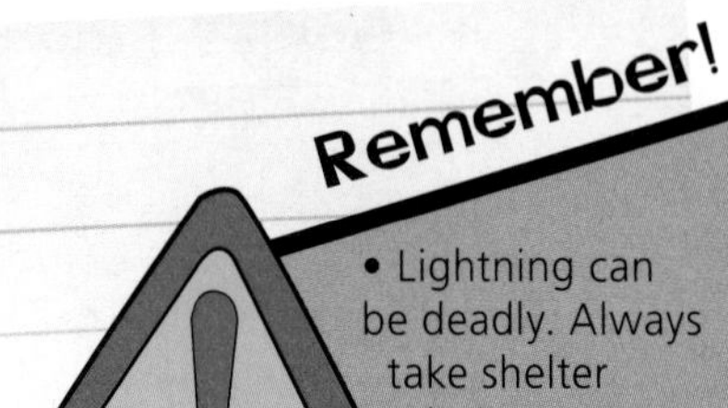

Remember!

- Lightning can be deadly. Always take shelter during a thunderstorm!
- Make sure balloons are not within reach of small children.

Electric Current

The constant flow of electric charge is called **current.** It is measured in **amperes**, or **amps**.

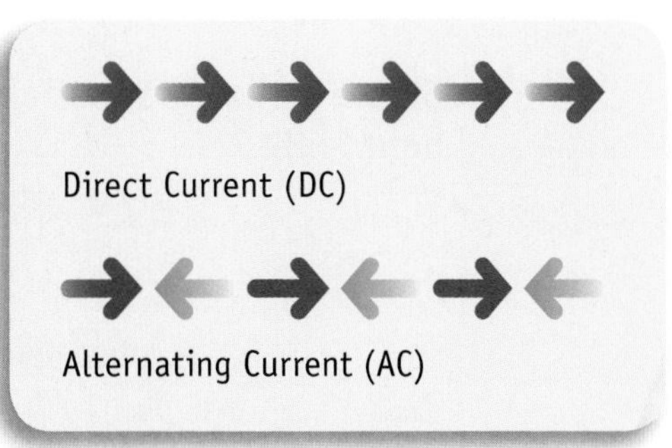

This illustration shows how electricity flows through direct and alternating currents.

There are two types of currents. **Direct current** (DC) moves in only one direction through a wire. An example of direct current is a flashlight battery. **Alternating current** (AC) changes direction within a wire. A power company delivers electricity to customers using alternating current.

Conductors are materials that carry electric current easily. Good conductors, such as copper and aluminum, are used in electric wiring. **Insulators** are materials that do not carry electric current easily. Good insulators, such as rubber and plastics, are wrapped around wires to contain the current.

See also: page 50
Atoms, Molecules, and Elements

Metallic wire conducts electricity easily. Thicker wire can handle more amps, or electric flow. Forcing too much electricity through a wire will make the wire hot and unsafe.

A light bulb has a black band at the base. The band is an insulator that prevents electricity from passing to parts of the lamp outside the bulb and wiring.

Electric Circuits

Before we learn about electric circuits, let's look at the parts of a light bulb.

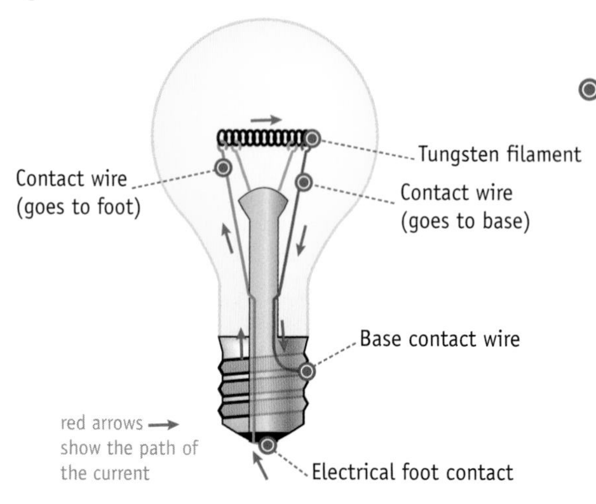

Electric current enters the light bulb through the foot contact, travels up through the contact wire, and into the tungsten filament. The current heats the filament to more than 3000° C (5400° F)—so hot that it glows. The super-hot filament gives off heat and light energy.

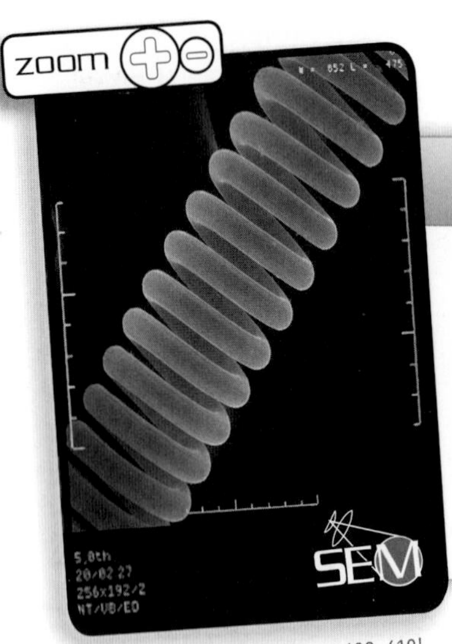

Answers on pages 408–419!

Thinking BIG™

www.carolinacurriculum.com/ThinkingBig

Record your ideas in your science notebook.

- **This is part of a system.**
- **What state of matter is it?**
- **How might this be used?**
- **How might the shape be important?**
- **What might the whole system look like? Draw your ideas.**

Solve IT

1) I'm part of a system.
2) You see me only through glass.
2) Metal conductors keep the flow going.
3) My shape might be a surprise.
4) Most don't know I'm so long.
5) Connect a battery. Close the circuit. Light.
6) What am I?

A circuit is a system along which electric current flows. Every circuit has four parts: (1) the energy source; (2) the load, or receiver; (3) connecting wires; and (4) a switch. Let's consider a simple circuit with a battery and light bulb.

There are two types of circuits that power multiple loads: series and parallel. What's the difference? Well, let's say that the loads in both circuits are light bulbs.

See also: page 292 Heat

In a **series circuit**, the loads are connected in a line. If one load malfunctions, the current cannot continue to the next. So, if one bulb burns out, all the loads in the circuit go dark.

See also: page 302 Light

In a **parallel circuit**, each load is connected in a closed circuit. If one load malfunctions, the other circuits will still work. So, if one bulb burns out, the others will stay lit.

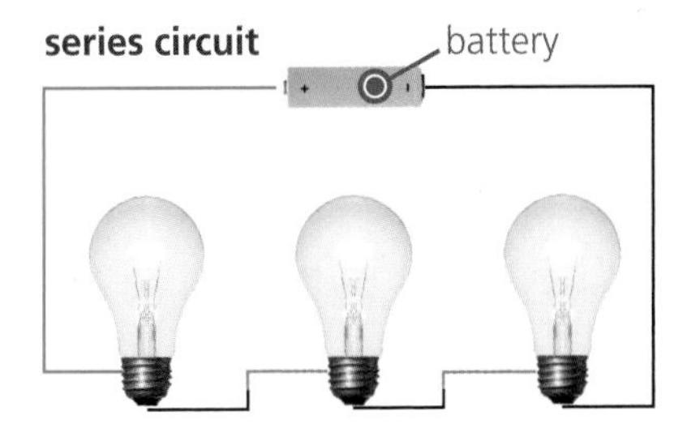

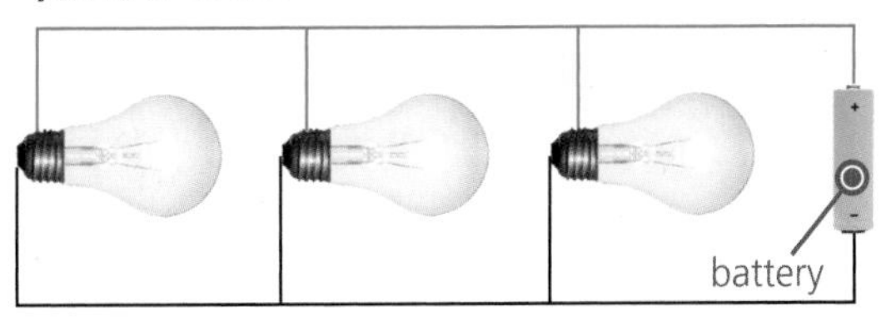

Do you see the difference between a series circuit and a parallel circuit?

Measuring Electricity

Light bulbs come in different shapes and sizes. All of them have one thing in common, though. They are defined by how many watts they put out. You have probably seen a light bulb that says "60 Watts" on it. A **watt** is a unit of measurement. It measures the electric power of an object. Light bulbs and small appliances put out electric power in watts. The amount of electric power that a building uses is so large that it is measured in **kilowatts**, or thousands of watts.

Until recently, most light bulbs have been incandescent bulbs. When something is **incandescent**, it can give off light energy if it is heated to a high enough temperature. This is the kind of light bulb described earlier in this module. In the last decade or so, a new energy-efficient light bulb has come on the market. This kind of bulb is called a CFL, which stands for compact fluorescent light bulb. CFLs are more expensive than incandescent light bulbs, but they give off the same amount of light, use less energy, and last longer.

What Is Magnetism?

The compass has been used for centuries to direct travelers on their journeys around the world—or just back to camp!

A **magnet** is an object that attracts most metals. A magnet has two poles, or areas where opposing electric charges are concentrated. These poles, commonly called the north and south poles, exert the most magnetic force on an object. You can think of magnetic force as a series of pushes and pulls. Poles with the same charge push against each other. Poles with different charges pull toward each other. A magnetic field is that area around a magnet that is acted upon by magnetic force. The way that objects behave in the presence of a magnetic force is called **magnetism.**

Why does a compass always point in the direction of the North Pole? The Earth acts like a huge magnet because it has a magnetic core. The south end of this magnet is at the North Pole, and the north end of the magnet is at the South Pole. A compass needle is magnetized so that its point is the north end of the magnet. It always seeks out the south end of the Earth's magnetic core, which is the North Pole.

Remember!

• The nail used in this experiment probably has a sharp point. Be careful when handling it.

BE A NAVIGATOR!

Make your own compass. Test it to see if it can find the Earth's north magnetic pole.

1. Rub a small steel nail with one end of a magnet. Be careful to rub in the same direction, not back and forth.
2. Have your teacher cut a cork in half. Tape your nail to the flat surface of the cork. The two ends of the nail should extend over the edges of the cork and should not be covered by the tape.
3. Fill a large bowl with water. Gently float your cork compass in the bowl of water. When the cork stops moving, the point of the nail will be directly north or south. Label the sides of the bowl with labels indicating north, south, east, and west. Now you have made your own compass!

Electromagnetism

This tiny electromagnet was used in early digital cameras.

Electromagnetism (eh LEK tro MAG nuh tiz em) is a specific type of magnetism that deals with electric charges that are in motion. An electric current produces a magnetic field. This field is weak around a single wire, but many wires coiled together and wrapped around an iron core create a powerful **electromagnet**. The electromagnetic field increases as the number of bends and loops in the wire increases.

Remember!

- The nail used in this experiment probably has a sharp point. Be careful when handling it.
- Use this battery only for electrical experiments.

EXPLORE MORE

CHECK YOUR ANSWERS, PAGES 408–419!

POWER UP!

Now you will make your electromagnet. You will need a 9-volt battery, a long iron nail, and about 1 m (3 ft) of copper wire.

1. Wrap the wire tightly around the nail. Leave about 15 cm (6 in) of wire free at each end.
2. Wind one end of the wire around the positive terminal of the battery. Wind the other end around the negative terminal. This is your electromagnet.
3. Attempt to pick up several metal objects, such as paper clips, staples, safety pins, straight pins, small nails, and jewelry findings.
4. Record the results of your attempts in your science notebook.
5. Compare your results with those of your classmates. If your results were different, try to find out why. Did you wind the copper coils in the same way? Did you use different lengths of wire? Did you use batteries with different voltages? Discuss how these differences might affect your outcome.

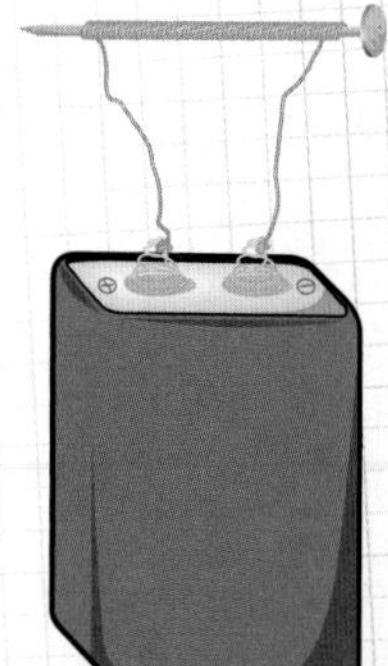

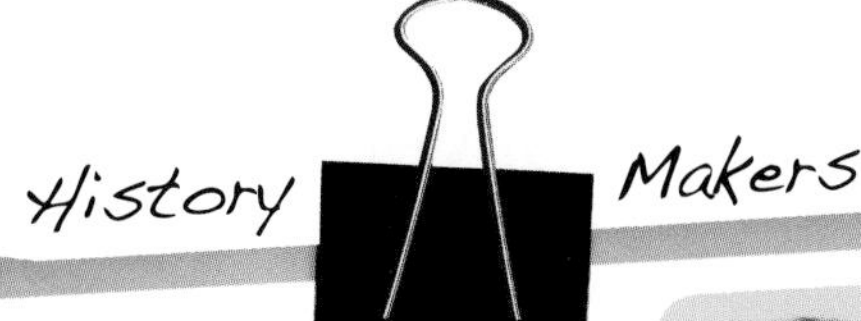

Hans Christian Ørsted (1777–1851)

In 1820, Hans Christian Ørsted, a Danish physicist, was the first to discover that an electric current creates a magnetic field around it. In the same year and immediately after, other scientists also made important contributions to the new field of electromagnetism.

See also: page 266 Energy

See also: page 330 Forces and Motion

The story of electromagnetism has not come to an end. There is continuing research into new and better energy-saving uses for this device. For example, the number of cars using electric motors continues to increase every year. Electromagnetic energy storage systems hold hope for environmentally safe energy use. Electromagnets can be made so small that they might be used in microchips in computers, in medical research, and in other scientific ventures that are limited only by the imagination!

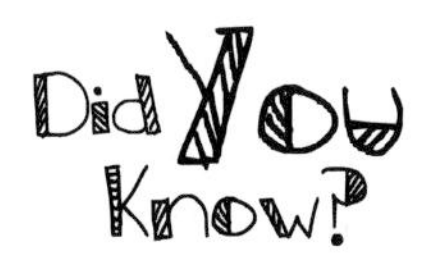

Magnetic Levitation (maglev)

The vehicle looks like a streamlined train. It travels above a guard rail-type track, but it does not touch the track. Instead, it hovers just above it with the use of powerful magnets. Maglev trains can travel more quietly, more smoothly, and faster than regular trains. This maglev train operates at Pudong International Airport in Shanghai, China.

Motors and Generators

An electric motor converts electric energy to motion energy. At the heart of every electric motor is an electromagnet. A black bar, called the axle, passes through the top of the electromagnet. The push and pull of the magnetic field on the coil causes the axle to spin. The spinning axle is the circular motion of the electric motor. There are other parts in an electric motor, but they all help to keep the electromagnet and power source operating that axle smoothly.

See also: page 266 Energy

This electric motor can fit in the palm of your hand.

An electric generator works in a way that is completely opposite to an electric motor. A generator uses motion energy to create electric energy. Just like a motor, an electromagnet is at the heart of a generator, and the parts of each are basically the same.

An electric generator can be used while camping.

Is That a Fact?

Electric Cars

Electric motors are already being used in some cars in addition to gasoline engines. Electric motors are quiet and fuel-efficient. The electric motors in cars rely on electromagnets, just like those you have been studying.

Conservation of Electricity

Energy conservation is important, and using electricity wisely is no exception. How can you conserve electricity? Well, every little bit helps! Here are some ideas you can use at home and at school:

1. Turn off lights and electronic devices when you leave a room.

2. Dress for the season, so you use less heat and less air conditioning.

3. Reduce the temperature of your hot water heater to 120° F.

4. Take a shower rather than a bath, to cut down on the energy needed to heat the water. While you're at it, take a shorter shower!

5. Know what you want inside the refrigerator before you open the door.

6. Use shades in your house to block hot summer sun. Open them in winter to let the sun in.

Go Online

To learn more about electricity, check out these Web sites!

- **Electricity and Conservation**
  http://www.tvakids.com/electricity/conservation.htm
- **Where Lightning Strikes**
  http://science.nasa.gov/headlines/y2001/ast05dec_1.htm

Forces and Motion

What Is Force?

Have you ever tried to pack a suitcase only to find that your clothes will not all fit? You might have needed to force that last pair of shorts inside by pushing on the suitcase. Any push or pull on an object is a **force**. Sometimes a force makes something move. You pull on a door handle to make the door open. Or you might push on a chair to move it.

Forces on this tire swing cause it to move.

Forces have direction. If you pull a wagon toward you, it will move toward you. If you push on a tire swing, it will move away from you. The force will move the swing forward. The chain of the swing pulls the swing upward. Eventually, the swing will move back toward you. That is because a force called gravity pulled the swing down.

Force can affect objects by setting them in motion. Forces can change the speed or direction of an object. Forces can also act on objects that are not moving.

All motion is caused by forces, but not all forces cause motion. A basketball sitting on a gym floor exerts force on the floor. The size of the force is equal to the weight of the basketball. The direction of the force is downward. The basketball does not move.

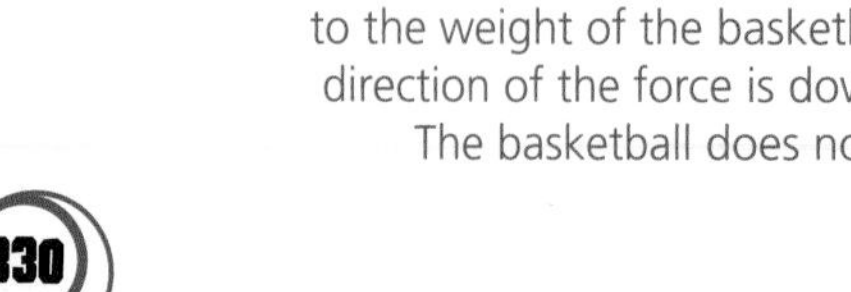

Measuring Force

Forces have **magnitude**, or size. Forces can be large or small. Some forces are gentle, like the force of a light breeze pushing on a flag. Other forces are stronger, like the force of a bat hitting a baseball.

A force can be very large or very small. A gentle breeze releases dandelion seeds. The high winds of a hurricane batter these palm trees.

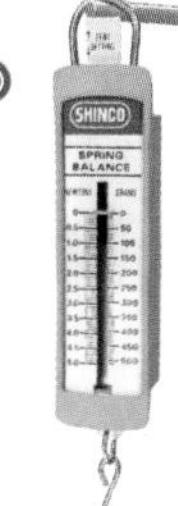

You may use a spring scale to measure pulling forces. One side of this spring scale measures force in grams. The other side measures force in newtons.

You can measure force the same way you measure weight. In the United States, the unit for weight is the pound. In other countries, weight is measured in kilograms. Scientists measure force using units called **newtons**.

Science Tools

A bathroom scale measures the force of your body pushing down on it.

History 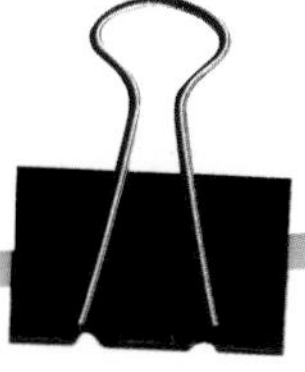 Makers

Isaac Newton (1642–1727)

Isaac Newton was an English scientist. He also studied mathematics. He was the first scientist to use mathematics to show how the force of gravity pulls objects on Earth and in space. Newton's three Laws of Motion are still used to describe how forces make objects move. Scientists gave his name to the unit used to measure force, the *newton*.

Contact Forces

Forces may be described as contact forces and non-contact forces. Many of the forces that you apply to things every day are contact forces. A **contact force** is a push or a pull that happens between objects that are touching.

Think about all the contact forces that happen during a game of baseball. You touch a baseball to make it move. You push it with your hand when you throw it. You also apply a force that stops the ball when you catch it. You push the ball with a bat when you hit it. In each case, a contact force makes the ball move or stop moving.

A dog pulls forward on the leash. The girl pulls back on the leash. Pulling the leash at both ends creates tension. Tension is a contact force.

A boy pushes the bowling ball. The ball hits the pins. Both are contact forces.

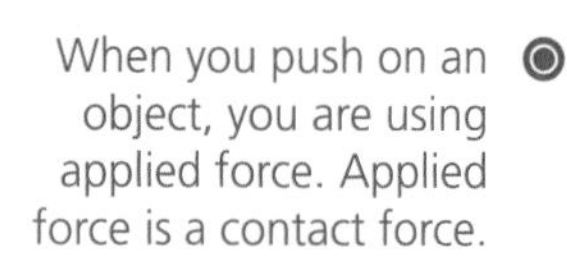

When you push on an object, you are using applied force. Applied force is a contact force.

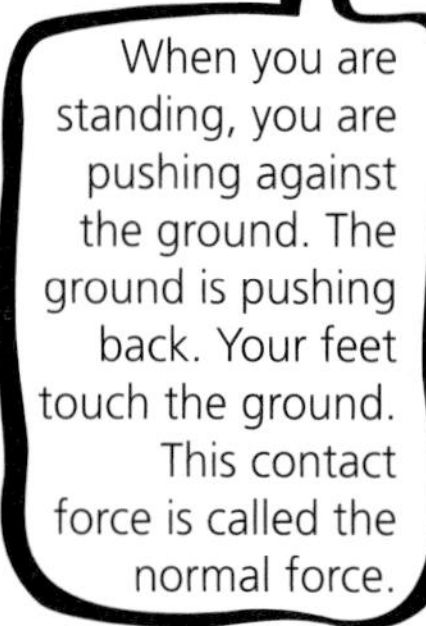

The boy pushes the weights up to lift them. The boy and the weights are touching.

Friction Did you ever slide on a smooth floor in your socks? You can slide pretty fast on a smooth floor. But you cannot slide very much at all on a carpet or on rough concrete. A rough surface will slow your movement. The difference in your movement on each type of floor has to do with friction.

When you rub any two things together, they both slow down. **Friction** is the force that slows down movement between two surfaces that are touching. Friction works in the opposite direction of motion. Rough surfaces have a lot of friction. When you rub rough surfaces together, they snag and pull on each other. This happens when you try to rub two pieces of sandpaper together.

A hockey player can glide easily on ice because ice has a very smooth surface. There is very little friction acting between his ice skate and the ice.

A baseball player slides more easily on loose dirt than he would on rough cement. There is more friction between the player and the cement than between the player and the loose dirt.

GO ONLINE

To learn more about force, check out this Web site!

- **Forces and Falling Towers**
  http://www.physicscentral.com/experiment/physicsathome/towers.cfm

Non-Contact Forces

Sometimes objects can apply forces on one another without touching. A **non-contact force** is a push or a pull between objects that are not touching.

See also: page 320 Electricity and Magnetism

Electricity Have you ever pulled socks out of a dryer and had them cling to each other? Sometimes if you hold them near a loose thread, you can make the loose thread stand up. The socks do not have to touch the thread, but they can still pull on it. This force, called **electric force**, is a non-contact force. A non-contact force works between things that do not touch.

Magnetism Magnets also apply non-contact forces. **Magnets** are objects that attract things that have iron in them. When you hold a magnet near a steel paper clip, it will pull the paper clip toward it. It does not have to touch the paper clip to apply a force. The area around a magnet that applies a force is called the **magnetic field**. The paper clip must be inside the magnet's magnetic field to be pulled by the force.

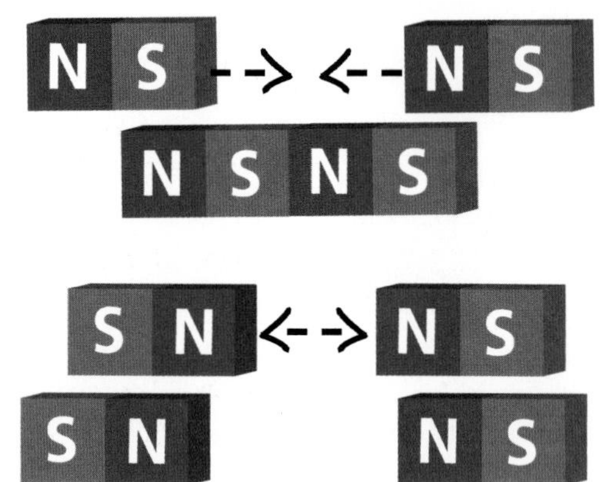

Every magnet has a north pole and a south pole. North poles pull toward south poles and vice versa. But two north or two south poles push away from each other.

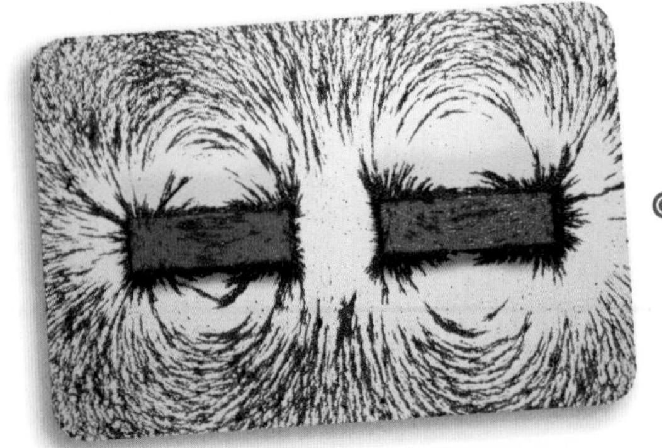

These bits of iron show the curved areas of the magnetic fields around these two magnets.

Satellites in space stay in orbit around Earth because of gravity.

Gravity Bump a soccer ball. The contact force sends the ball upward. It soon comes down again. A non-contact force called gravity pulls the ball to Earth. **Gravity** is a pulling force between objects that have **mass**. Gravity acts between Earth and all objects near Earth. It pulls objects down toward Earth.

All things that are made of matter have gravity. The bigger something is, and the closer you are to it, the stronger its gravity. Earth is very large, and the pull of its gravity is very strong.

The area around objects where the force of gravity acts is called the **gravitational field**. Some things in space get close enough to Earth's gravitational field that they are pulled toward Earth. The Moon is such an object. Its orbit, or circular path around Earth, is due in part to gravity.

The Sun is much larger than Earth. It has more mass than any planet in the solar system has. The gravity from the Sun holds all of the planets in orbit.

The ball and the soccer player are pulled to Earth by gravity. To bump the ball up, the soccer player has to use a force greater than the force of gravity pulling the ball downward.

Which-Is-Which?

FRICTION

THROW

MAGNETIC

Types of Force Which of the words above describe a contact force? Which of the words describe a non-contact force? Write your answers in your science notebook.

FIND ANSWERS ON 408–419

Balanced Forces

More than one force can act on an object at one time. If you are standing still, gravity is pulling you down to Earth. The ground is pushing up on your feet. The weight of your backpack or any object that you are holding, even your clothes, pushes on you, too. Even the air around you is pushing on you. How can you stand still with all these forces acting on you? The reason has to do with the sizes and directions of these forces.

When the forces that are pushing are equal to the forces that are pulling, the forces are **balanced**. When the forces on an object are balanced, the object's motion does not change. A moving object will keep moving at the same speed, in a straight line, in the same direction. An object that is standing still will remain still. It will move only when the forces on it are no longer balanced.

If both sides of a seesaw have the same weight on them, the seesaw will be balanced. It will not move until someone applies a new force.

Is That a Fact?

Balanced Yet Falling

Gravity is not the only force acting on the skydiver. As she falls, air pushes up against her. The force of the air is called **air resistance**. The faster the skydiver falls, the harder the air pushes. Most of the time she is falling, the force of gravity is greater than the force of air resistance. But she will reach a point during her dive when the force of gravity is equal to the force of air resistance. The forces will be balanced. The skydiver will not stop falling. She will not slow down. She will not fall any faster—she will fall at the same steady speed until she opens her parachute.

Unbalanced Forces

Motion changes only when forces are **unbalanced**. Think about a billiard ball. When it is in motion, it will keep moving in a straight line until a new force acts on it. That new force causes the balanced forces acting on the ball to become unbalanced. The ball could bump the side of the table. It could hit another ball. The ball may change direction or it may stop moving because of friction.

See also: page 252 Matter

See also: page 218 Air

The same is true of a billiard ball that is standing still. Gravity pulls it downward. This force is balanced by the push of the table that holds it up. It will not move until another force pushes against it. You may hit the ball with a pool cue. You may roll another ball against it.

When the forces on both sides of a game of tug-of-war are even, they are balanced. The rope does not move. If there is more force on one side, the forces are unbalanced. The rope moves to the side with greater force.

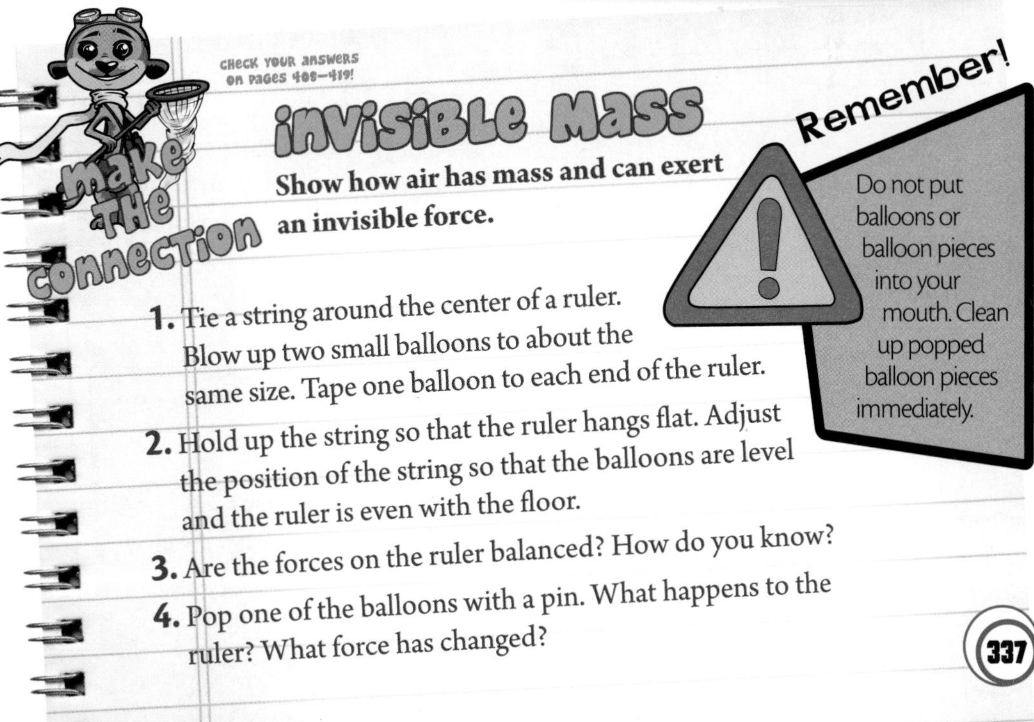

Check your answers on pages 408–419!

Invisible Mass

Show how air has mass and can exert an invisible force.

1. Tie a string around the center of a ruler. Blow up two small balloons to about the same size. Tape one balloon to each end of the ruler.
2. Hold up the string so that the ruler hangs flat. Adjust the position of the string so that the balloons are level and the ruler is even with the floor.
3. Are the forces on the ruler balanced? How do you know?
4. Pop one of the balloons with a pin. What happens to the ruler? What force has changed?

Remember! Do not put balloons or balloon pieces into your mouth. Clean up popped balloon pieces immediately.

Forces and Motion

The motion of objects depends on forces. Things do not move or stop moving, or change their motion by themselves. The scientist Isaac Newton came up with three rules to describe how motion and forces are related.

Newton's First Law of Motion An object that is moving will keep moving until an unbalanced force acts on it. It will move at the same speed and in the same direction. An object that is not moving will not move until an unbalanced force acts on it.

In other words, an object in motion will stay in motion. An object at rest will stay at rest.

Many games require you to know how a force will affect the motion of a ball.

An unbalanced force is a force strong enough to cause a change in speed or direction. All objects resist changes in motion. This is called **inertia**. The more mass an object has, the more inertia it has. An object with a large mass is harder to move than an object with a small mass.

Newton's Second Law of Motion The acceleration of an object depends on the mass of the object. Acceleration also depends on the amount of force applied.

Acceleration is any change in the speed or direction of a moving object. The more mass an object has, the harder it is to move it. A large force is needed to move a large object.

Think about two shopping carts in a grocery store. One cart has just a few items inside. The other cart is full of heavy items. It takes force to move the carts. The size of the force needed to move the carts depends on how heavy the load is. It takes less force to move the cart that has fewer items. It is carrying less mass. It takes more force to push the cart full of stuff.

Newton's Third Law of Motion Forces act in pairs. When one object applies a force to a second object, the second object applies a force back.

You cannot always see or feel both forces. Think about pushing a heavy box. If you push against the box to move it, you will feel it pushing back on you. The push it applies to your hands is equal and opposite to the force your hands apply to the box.

What Is Motion?

A baseball speeds over home plate. A hummingbird flashes from one flower to another. A gymnast dismounts from the balance beam. The baseball, the hummingbird, and the gymnast have all changed position. **Motion** is a change of position of an object.

How can you tell that an object is moving? You cannot tell if an object is moving just by looking at it. You see the movement of an object by comparing it to an object that stays in place. The object that stays in place is called a **reference point**. The baseball moves compared to home plate. The hummingbird moves compared to the flowers. The gymnast moves compared to the balance beam.

Motion depends on energy and forces. **Energy** is the ability to do work. A force is needed to push or pull an object. In science, **work** is done when forces cause something to move.

Objects that move use two kinds of energy. An object that is waiting to move has potential energy. **Potential energy** is stored energy. An object that is moving has kinetic energy. **Kinetic energy** is energy in motion.

While this girl is reading, she is not doing work. She is not using energy and force to move an object.

These athletes are doing work. They are using energy and force to run and move the ball.

In general, we want to know three things about a moving object:

- How far did it go?
- How fast did it go?
- In what direction did it go?

Distance is the amount of ground covered by a moving object. **Speed** measures how far an object has moved in a certain amount of time. Speed is found by dividing the distance traveled by the total time it took to travel the distance.

$$\text{speed} = \frac{\text{distance}}{\text{time}}$$

Most objects do not travel at a constant speed. Most often, when we talk about speed we mean **average speed**.

Speed and velocity are not the same things. **Velocity** measures the speed of an object as it moves in a certain direction. Direction is usually given as a direction on a compass, such as north, south, east, or west.

Velocity always includes direction. The speed of this truck is 55 mph. The velocity of this truck is 55 mph west.

speed = 45 mph (miles per hour)
velocity = 45 mph east

Acceleration is a change in the speed or direction of a moving object. So acceleration is a change in velocity.

THERE ARE SOME REALLY INTERESTING PEOPLE BACK HERE! I REALLY LIKE LEARNING ABOUT ALL THE SCIENTISTS AND INVENTORS WHO MADE REALLY COOL DISCOVERIES. WHAT ARE *YOU* INTERESTED IN?

Study Guide

Don't stop now—there is still a lot more to learn! Check out this section for great science facts and tips that will help you right now and in the future. Are you making a map? Just how big is a millimeter? How about figuring out a fancy new science word? All this, plus answers to the questions you've seen throughout the book are right here.

Reading Maps

What Is a Map?

You may think that a map is just a drawing. You could draw a map yourself! You could make a picture or a diagram of the streets around your house. You could add a picture of your house. That would be a pretty good map. Wouldn't it?

A map is not really a drawing at all. It is a collection of data.

A **map** shows the locations of things on Earth. Maps can show where things are physically located. You can find your city on some maps. Maps also show locations compared to other places.

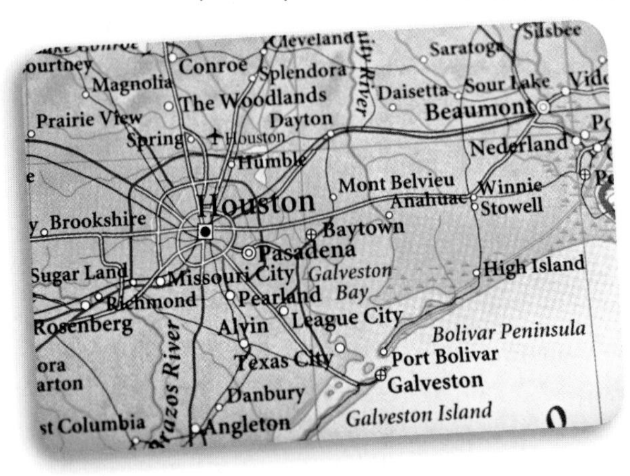

This map shows the location of Houston, Texas. It compares the location of Houston to the location of other cities. It also shows how close it is to Galveston Bay.

Maps can show Earth's landforms. Some maps show the types of plants and animals that live in an area. Maps can show where people live and the streets and highways that link them together.

Geographers, map makers, and other scientists work together to collect the information shown on maps. Satellites above Earth and researchers on the ground collect data for maps.

How to Read a Map

Scale Locations and features on a map cannot be shown at the same size they are on Earth. Everything on a map is reduced by the same proportion. The reduction is called the **scale**. A small-scale map shows a lot of area. A large-scale map shows more detail.

miles 5 0 5 10 15

1:250,000

Graphic Scale or Scale Bar The **scale bar** on a map shows how many miles or kilometers are shown per unit of measurement.

Fractional Scale A scale can be written as a ratio of map distance to distance on the ground. If the unit you use is a centimeter, this scale means that 1 cm equals 250,000 cm, or 2.5 km. If you use inches, it means that 1 in equals 250,000 in, or about 4 mi.

Compass Rose Most maps have a symbol to indicate the direction north. Other directions may be shown. The symbol on the map that indicates direction is called the **compass rose**.

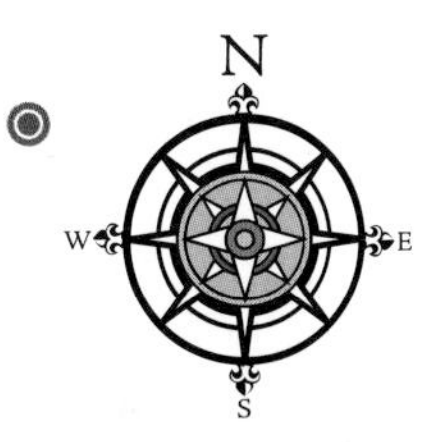

A compass rose shows the directions north, south, east, and west as they appear on the map.

Grid A map index shows locations using letters and numbers that mark a **grid**.

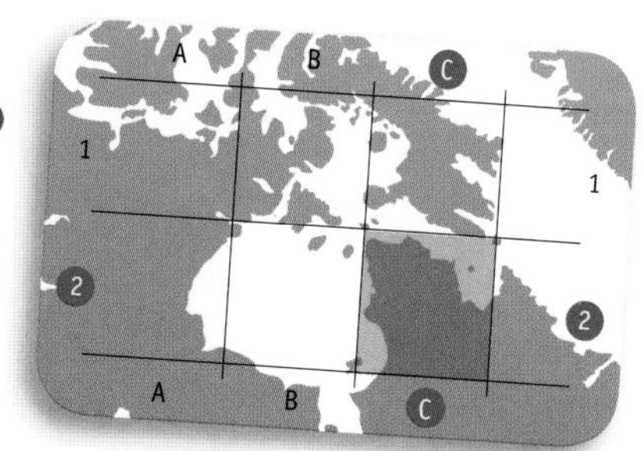

Find the grid coordinates C-2. First, read the letters across the top of the map. Then, read the numbers down the side of the map. The location is inside the square where the letters and numbers meet.

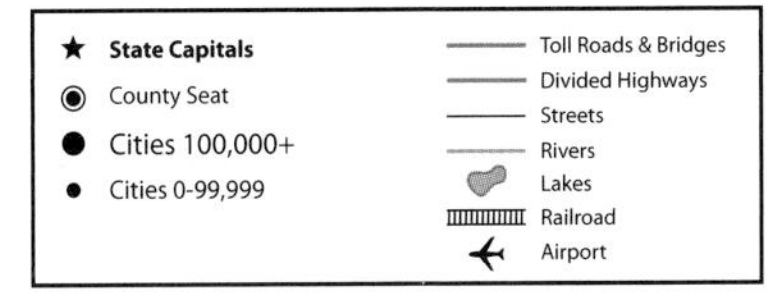

Features such as rivers and railroad tracks, schools and hospitals, cities and roads are shown on maps with symbols.

Legend Symbols are used to show special types of information on a map. The symbols are usually explained inside a map **legend**. The legend is a box shown in a corner of the map.

How Scientists Use Maps

Scientists use many kinds of maps in their work. Like every other traveler, scientists rely on maps to find locations. Sometimes scientists need to work in areas that are hard to find. They may use a special kind of map called a **topographic map**.

The lines on a topographic map are called contour lines. They show elevation. The closer the lines are, the steeper the climb.

Some scientists study the movement of Earth's surface. They use maps that show undersea ridges, mountain ranges, and volcanoes. Some maps show the locations of tectonic plates. Data for these maps is collected by sensors on the ground, and by satellites and aircraft in the air.

Scientists need to know how deep the water is at different spots. What does the land look like underneath the water? There are maps of the sea floor that can show where different underwater objects are located.

Arctic National Wildlife Refuge Bird Migration Routes

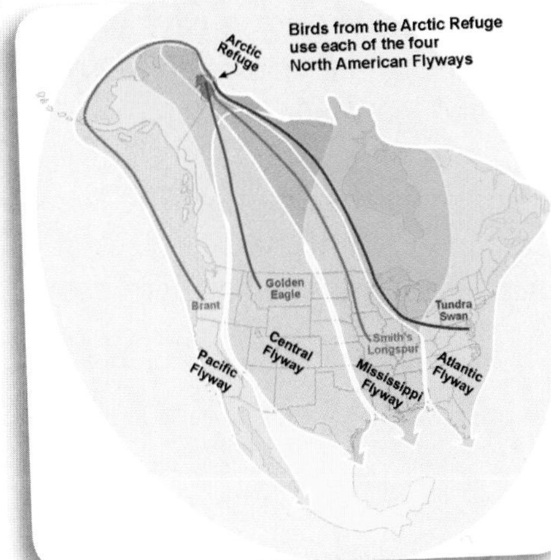

Scientists use maps to show the movement of animals and people. Many North American birds nest in the cold northern regions. They migrate south to warmer climates in the winter. This map shows the four major routes, or flyways.

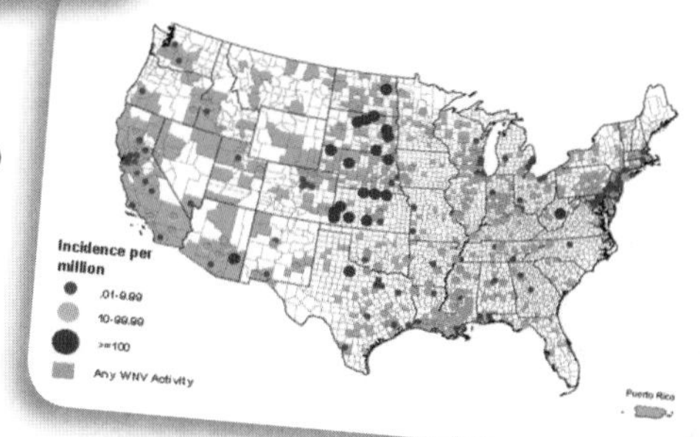

Scientists use maps to track diseases. This map shows where cases of West Nile Virus were found in the United States in 2008.

Environmental scientists use many kinds of maps. Sometimes scientists compare old maps with newer maps. They can see how weather patterns have changed over time. Maps may help them measure the rate at which glaciers melt. Scientists may also use the maps to track the spread of diseases.

Latitude and Longitude

Have you ever seen a map of all the continents? Have you seen a globe representing the entire Earth? If so, you have probably noticed a grid covering the surface of the map or globe. The grid is made up of lines of latitude and longitude.

Latitude All **latitude** lines begin from the **Equator.** The Equator is an imaginary line that runs around Earth like a belt. It is at 0 degrees latitude. The equator separates the northern and southern hemispheres.

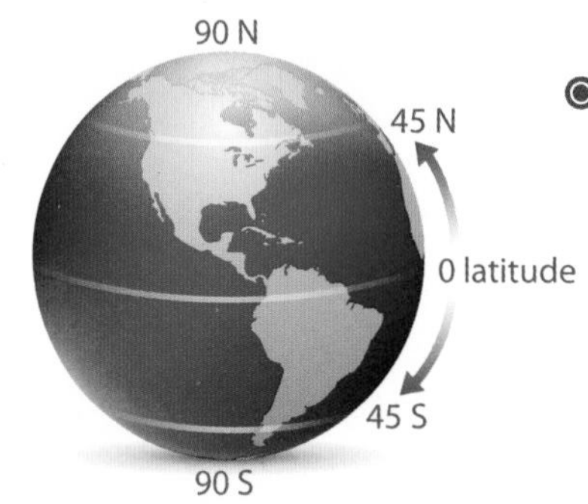

Latitude lines, or *parallels*, run parallel to the equator. They circle Earth from east to west. They run north from 0° to 90° to the North Pole, and south from 0° to 90° to the South Pole.

Longitude The **prime meridian** is an imaginary line that begins at the North Pole and runs south to the South Pole. All **longitude** lines begin from the prime meridian.

Latitudes and longitudes are divided into degrees (°), minutes ('), and seconds ("). Any place on Earth can be located by specifying the degrees, minutes, and seconds of latitude and longitude. For example, the location of Chicago, Illinois, is 41°52'55" north (latitude), and 87°37'40" west (longitude).

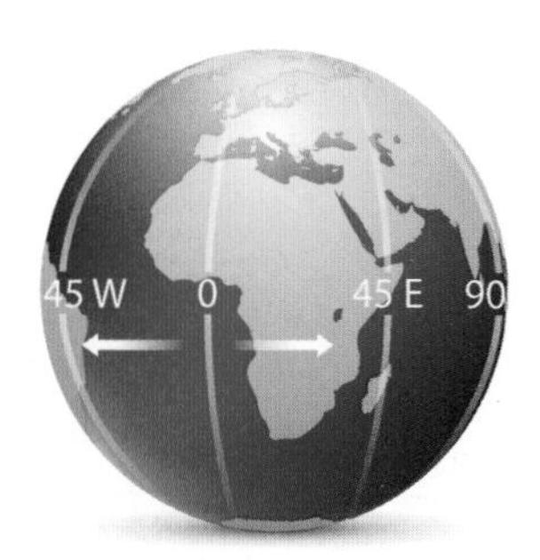

The prime meridian is at 0° longitude. It separates the eastern and western hemispheres. Longitude lines, or *meridians*, run from north to south. They are measured in degrees east or west of this line.

The Global Positioning System

The **Global Positioning System**, or **GPS**, can help you find your position anywhere on Earth. GPS depends on a system of 24–32 satellites that orbit Earth. At various times, at least 24 of these satellites transmit radio signals to GPS receivers. Drivers, hikers, pilots, and sailors depend on GPS.

GPS satellites orbit Earth twice a day. Signals from the satellites can pass through any kind of weather. The signals cannot pass through most solid objects. But they can pass through plastic and glass, including windshields. GPS works best in open areas with no trees or buildings.

Each satellite constantly sends information about its location and the current time. The signals move at the speed of light. GPS receivers find the distance to the satellites by estimating the amount of time it takes for their signals to reach the receiver. When the receiver estimates the distance to at least four GPS satellites, it can calculate its latitude, longitude, and altitude precisely.

Think of GPS satellites this way:

Satellite 1 tells the receiver "you are 200 miles from Denver."
Satellite 2 tells the receiver "you are 90 miles from Boulder."
Satellite 3 tells the receiver "you are 90 miles from Grand Junction."

Your receiver can now determine a location somewhere near Aspen. A fourth satellite can help narrow down the location. Your receiver will have a latitude, longitude, and altitude. You might even get a street address!

Science and Math Together

Why Do Scientists Use Math?

Scientists use the **scientific method** to answer scientific questions. An important step in the scientific method is **collecting data**. To a scientist, data are information that can be observed and measured. Math is often very useful in gathering scientific data.

Another important step in the scientific method is **interpreting data**. The measurements scientists take must be put into a form that is easy to understand. Math can also help scientists see patterns in data. Sometimes scientists make calculations so that they can make graphs or charts. Scientists can also use math to make comparisons between different trials, or tests, of their experiments.

When scientists complete their investigation, they **share their results**. Often, they will make a graph or chart of the data they collect. Then, they **think of new questions**. They may use math to help predict events in the future. Math is often used to help scientists form a new hypothesis and to plan new experiments.

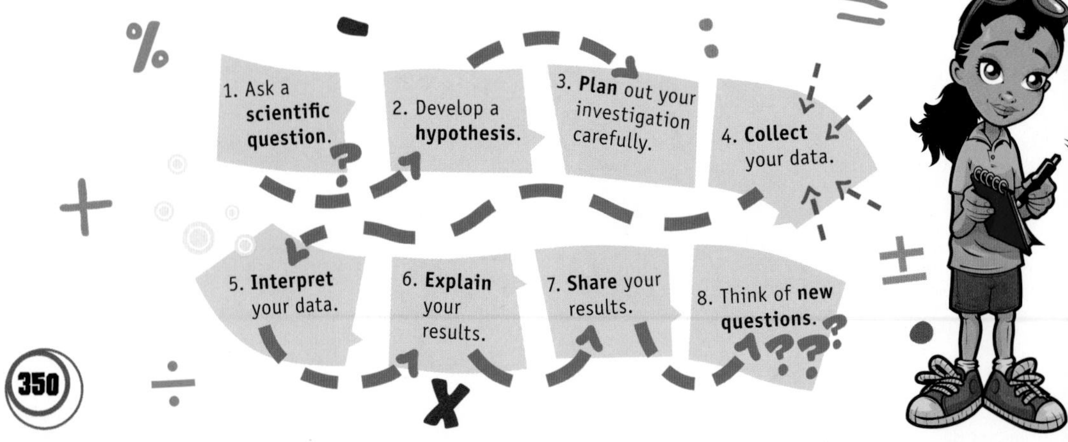

Using the Metric System

The first measurement called an inch was about the width of a human thumb! How wide is your thumb? Is it an inch wide? Is it less?

Scientists must use accurate measurements. All scientists must agree on what a measurement means. Most people in the United States use the English system of measurement. Length and distance is measured by the inch (in), foot (ft), yard (yd), and mile (mi). The ounce (oz) and pound (lb) are measures of mass and weight. Fluid volume is measured by the cup (c), pint (pt), quart (qt), and gallon (gal).

Scientists around the world use the metric system. The metric system is also called the International System of Units (SI).

In the metric system the **meter** (m) is the base unit used to measure length and distance. The **gram** (g) is the base unit for measuring mass. The volume of a liquid, a gas, or an irregular object is measured in **liters** (L).

Length and Distance

1 centimeter (cm) = 10 millimeters (mm)
1 decimeter (dm) = 10 centimeters (cm)
1 meter (m) = 10 decimeters (dm) = 1,000 millimeters (mm)
1 kilometer (km) = 1,000 meters (m)

Mass

1 centigram (cg) = 10 milligrams (mg)
1 decigram (dg) = 10 centigrams (cg)
1 gram (g) = 10 decigrams = 1,000 milligrams (mg)
1 kilogram (kg) = 1,000 grams

Volume

1 centiliter (cL) = 10 milliliters (mL)
1 deciliter (dL) = 10 centiliters (cL)
1 liter (L) = 10 deciliters (dL) = 1,000 milliliters (mL)

Using Comparisons to Measure

If you have trouble visualizing measurements in the metric system, you can use the size of everyday objects to make comparisons. Remember, when you are estimating, you are not making an exact measurement.

1 mm

1 cm

Estimating Length and Distance

1 mm = about the thickness of a dime
1 cm = about the width of a dime
1 dm = about the length of a whole piece of chalk
1 m = about the height from the floor to a doorknob

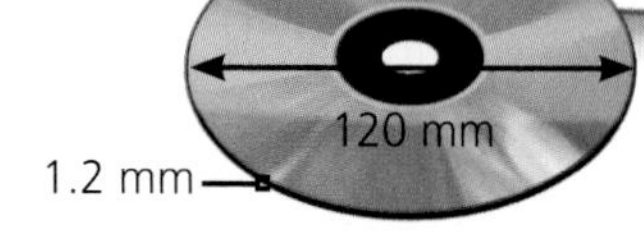

Estimating Mass

1 g = a dollar bill
1 kg = about six medium potatoes

Estimating Liquid Volume

1 mL = about 1/5 of a teaspoon
1 cL = about 2 teaspoons
1 dL = about one small scoop of ice cream
1 L = the liquid in a sports drink bottle

2L

+

= 1cL

How do English system measurements compare to metric measurements?

1 in = about 2.5 cm = about as long as a paper clip
1 yd = about 0.9 m = about as long as a baseball bat
1 mi = about 1.6 km = the elevation of Denver, CO

1 oz = about 28 g = about one slice of cheese
1 lb = about 0.4 kg = about one basketball

1 cup = about 240 mL = school lunch carton of milk
1 pt = about 473 mL = a basket of blueberries
1 qt = about 0.9 L = a bottle of motor oil
1 gal = about 3.8 L = large container of milk

Measuring Temperature

Most thermometers show both the Fahrenheit and Celsius scales.

In the United States, we frequently measure temperature in degrees Fahrenheit (F). On the Fahrenheit scale, water freezes at 32° and boils at 212°.

In most of the world, temperature is measured using the Celsius (C) scale. The Celsius scale is part of the metric system. On the Celsius scale, water freezes at 0° and boils at 100°.

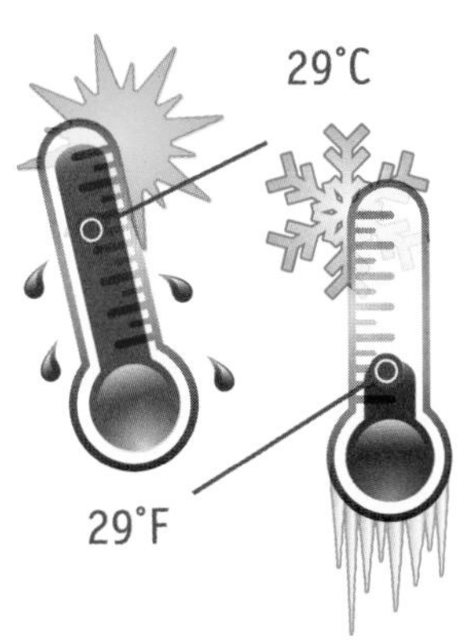

It's 29°F outside—I'll need my coat and gloves!

It's 29°C outside—I'll wear shorts and sandals!

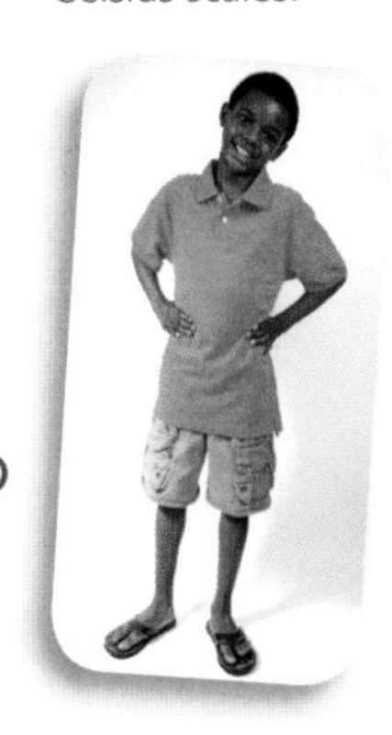

The average human body temperature is 98.6°F or 37°C. You are probably comfortable at a temperature of 68°F or 20°C when you are inside your home or classroom. In the winter you may put on your winter coat when the temperature drops to 45°F or 7°C.

Ice cream must be stored in the freezer at a temperature of -5°F or -21°C or it will begin to melt.

How hot is hot chocolate? Hot chocolate is hotter than your body temperature and colder than the temperature of boiling water. If you don't want to burn your tongue, you should drink hot chocolate at a temperature of about 136°F or 58°C.

Equation Review

As you collect scientific data, you will perform math operations over and over. Here are a few useful equations used in science.

Area The measurement of the surface of an object is called its **area**. Area is measured in square units.

area of a rectangle = length × width

area = 4 cm × 5 cm = 20 cm^2

4 cm

5 cm

area of a triangle = $\frac{1}{2}$ base × height

$\frac{1}{2}$ × 4 cm × 5 cm = 10 cm^2

5 cm

4 cm

Volume The amount of three-dimensional space an object takes up is called its **volume.** Volume is measured in cubic units.

volume of a cube = side × side × side

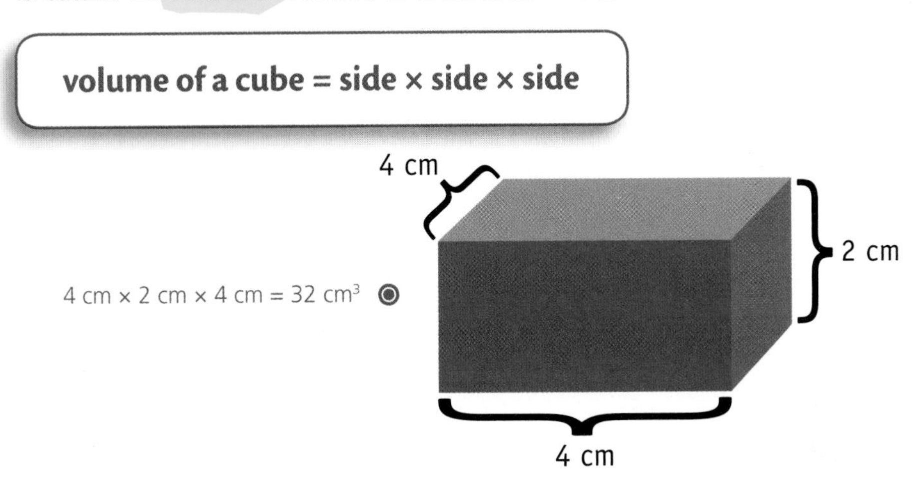

Sometimes you will need to compare two numbers that are not measured in the same units. The result is called the **rate.** When you compare two numbers you are **dividing**. The fraction bar in the result is read as "per."

Density The mass an object has per unit of volume is its **density.**

density = mass ÷ volume

Find the density of a cube that has a mass of 2 kilograms and a volume of 32 cubic meters.

$$2 \text{ kg} \div 32 \text{ m}^3 = 0.0625 \text{ kg/m}^3$$

Speed The distance an object moves per unit of time is its **speed.**

speed = distance ÷ time

Find the speed of a train that travels 30 kilometers in 50 minutes.

$$30 \text{ km} \div 50 \text{ min} = 6 \text{ km/min}$$

Math Skills Review

When you collect scientific data, you will make measurements. Sometimes you will repeat a step in an experiment. The measurements you take will not always be the same. You can use the **average** of your measurements as a typical measurement. Sometimes the average is called the **mean.**

To find the average of a set of measurements, first add all of the numbers together. There are four numbers in the example:

Your teacher can help you decide what data to collect during your investigation. The teacher may also suggest ways to organize the data.

$$2 \text{ m} + 2.2 \text{ m} + 3 \text{ m} + 1.8 \text{ m} = 9 \text{ m}$$

Then, divide the sum by the number of measurements in the set.

$$9 \text{ m} \div 4 = 2.25 \text{ m}$$

You can compare two numbers using a **ratio.** A ratio is usually written as a **fraction**. Suppose you needed to compare the number of rainy days to the number of sunny days in November.

10 rainy days
20 sunny days
Total: 30 days

The ratio for rainy days to total days is 10/30, which can be simplified to 1/3. One-third of the days in November were rainy days. The ratio for sunny days is 20/30. Two-thirds of the days in November were sunny.

A **percentage** is the ratio of a given number to 100. The symbol for **percent** is %.

Suppose you know that 5 of the students in your classroom have dogs as pets. There are 25 students in your classroom. What percentage of the students has dogs?

First, write the ratio of dog owners to the number of students:

5 dog owners / 25 students = $\frac{1}{5}$

Then, multiply the ratio by 100.

$$\left(\frac{1}{5}\right) \times 100 = 20$$

The answer is the percentage of dog owners to students in the classroom. In this case, 20% of the students in your classroom have a dog.

Studying Science

Study Tips

Here are some study tips that can help you become more organized and better prepared for your science class.

Classroom Learning Skills

- Listen.
- Pay attention.
- Ignore distractions.
- Stay focused.
- Take good notes.
- Be organized.
- Read actively.

Time Management

- Plan your time wisely. Complete tasks in order of importance.
- Plan ahead for long-term projects. Break a big project into smaller parts. Use a calendar to write down the due dates of the smaller parts to be completed.
- Plan ahead for any tests or quizzes.

References and Resources

- Libraries are a great source for science information. Most of the library's catalog will be found on the library's computer database. You can search by subject, author, or title. Try to be as specific as possible. For example, searching for "alligators" will give you more specific choices to research than searching for "animals" will give you.

- Searching the Internet is also a good way to find information. You can search by topic, or you can type in the address for a specific Web site. Know the difference between sites ending in .org (nonprofit organization), .edu (educational), and .gov (government).

Internet Safety

- Never give your name, address, or phone number over the Internet.
- Do not share your password with anyone.
- Do not say anything on the Internet that you don't want someone else to know. Your e-mail is not private.
- Tell an adult if someone is talking about inappropriate topics online.
- Use appropriate language. Do not swear or use insults.

Listing Sources

- Some sources will be helpful, and some will not. Keep a list of your sources.
- Write down the titles of any books, magazines, or other print references you use. Also, list any Internet sites that are helpful.
- Check with your teacher to find out what information to include and how to list your sources.
- *Plagiarism* (PLAY juh ri zuhm) means to take credit for someone else's words or ideas. Be sure that any information, notes, or ideas you get from outside sources are written down in your own words, and be sure that credit is given to the source.

How to Study

Good study habits can help you develop the skills you need to complete assignments on time. They can also help you be a successful student.

- Gather all notes, worksheets, and study guides together.
- Find a quiet place to study. Keep a regular study schedule. Be sure you have enough room to spread out books and papers. Keep necessary school supplies nearby.

- Use index cards to make flash cards of vocabulary words. Write the definitions on the back. You can also use index cards to review main concepts and questions from the study guide. Write the answers on the back. Have someone quiz you.
- Study with a friend or with a small group.
- Use memory devices like rhymes or silly sentences to help you remember things in a sequence. For example: **M**y **V**ery **E**xcited **M**other **J**ust **S**ent **U**s **N**oodles is a way to remember the planets in order (*Mercury, Venus, Earth, Mars, Jupiter, Saturn, Uranus, Neptune*).

Make a flow chart of the material to show how things relate to each other.

The Life Cycle of a Monarch Butterfly

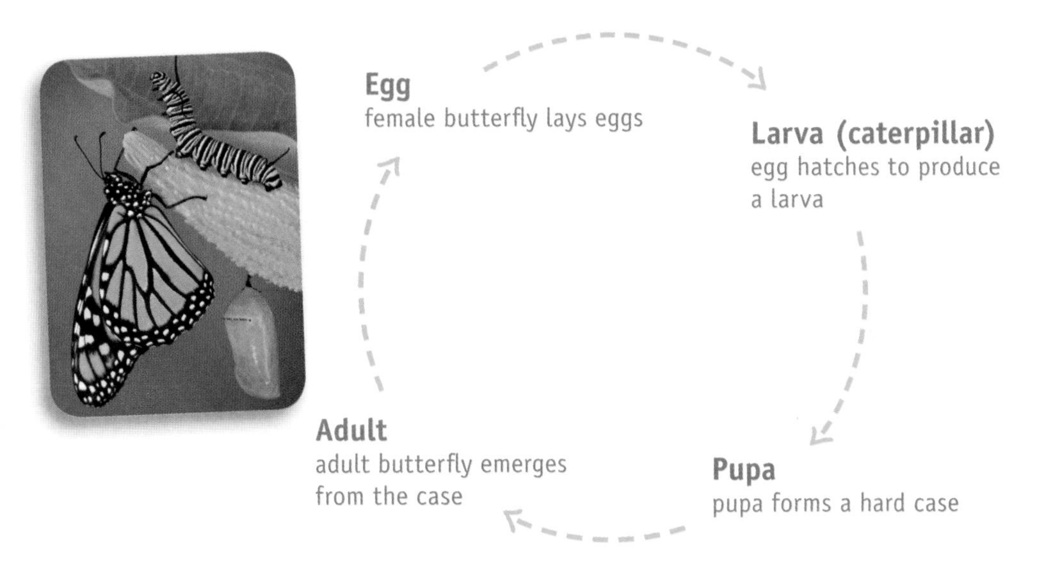

When working with how things are similar and different, make a Venn diagram.

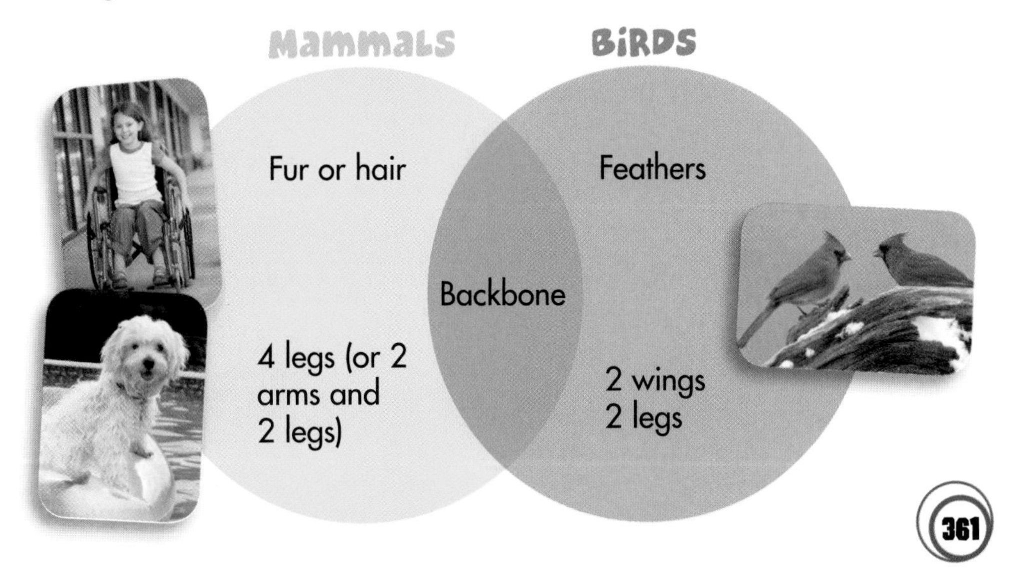

Test-Taking Skills

To be good at anything takes practice. Following these test-taking skills can help you improve your performance on science tests.

Before the Test

- Know what material will be covered on the test.
- Look over notes and homework. Be sure you have any notes or handouts you may have missed.
- Write down any questions that you need to ask the teacher before the day of the test.
- Get a good night's sleep the night before the test. You will concentrate better if you are not tired.

The Day of the Test

- Eat a healthy breakfast. Food provides energy for your brain.
- Go to class with all necessary materials. Relax.

When You Get the Test

- Write your name on the test.
- Read all directions carefully.
- Check out how many questions there are and what types of questions are being asked.

Taking the Test

- If you come to a hard question, skip it and come back to it later. Go through and answer all the questions you are sure of first.
- Read each question carefully.
- For True/False questions, write out the word *true* or the word *false*. Read the directions carefully. Sometimes the directions tell you to change a word in a false statement to make it true.
- For Short Answer questions, the directions may tell you to use complete sentences, or it may be acceptable to write two- or three-word phrases.
- If the question has a chart or a graph, remember to read all information that may be part of it.
- For Multiple Choice questions, notice words like *all, always, never, most,* and *best* in the answers. They frequently are part of a wrong answer.
- Review your answers before handing in your test.

After the Test

- After the test is graded and returned, review any wrong answers. Be sure to ask questions if you are unsure why one of your answers is wrong.
- Think about what you might do better to be even more prepared next time.

Cool Scientists

Notable Scientists and Inventors

Luis Alvarez (1911–1988) and **Walter Alvarez** (1940–present)
Walter Alvarez is an American geologist. His father, Luis Alvarez was an American physicist who won a Nobel Prize for physics in 1968. They were famous for discovering a layer of rock that formed around the time the dinosaurs went extinct. The layer contained iridium, an element that is rare on Earth but found in space objects. The Alvarezes thought the layer formed when a space rock hit the Earth. Dust from the space rock blocked the Sun. Without the Sun, food chains were disrupted. Many species went extinct. The Alvarez theory is widely accepted today as the most likely reason that the dinosaurs died.

Mary Anning (1799–1847)
See the History Makers entry in Fossils, p. 180.

Archimedes (287–212 B.C.E.)
See the History Makers entry in Simple Machines, p. 317.

Aristotle (384–322 B.C.E.)
Aristotle was a Greek scientist and philosopher. He studied many subjects, including astronomy, geology, meteorology, physics, and zoology. Aristotle invented the first system to classify living things. Some of his ideas were wrong, however. For example, he thought that the planets revolved around the Earth. This idea was widely accepted until the 1500s.

Neil A. Armstrong, Michael Collins and Edwin E. "Buzz" Aldrin.

Neil Armstrong (1930–present); **Edwin "Buzz" Aldrin, Jr.** (1930–present); and **Michael Collins** (1930–present)
American astronauts Neil Armstrong, Edwin "Buzz" Aldrin, and Michael Collins became world famous as the crew of Apollo 11, the first successful mission to the surface of the Moon. On July 20, 1969, Armstrong stepped onto the Moon's surface, the first person ever to do so. Aldrin followed Armstrong, becoming the second person to walk on the Moon. While Armstrong and Aldrin explored the lunar surface, Collins kept the command module in orbit around the Moon.

Alexander Graham Bell

Alexander Graham Bell (1847–1922)
Alexander Graham Bell was a scientist, engineer, and inventor. He was best known for inventing the telephone. He also invented the microphone and helped develop many other devices. Being able to hear and to communicate was important to Bell. As a young man, he taught speech to deaf children. He later founded a school for the deaf.

Cai Lun (62?–121)
Cai Lun was a Chinese court official. He was the first person to invent the form of paper that we use today. Before then, most writing had been done on materials such as bamboo and silk. Cai invented a process that mixed pieces of tree bark with plant fibers and water. The process was easy, and the product did not cost too much to make. And it could be made in large amounts. Within a few hundred years, Cai's method of papermaking was being used in other parts of Asia. It spread from there to the Middle East, and later to Europe.

Annie Jump Cannon (1863–1941)
Annie Jump Cannon was an American astronomer. She was best known for counting and classifying the stars. She created a system to classify them based on certain properties of the light they emitted. She also discovered 300 stars. Cannon was the first person to receive an honorary doctorate degree from Oxford.

Annie Jump Cannon

Rachel Carson (1907–1964)
Rachel Carson was an American biologist and writer. She was famous for her book, Silent Spring. It warned about the effect of chemical pesticides on wildlife. Her book made people aware of taking care of the environment.

Rachel Carson

George Washington Carver (1864–1943)
See the History Makers entry in Matter, p. 260.

Anders Celsius (1701–1744)
Anders Celsius was a Swedish astronomer and physicist. He was best known for developing a temperature scale based on the boiling point and melting point of water. His scale defined 0° as the freezing point of water, and 100° degrees as the boiling point of water. The Celsius scale is still used today in almost all scientific work.

Nicolaus Copernicus (1473–1543)
Nicolaus Copernicus was a Polish astronomer. He was the first to recognize that the Earth and the other planets moved around the Sun. Before that, people had believed that everything revolved around the Earth. Copernicus was also the first to explain that the Earth rotates on its axis. His ideas were not accepted until many years after his death. Today, Copernicus is considered the founder of modern astronomy.

Marie Curie (1867–1934) and **Pierre Curie** (1859–1906)
Marie Curie was born in Poland but later moved to France and became a French citizen. Marie was a physicist and chemist. She was the first person to be awarded two Nobel Prizes. She shared the 1903 Nobel Prize in Physics with her husband Pierre Curie, a physicist. This honored their studies of radioactivity. Marie won the 1911 Nobel Prize in Chemistry for her discovery of two radioactive elements: radium and polonium. Marie Curie became ill and died from working with radioactive elements. At that time, no one knew that radioactivity could cause serious illness.

John Dalton

John Dalton (1766–1844)
John Dalton was an English chemist. He was famous for developing the atomic theory of matter. Dalton explained that matter is made up of atoms, and that each atom of an element has the same properties. He also explained that the elements are different because they have different properties.

Charles Darwin (1809–1882)
Charles Darwin was an English scientist who studied plants and animals. He became famous for his theory of natural selection. It states that all species have evolved over time from common ancestors. The theory describes how traits that help an organism survive in its surroundings become more common in future generations. Traits that hurt an organism's chances of survival become less common over the same time.

Charles Darwin

Rudolf Diesel (1858–1913)
See the History Makers entry in Alternative Energy, p. 280.

Charles Drew (1904–1950)
Charles Drew was an American doctor and inventor. He was best known for developing a way to store blood plasma (the liquid part of blood). He helped establish a large blood bank (blood storage) system for soldiers during World War II. This made it easier for military doctors to get blood and plasma for transfusions, saving the lives of thousands of British and American soldiers.

Thomas Edison

Thomas Edison (1847–1931)
Thomas Edison was an American scientist and inventor. He invented more than a thousand devices during his lifetime. He is most famous for inventing the first light bulb that could be used in the home. He invented the phonograph and the motion picture camera. He also helped improve the telephone and the telegraph.

Albert Einstein (1879–1955)
Albert Einstein was a physicist, and one of the most famous scientists of modern times. Einstein was born in Germany, but moved to the United States in the 1930s. He became best known for his theory of relativity. It describes the connections between time, space, matter, and energy. The theory changed the way scientists viewed the universe and earned Einstein the 1921 Nobel Prize in Physics.

Albert Einstein

Daniel Fahrenheit (1686–1736)
Daniel Fahrenheit was a German physicist and engineer. He became famous for inventing the first mercury thermometer. It was the first thermometer that measured temperature correctly each time it was used. Fahrenheit used his invention to develop the Fahrenheit temperature scale.

Michael Faraday (1791–1867)
Michael Faraday was an English chemist and physicist. He became famous for discovering that a magnetic field can produce an electric current.

Philo T. Farnsworth (1906–1971)
Philo T. Farnsworth was an American engineer and inventor. He was most famous for inventing the television set. He also helped develop radar, infrared night lights, electron microscopes, and many other machines.

Enrico Fermi

Enrico Fermi (1901–1954)
Enrico Fermi was an Italian physicist and mathematician. He was best known for his work in nuclear energy.

Alexander Fleming (1881–1955)
Alexander Fleming was a Scottish scientist. He studied bacteria and other single-celled organisms. Fleming was famous for discovering the first antibiotic: penicillin. He made his discovery when he saw that bacteria didn't grow on lab dishes that contained mold. Fleming shared the 1945 Nobel Prize in Medicine with Howard Florey and Ernst Boris Chain, who had developed a way to mass-produce penicillin for the general public.

Walther Flemming (1843–1905)
See the History Makers entry in Cycles, p. 129.

Dian Fossey (1932–1985)
Dian Fossey was an American biologist. She was best known for her studies of the endangered mountain gorillas of central Africa. She moved to Africa in order to study the gorillas in their natural habitat in the Virunga Mountains. She observed them every day for 18 years and became the world's leading authority on mountain gorillas.

Benjamin Franklin

Benjamin Franklin (1706–1790)
Benjamin Franklin was an American inventor, politician, and writer. He was famous for his kite experiment, which showed that lightning is a form of electricity. He invented many things, including specialized eyeglasses, swimming fins, and a heat-efficient stove. Learn more about Benjamin Franklin in Doing the Work of Scientists, p. 17.

Biruté Galdikas (1946–present)
Biruté Galdikas is a Canadian biologist who became the worlds leading authority on orangutans. She spent more than 30 years in Indonesian Borneo studying orangutans in their natural habitat.

Galileo Galilei (1564–1642)
Galileo Galilei was an Italian physicist, mathematician, and astronomer. He used lenses to improve the early telescope, and then used it to study the skies. He discovered sunspots on the Sun and mountains on the Moon. He also discovered four of Jupiter's moons. These discoveries helped support Copernicus's theory that the planets revolve around the Sun. Galileo also studied how objects move and fall. His findings later helped Isaac Newton develop the laws of motion.

John Glenn (1920–present)
John Glenn was an American astronaut. He was the first American to circle the Earth from space. As part of the Mercury 6 space mission in 1962, he was the first American to make a complete orbit of the Earth. He left the space program and in 1974 was elected senator of his home state Ohio. In 1998, Glenn returned to space as part of the shuttle Discovery. At age 77, he was the oldest person to travel in space.

John Glenn

Robert Goddard (1882–1945)
Robert Goddard was an American physicist. He became famous for his work on missiles and space flight. In 1926, he designed and successfully launched the first rocket that used liquid fuel. Today he is often called the father of rocketry.

Maria Goeppert-Mayer (1906–1972)
Maria Goeppert-Mayer was a German physicist. She was best known for her model of atomic nuclei. This model was the first to show that electrons orbit around the nucleus of an atom. In 1963, she received the Nobel Prize for Physics.

Jane Goodall (1934–present)
Jane Goodall is an English scientist who became the world's leading authority on chimpanzees. She studied chimpanzees for 45 years in their natural habitat: the tropical rainforest near Gombe Stream in Tanzania. Her work helped people around the world learn more about the ways that chimpanzees act with each other every day.

William Harvey (1578–1657)
See the History Makers entry in The Human Body, p. 92.

Stephen Hawking (1942–present)
Stephen Hawking is a British physicist. He was famous for his theories on black holes. Hawking also wrote books such as A Brief History of Time. His books helped many people outside of science to appreciate his ideas.

Ibn al-Haytham

Ibn al-Haytham (965–1040)
Ibn al-Haytham was born in Persia (now Iran) but moved to Egypt as an adult. He was an astronomer and mathematician. He was famous for his studies of the way light reflects from objects. His theories about light and vision were widely accepted and had an impact on the work of later scientists, such as Johannes Kepler.

Robert Hooke (1635–1703)
See the History Makers entry in Cells, Tissues, and Organs, p. 58.

Edwin Hubble (1889–1953)
Edwin Hubble was an American astronomer. He was best known for demonstrating that other galaxies besides our own Milky Way exist. To honor his contributions to astronomy, NASA named the famous Hubble Space Telescope after him.

NASA's Hubble Space Telescope

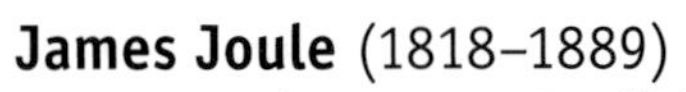

James Joule (1818–1889)
James Joule was an English physicist. He was famous for developing the law of conservation of energy. This is called Joule's Law. It states that energy can change forms but cannot be lost. Joule based this on his discovery that heat is produced in an electrical conductor. Scientists honored his work by naming the international unit of energy—the joule—after him.

Percy Julian (1899–1975)
Percy Julian was an American chemist. He was famous for finding a way to produce artificial versions of certain natural plant chemicals. These artificial chemicals were then used to produce different types of medicines at a low cost. This meant that more people were able to afford medicines to treat ailments. Julian's research led to mass production of a drug used to treat glaucoma, a serious eye disease. The drug also is used today to treat patients with Alzheimer's disease.

Johannes Kepler (1571–1630)
Johannes Kepler was a German astronomer. He was the first person to correctly describe how planets move. He explained that a planet's orbit is shaped like an ellipse (oval) and not a circle. Kepler used this to develop the Three Laws of Planetary Motion. He also designed a telescope and was the first person to correctly describe how humans see.

Lewis Latimer (1848–1928)
Lewis Latimer was an American scientist and inventor. He was best known for the work he did with Thomas Edison. In Edison's lab, Latimer designed important features that improved the light bulb. He also worked with Alexander Graham Bell, and made the technical drawings for Bell's telephone.

Lewis Latimer

Mary Leakey (1913–1996) and **Louis Leakey** (1903–1972)
Mary and Louis Leakey were British scientists. They studied the fossil remains of very early humans in East Africa. In 1948, they discovered the first skull of a fossil ape ever to be found. Mary Leakey became famous for discovering the Laetoli footprints, a trail of fossilized footprints that were made 3.6 million years ago.

Carl Linnaeus (1707–1778)
See the History Makers entry in Organisms, p. 76.

Anton von Leeuwenhoek (1632–1723)
Anton von Leeuwenhoek was a Dutch scientist. He was best known for his improvements of the microscope. He also was the first person to view and describe bacteria and other one-celled organisms. He called these organisms "animalcules" (tiny animals).

Wangari Maathai

Wangari Maathai (1940–present)
Wangari Maathai was born in Kenya, in East Africa. She is an environmental scientist who organized the planting of trees in towns and villages across Africa. By the turn of the century, more than 40 million trees had been planted. This reduced soil erosion, preserved water resources, and helped to restore rainforests. In 2004, Maathai became the first woman from Africa to receive the Nobel Peace Prize.

Guglielmo Marconi (1874–1937)
Guglielmo Marconi was an Italian inventor and entrepreneur. He helped develop the wireless telegraph. For his work, Marconi shared the 1909 Nobel Prize in Physics with German inventor Karl Ferdinand Braun.

Gugliemo Marconi

Gregor Mendel (1822–1884)
See the History Makers entry in Organisms, p. 72.

Dmitri Mendeleev (1834–1907)
Dmitri Mendeleev was a Russian chemist. He was famous for organizing the chemical elements into a chart called the Periodic Table. The table shows the elements arranged by atomic number.

Samuel Morse

Samuel Morse (1791–1872)
Samuel Morse was an American scientist and inventor. He was famous for his inventions of the telegraph and Morse code. These inventions allowed people in distant parts of the United States to communicate with each other.

Isaac Newton (1642–1727)
See the History Makers entry in Forces and Motion, p. 331.

Hans Christian Øersted (1777–1851)
See the History Makers entry in Electricity, p. 327.

Georg Ohm

Georg Simon Ohm (1787–1854)
Georg Simon Ohm was a German physicist. He was the first person to explain the relationship between voltage, current, and resistance. His findings led to a better understanding of how electrical circuits work.

Louis Pasteur (1822–1895)
Louis Pasteur was a French chemist and biologist. He invented a process to stop foods and liquids from spoiling. He also made discoveries about the immune system, vaccinations, and diseases. Learn more about Pasteur in Technology, p. 34.

Ivan Pavlov (1849–1936)
See the History Makers entry in Animals, p. 86.

Ptolemy (100?–170?)
Ptolemy was born in Greece but later lived in Egypt. He was an astronomer and mathematician. Ptolemy believed that the planets revolved around the Earth. Ptolemy's theory was accepted as correct until Copernicus proved it was wrong in the 1500s.

Charles Richter (1900–1985)
Charles Richter was an American scientist who studied earthquakes. He was famous for developing the Richter scale, which measures the strength of earthquakes.

Sally Ride (1951–present)
Sally Ride is an American astronomer, physicist, and astronaut. In 1983, she became the first American woman to travel in space.

Sally Ride

Ernst Ruska (1906–1988) and **Max Knoll** (1897–1969)
Ernst Ruska was a German physicist. Max Knoll was a German engineer. They are best known as the inventors of the electron microscope.

Ernest Rutherford (1871–1937)
Ernest Rutherford was a New Zealand-born physicist and chemist. He was best known for developing a new model of the atom. Rutherford's findings earned him the 1908 Nobel Prize in Chemistry. The chemical element rutherfordium (atomic number 104) is named after him.

Jonas Salk (1914–1995)
Jonas Salk was an American medical scientist. He was best known for his discovery of the first safe vaccine against polio.

Alan B. Shepard, Jr. (1923–1998)
Alan Shepard was an American astronaut. He was the first American astronaut to travel into space.

Alan B. Shepard, Jr.

Nikola Tesla

Nikola Tesla (1856–1943)
Nikola Tesla was a physicist and inventor. He was born in Serbia, but did most of his work in the United States. Tesla was best known for developing the use of alternating current (AC) for electrical products. Tesla also helped improve radio communications and electrical lighting for the public. During his lifetime, he invented and patented more than 700 devices.

Evangelista Torricelli (1608–1647)
Evangelista Torricelli was an Italian physicist and mathematician. He was famous for inventing the barometer, a tool that measures the amount of pressure in air and in water.

James Watson (1928–present); **Francis Crick** (1916–2004); **Rosalind Franklin** (1920–1958); and **Maurice Wilkins** (1916–2004)
Francis Crick, Maurice Wilkins, and Rosalind Franklin were British scientists; James Watson was an American scientist. Together, this team of scientists determined the structure of DNA, the chemical that controls inheritance in all living cells.

James Watt (1736–1819)
James Watt was a Scottish inventor and engineer. He was best known for his improvements of the steam engine. Watt's work helped make the engine a more useful tool for running machinery.

Eli Whitney

Eli Whitney (1765–1825)
Eli Whitney was an American inventor, engineer, and manufacturer. His best-known invention was the cotton gin, a machine that separates cotton seeds from raw cotton fibers.

Wilbur Wright (on ground) and Orville Wright (controlling machine)

Orville Wright (1871–1948) and **Wilbur Wright** (1867–1912)
The Wright brothers were American inventors. They became famous for building and flying the first airplane that could remain in the air for more than a few seconds. The first flight took place on December 17, 1903. It lasted only 12 seconds, but forever changed the way humans travel.

Today's Scientists

What does a "real" scientist do all day? That depends on the scientist! Some scientists chase tornadoes across grasslands. Some chase monkeys through rainforests. Some scientists build robots that can explore the planets. Others design video games. Science is hard work. But new technologies and discoveries make it more fun and exciting than ever.

Pardis Sabeti

Can a biologist also be a rock star? Pardis Sabeti can! Dr. Sabeti studies how human genes have changed over time to resist certain diseases. Her work has earned many awards and honors. In her spare time, Dr. Sabeti sings and plays bass with her rock band, Thousand Days.

Ever wonder what's it like to be a National Geographic explorer? Ask Sylvia Earle! She's a marine biologist who studies plant life in ocean ecosystems. Dr. Earle has led more than 50 expeditions to study the world's oceans. Along the way, she discovered new species, walked on the ocean floor, and wrote many books and papers about the ocean.

Want to build things? Physical science and math are your main tools. Paul Schmitt has built more than ten million skateboards over the past 30 years. He started an organization to teach young people to build skateboards and learn about science—especially physics, biology, and math. Of course, it helps to be artistic! In fact, mixing art and science is behind designing and building almost anything—from action toys to video games, sports equipment, and roller coasters.

Paul Schmitt

Terry Fong (far left)

Science, art, fun—that's a good way to describe Terry Fong's job. Dr. Fong designs and builds robots for NASA's Intelligent Robotics Group (IRG). He uses physics, math, biology, Earth science, and space science to develop robots for exploring extreme environments—like the Moon!

What if you like more than one area of science? No problem. Some areas of science combine information from different fields!

An **astrobiologist** looks for other forms of life (biology) in the universe (astronomy). NASA astrobiologist Victoria Meadows makes computer models to study what types of planets might have the right resources for living things.

An **astrophysicist** studies physical properties of planets, stars, and other parts of the universe. Astrophysicist Neil DeGrasse Tyson is the director of the Hayden Planetarium in New York City. He also hosts a public television show about science.

Neil DeGrasse Tyson

A **biometeorologist** like Nancy Kiang studies how living things (biology) and the atmosphere (meteorology) interact.

A **marine geophysicist** studies the ocean floor. A **marine geographer** maps it. Can't decide which to do? Neither could Dawn Wright—so she does both! Dr. Wright explores underwater mountains and ridges. Then, she uses her computer to analyze and map what she finds on the ocean floor.

Understanding Science Words

Understanding Special Terms

Science words become easier to understand when you break the words down into their different parts. Then, when you understand each word part, you can figure out the meaning of an unfamiliar science word. The different parts are root, prefix, and suffix.

- A **root** is the main part of a word.
- A **prefix** is a word part that you can add to the beginning of a root to change its meaning.
- A **suffix** is a word part that you can add to the ending of a root to change its meaning.

Many science word parts can be traced back to the ancient languages of Latin and Greek. For example, the root word micro is from the Greek word that means "very small." The suffix scope is from the Greek word that means "examine." If you add micro to the beginning of scope, the word becomes microscope, which is an instrument that examines very small objects.

Check out the following tables to help you understand science words you don't know.

Root	Language Source	Roots Meaning	Example
astr	Greek	star	astronomy
aud	Latin	hear	auditory
bio	Greek	life	biology
centi	Latin	hundred	centipede
chloro	Greek	green	chlorophyll
derm	Greek	skin	dermatology
gen	Greek	birth, produce	genetic
geo	Greek	earth	geology
hydr	Greek	water	dehydrate
luc	Latin	light, shine	translucent
luna	Latin	moon	lunar
magn	Latin	large	magnify
ped	Greek	child	pediatrician
sci	Latin	know	scientific
scope	Greek	examine	telescope
sol	Latin	sun	solar
sphere	Greek	ball, globe	atmosphere
techne	Greek	art or skill	technology
vis	Latin	see, view	visible
vor	Latin	eat, devour	carnivore
zoo	Greek	animal	zoology

Prefix	Source Language	Meaning	Example
a-, an-	Greek	not, without	anesthetic
amphi-	Greek	both	amphibian
anti-	Greek	against, opposite	antibody
cyto-	Greek	hollow vessel, cell	cytoskeleton
de-	Latin	concerning	decompose
dis-	Latin	apart	dissolve
ecto-	Latin	outside	ectoderm
endo-	Latin	inside, within	endoderm
epi-	Latin	upon, at	epidermis
ex-	Latin and Greek	out of, from	exoskeleton
hyper-	Greek	over, above	hyperactive
im-, in-	Latin	not	inexhaustible
meta-	Greek	after, behind	metamorphic
micro-	Greek	very small	microscope
photo-	Greek	light	photosynthesis
post-	Latin	after	posterior
pre-	Latin	before	predict
re-	Latin	again	recycle
super-	Latin	more than, excessive	supernova
ultra-	Latin	beyond	ultraviolet

Suffix	Source Language	Meaning	Example
-en	Greek	made of	frozen
-er, -or	Latin	one who takes part in	inventor
-ful	Middle English	full of	harmful
-ic	Latin	nature of	metallic
-ide	German, French	indicating binary compound	oxide
-ist	Greek, Latin	one who specializes in	paleontologist
-ite	Latin	a division, part	meteorite
-itis	Greek	inflammation	tonsillitis
-ology	Greek	field of study	pathology
-meter	Greek	measurement	thermometer
-oid	Greek	like, similar	asteroid
-phobia	Greek	exaggerated fear	claustrophobia
-tion	Latin	state of	evaporation
-tomy	Greek	cutting	anatomy

On the pages that follow is a glossary of the terms that are used throughout this book.

Glossary

A

absolute age the actual age of a rock (181)

absorb the ability to take something in, such as nutrients or lightwaves (117, 306)

acceleration a change in the speed or direction of a moving object (338)

acid rain formed when water is mixed with gases from burning fuels and chemicals; can wear away or dissolve hard materials like metal and stone (227)

adaptation a trait that helps an organism survive in its surroundings (73, 83)

adult an organism that is completely developed (130)

air a mixture of gases that surrounds Earth (218)

air mass a large volume of air that has the same temperature and moisture levels (234)

air pollution harmful substances in the air (226)

air pressure the weight of the atmosphere in a particular place on Earth's surface (232)

air resistance force that pushes against objects as they move through air (222, 336)

alternating current (AC) current that changes direction through a wire (321)

alternative energy an energy source that does not use fossil fuels (272)

alveoli air sacs in the lungs where the exchange of oxygen and carbon dioxide takes place (93)

amber hardened tree resin (178)

amino acids the building blocks of protein (106)

amperes, or **amps** the measure of electric flow (321)

amphibians animals that spend part of their life cycle in water and the other part on land; the only vertebrate that goes through metamorphosis (134)

amplitude the distance in height that a sound wave travels from its starting point; also known as volume, or loudness (284)

anemometer an instrument that measures wind speed (236)

anesthetic a gas used to put patients to sleep (225)

animal a living thing that moves, reproduces by laying eggs or giving birth, and eats plants or other animals (78, 248)

Animalia the kingdom that includes all animals (78)

annual plant a plant that goes through its entire life cycle in one growing season (122)

antibiotic a chemical compound that slows the growth of bacteria (192)

anus the organ in the digestive system that transports solid waste to the outside of the body (95)

artery a blood vessel that carries blood away from the heart (92)

arthropod an animal with an exoskeleton and jointed legs (80)

asteroid a chunk of rock that orbits the Sun (197)

atmosphere air, moisture, and small particles of matter such as dust that surround Earth in a thick layer (154, 218, 228)

atom the smallest part of a substance that has the same properties as the substance (50, 252)

atomic number the total number of protons in an atom (51)

average speed the average rate at which an object travels (341)

axis the imaginary line around which a planet spins (135, 202)

B

balanced forces a pair of forces in which the forces that are pushing are equal to the forces that are pulling (336)

bar graph a graph that lets you compare numbers quickly (23)

barometer an instrument that measures air pressure (236)

beach a landform that forms when sediment is deposited by waves or currents (158)

bedrock the bottom layer of soil made up of solid rock (187)

behavior how an animal acts (86)

behavioral adaptation an activity that helps an organism survive its surroundings (75, 86)

biofuel a fuel made from organisms that lived recently (280)

biosphere all living things on Earth (154)

bladder the part of the urinary system where urine is stored (94)

block and tackle a system of pulleys that lessens the force needed to lift a load (319)

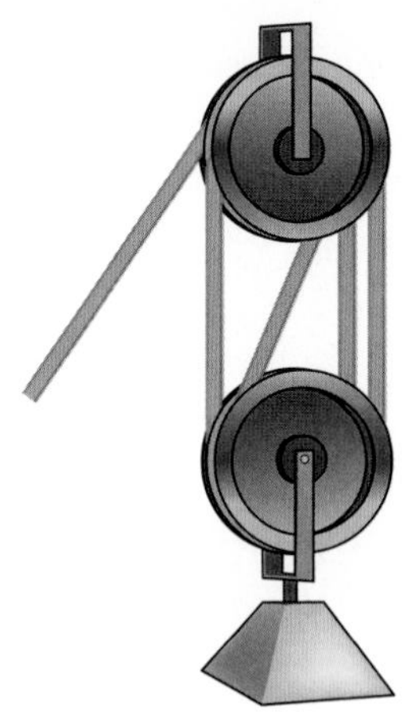

blood the fluid in the body made up of red blood cells, white blood cells, platelets, and plasma (92)

boiling the change of state from a liquid to a gas (259)

brain an organ that controls and regulates everything in the body (64)

brain stem the lower part of the brain that controls breathing, digestion, blood circulation, and heartbeat (91)

bronchi two tubes at the bottom of the trachea that bring air into the lungs (93)

C

calorie a measure of the amount of energy in a certain amount of food (113)

canyon a deep and narrow valley (158)

capillary a blood vessel that connects veins to arteries (92)

carbohydrate the substance plants create to store food energy; a nutrient in food that gives your cells energy (104, 120)

carbon and oxygen cycle the flow of carbon dioxide and oxygen through an ecosystem (137)

carbon dioxide one of the gases naturally found in air (137, 157, 242, 261)

cardiac muscle type of muscle found only in the heart (96)

carnivore an animal that eats only other animals (139)

cartilage tissue that lets bones move without grinding together; also forms structures like the tip of your nose or outer ear (96)

cast fossil a fossil that forms when sediment fills in the impression made by a mold (178)

catalytic converter a mechanism in cars, trucks, and buses that turns some harmful gasoline exhaust into carbon dioxide and water (227)

cell the smallest unit of life (58, 70, 128)

cell division when a cell can make new cells (58)

cell membrane a coating around each cell that controls what enters and leaves the cell (60)

cellular respiration the process in which cells burn sugar with oxygen to release energy, which produces carbon dioxide and water (104, 121)

cellulose the stiff, fibrous substance in plants (116)

cell wall the barrier of a plant cell that surrounds the cell membrane and gives the cell extra support (61)

cerebellum the part of the brain that controls movement and balance (91)

cerebrum the largest part of the brain; where thought and reason take place (91)

change of state a physical change in matter, such as freezing, melting, boiling, or condensing (258)

chemical change when matter changes from one type to another, taking on a new identity (264)

chemical energy a form of stored energy that can be converted to other forms of energy in order to do work (102, 266)

chemical property the ability of matter to go through a chemical change (265)

chemical symbol one or two letters that identify each element (54)

chemical weathering a process in which chemicals cause rock to dissolve or break down (157)

chlorophyll the part of a plant that gives the plant its green color; captures energy from sunlight for photosynthesis (61, 120)

chloroplast a special organelle in plant leaves where photosynthesis takes place (61, 76, 116)

circuit a system along which electric current flows (323)

circulatory system the organ system that supplies the body with nutrients and oxygen (92)

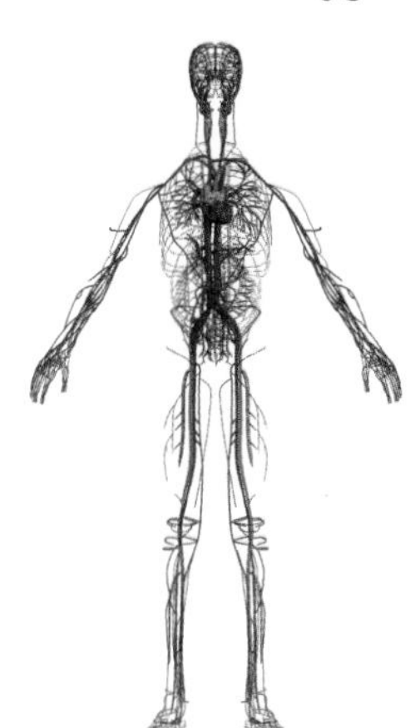

cirrus cloud a cloud formed of ice crystals found high in the atmosphere (233)

classification a system that groups organisms by how much they are alike (76)

clay a type of soil made up of very, very small pieces of rock (188)

cleavage the way in which a mineral splits when broken (168)

climate the average weather conditions of a place over a long period of time (228)

cloud millions of tiny water droplets suspended in the air (233)

coal a solid fossil fuel formed from prehistoric land plants (184, 240)

collaboration working together on a project or investigation (21)

comet a chunk of dust and ice that travels into our Solar System from another part of the galaxy (197)

commensalism a symbiosis where one organism benefits but the other is not affected (147)

community the populations in an area (136)

compass rose the symbol on a map that indicates direction (345)

competition what occurs when organisms need to use the same resource (146)

complete metamorphosis a process of change in animals that has four stages: egg, larva, pupa, and adult (133)

complex carbohydrates starches; contained in grains and vegetables (104)

compound a molecule made of two or more elements (53)

compressed air air that has been squeezed into a small space (224)

concave lens a lens that is "caved in" or curved inward (310)

conceptual model the model you make when you think about how something might work (48)

condensation a process in which water vapor cools and turns into drops of water (161)

condensing the change of state from a gas to a liquid (259)

conduction the flow of heat energy between things that are touching (297)

conductor a material that allows heat or electric current to flow through it easily (298, 321)

connective tissue tissue, such as ligaments, bones, and cartilage in an animal, that holds body parts together (62)

conservation ways to protect resources and use them without using them up (246)

consumer an organism that gets energy by eating other organisms (138)

contact force a push or a pull that happens between objects that are touching (332)

continental shelf the land around the edge of a continent that lies under the ocean (155)

convection the flow of heat energy through the movement of atoms in fluids (299)

convection current the cycling of air, which rises when heated by the Sun and sinks when it cools (232)

convex lens a lens that is rounded and curves outward (310)

coprolite a trace form fossil made from feces or dung (178)

crust the top of the lithosphere, the topmost layer of Earth (154)

cumulonimbus cloud a rain cloud (233)

cumulus cloud a fluffy white cloud that indicates fair weather (233)

current the constant flow of electric charge (321)

cycle a series of events that repeats; the last step leads to the first step (128, 170)

cytoplasm the fluid inside the cell membrane (60)

D

data information collected during a science investigation (22)

decay the breaking down of living things (186)

decibel a unit used to measure loudness (284)

decompose when an organism rots and breaks down into its smaller components (177)

decomposer an organism that gets energy by feeding on decaying organisms and wastes (138, 244)

delta a landform that forms when sediment builds where a river flows into an ocean or sea (158)

density a measure of how much mass there is in a given volume (255, 355)

dermal tissue the tissue that covers the outside of a plant (63, 118)

dichotomous key a tool that helps identify an object by answering a set of questions (77)

diffuse reflection when light hits a dull or rough surface and the rays bounce off in different directions, forming a soft or hazy pattern (304)

digestive system the organ system that breaks down food into chemicals cells can use; includes the mouth, esophagus, stomach, small intestine, large intestine, rectum, and anus (95)

direct current (DC) current that moves in only one direction through a wire (321)

distance the amount of ground covered by a moving object (341)

dune a landform that forms when a lot of sand is deposited in one place (158)

dwarf planet an object in the sky that is smaller than a planet and shares its space with other same-sized objects (200)

E

Earth materials nonliving, natural materials, such as minerals and rocks, that are found in Earth (166)

Earth science the study of Earth and its place in the universe (36, 153)

echo a sound created when sound waves hit a surface and bounce back (288)

ecosystem a large natural system that includes a community of organisms and nonliving things (39, 136)

ectotherm an animal whose body temperature depends on its surroundings (81)

egg the part of an organism that is fertilized and grows into an offspring (124)

electrical energy energy delivered by electric charges (266)

electric force a non-contact force caused by electricity (334)

electricity energy that results from the movement interaction of positive and negatively charged atoms (320)

electromagnet an iron core with many wires coiled together and wrapped around it (326)

electromagnetic spectrum made up of visible and nonvisible light (303)

electromagnetism a specific type of magnetism that deals with electric charges that are in motion (326)

electron a particle that circles the atom's nucleus; has a negative charge (51)

electron cloud the area of the atom that contains the electrons (51)

element a pure substance that cannot be broken down into other substances (50)

endocrine system the organ system that includes a group of glands that make chemical messengers called hormones (97)

endotherm an animal whose temperature stays the same regardless of its environment (81)

energy the ability to make things change or move (102, 266, 340)

energy conversion the changing of energy from one form to another (270)

energy pyramid shows how much energy passes from one organism to another in a food chain (143)

energy resources natural resources that produce energy (238)

English system of measurement the system of measure used mostly in your daily life (30, 351)

epidermis the outer layer of skin made up mostly of dead cells (99)

epithelial tissue animal tissue that is thin and smooth and that protects the body and certain body parts (62)

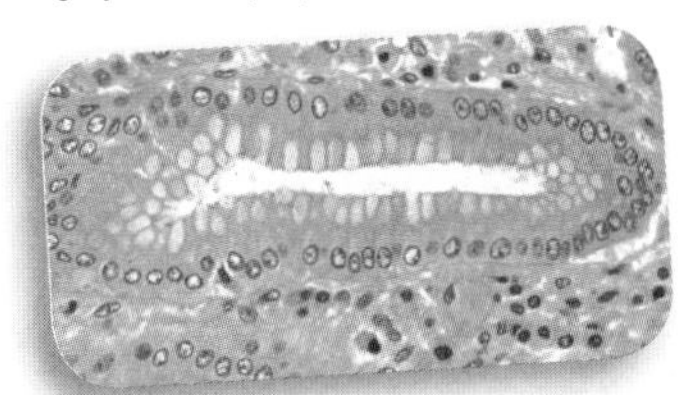

equator an imaginary line that runs around Earth like a belt (0° N/S) (348)

eroded mountains mountains formed when softer portions of Earth's surface are eroded by wind, rain, and ice leaving the harder rock behind (162)

erosion the movement of soft or loose weathered rock from one place to another; most often caused by water, ice, wind, or gravity (158)

esophagus an organ in the digestive system that pushes food into the stomach (95)

evaporation a process in which the Sun heats liquid water and changes it into a gas (161)

exosphere the atmosphere layer that starts above the thermosphere rising to about 10,000 kilometers (6,200 miles) above sea level (221)

experiment a scientific investigation (21)

fat a nutrient in food that stores energy (108)

fault a crack in the rocks of Earth's surface (162)

fertile soil that is able to hold nutrients and to make them available to plants (190)

fiber a complex carbohydrate that gives plants their structure (105)

fixed pulley a pulley that does not move (319)

flowering plant a plant that produces flowers and fruit (131)

fluids liquids and gases (299)

folded mountains mountains formed when layers of rock are pushed into each other from both ends (162)

food chain the path of energy through an ecosystem (when an organism eats another organism and food energy moves from the "food" organism to the "eater" organism) (140)

food web all of the paths that energy can take through an ecosystem (50, 142)

force a push or pull on an object (330)

fossil the trace or remains of a living thing that died a long time ago (176)

fossil fuel a fuel formed from organisms that lived millions of years ago; includes coal, petroleum, and natural gas (184, 240)

freezing the change of state from a liquid to a solid (258)

frequency the number of times that an object vibrates per second (286)

friction the force that slows down movement between two surfaces that are touching (295, 333)

front the border between one air mass and the other (234)

fruit the part of a flowering plant that develops around the seeds (131)

fulcrum the support required to make a lever effective (317)

G

galaxy an enormous grouping of matter in the universe held together by gravity (196)

gas a state of matter that does not have shape or volume (257)

gas giants another name for the four outer planets, which are made up of gas (200)

gasoline a liquid fuel that is made from oil (240)

geologist a scientist who studies minerals and rocks (167)

geosphere the rock and soil on Earth's surface and the layers under the surface (154)

geothermal energy energy that comes from heat stored in Earth (277)

germ something that causes infection and disease (100)

germination the process in which a seed splits and a plant begins to grow (122)

Global Positioning System (GPS) a system of satellites that can help you find your position anywhere on Earth (349)

glucose a sugar produced by plants using energy from sunlight to combine oxygen and carbon dioxide from the atmosphere; the basic building block of all carbohydrates (104, 218)

gravitational field the area around objects where the force of gravity acts (335)

gravitropism the tendency of a plant's roots to grow down in response to Earth's gravity (127)

gravity a pulling force between objects that have mass (220, 335)

grid the use of letters and numbers to show locations on a map (345)

ground tissue the tissue of a plant where photosynthesis takes place (63, 118)

groundwater freshwater that collects underground (241)

H

habitat the area where an animal lives (83)

hardness scale a scale to rank the hardness of minerals (169)

heart the organ of the circulatory system that pumps blood through the blood vessels (64, 92)

heat the movement of energy from one thing to another due to a difference in temperature (292)

heat energy energy that depends on the temperature of matter (266)

herbivore an animal that eats only plants or plant products (139)

hibernation a state of deep rest that lets an animal survive the winter without food (87)

horizon a layer of soil (187)

hormone a chemical messenger in the body that controls how your body grows and develops (97)

host an organism that is prey to a parasite in parasitism (147)

humus the layer of soil made up of dead plants and animals, and decayed animal wastes (187)

hydroelectric energy uses the energy in moving water to make electricity (275)

hydrosphere Earth's liquid water and ice, water in soil and rock, and water vapor in the air (154, 230)

hydrotropism when a plant's roots move toward moisture (127)

hypothesis an educated guess about how things will work that includes an explanation of why it does so (18)

I

igneous rock rock formed from lava that erupts from a volcano (170)

incandescent something that can give off light energy if it is heated to a high enough temperature (234)

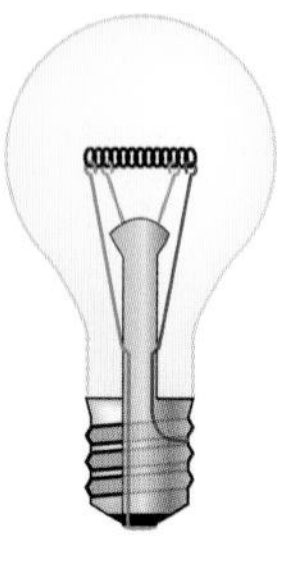

inclined plane a flat surface that slants (314)

incomplete metamorphosis a process of change in animals that has three stages: egg, nymph, and adult (132)

index fossil a fossil used to help find out the age of rock layers (181)

inertia when objects resist changes in motion (338)

inexhaustible energy energy from a source that can't be used up, such as the Sun or wind (274)

inherited trait a feature that is passed from parent to offspring, and from generation to generation (72)

inner planets the four planets closest to the Sun (198)

insulator a material that does not allow heat or electric current to flow through it easily (298, 321)

integumentary system the organ system that is made up of skin, hair, and nails (98)

invertebrate an animal without a backbone (80, 132)

J

joint the place where two or more bones come together (96)

K

kidneys two organs in the urinary system that filter blood (94)

kilowatts thousands of watts; a measure of electric power (234)

kinetic energy energy in motion (269, 340)

L

landform a natural geological feature on Earth's surface (155)

landslide rocks and soil that slide down a slope together (158)

large central vacuole found in plant cells; stores water, salts, and carbohydrates (61)

large intestine the organ in the digestive system where water from liquid waste is absorbed (95)

larva a stage during complete metamorphosis that is hatched from an egg (133)

larynx the part of the throat that contains the vocal cords (93)

laser **L**ightwave **A**mplification by **S**timulated **E**mission of **R**adiation; a special kind of light that contains light rays that are all the same wavelength and travel parallel to each other (305)

latitude lines on a globe that run parallel to the equator (348)

lava magma (molten rock) that reaches Earth's surface (170)

law of electric charges the law that says that positive charges push away from each other and negative charges push away from each other, and a positive charge and a negative charge pull toward each other (52, 320)

leaf the part of a plant where photosynthesis takes place (65, 119)

legend a box in the corner of a map that explains symbols used on the map (345)

lever a board and a support used to lift heavy objects (317)

life cycle the way an organism grows and reproduces (128)

life science the study of all living things (35 ,57)

light a form of energy that travels in waves; the visible part of the electromagnetic spectrum (302)

light energy energy that is transferred by electromagnetic waves (including light waves) (266)

line graph a graph that shows how data change over time (25)

liquid a state of matter that has volume but not shape and can be poured from one container to another (257)

lithosphere the rocky parts of Earth's surface; the top part of the geosphere (154)

loam soil made up of about equal parts of humus, sand, silt, and clay (189)

longitude lines on a globe that run north and south (348)

lungs two organs in the respiratory system that allow you to breathe in and out (64, 93)

luster the way the surface of a mineral reflects light (168)

lymphatic system the organ system that carries away excess fluids, dead cells, germs, and white blood cells; also helps make antibodies to fight disease (100)

M

machine a tool that uses energy to do work (312)

magma molten rock below Earth's surface (162, 170)

magnet an object that attracts certain types of metals (325, 334)

magnetic field the area around a magnet that applies a force (334)

magnetism the way that objects behave in the presence of a magnetic force (325)

magnitude the size of a force (331)

map shows the location of things on Earth (344)

mass a measure of how much matter there is in an object (33, 51, 220, 254, 335, 352)

material resources natural resources used to produce other products (238)

mathematical model a model that scientists make to make predictions or demonstrate how something works (48)

matter anything that has mass and takes up space (252)

mechanical energy the energy in matter; has two forms: energy that is stored and energy that is moving (267)

medium a gas, a liquid, or a solid that can transmit or hold particles or waves (283)

melting the change of state from a solid to a liquid (258)

mesosphere the atmosphere layer that starts above the stratosphere rising to about 85 kilometers (53 miles) above sea level (221)

metamorphic rock rock made when heat and pressure melts and squeezes igneous or sedimentary rock for a very long time (171)

metamorphosis a process in which an animal changes form as it grows (132)

meteor a space object that burns and lights up from friction as it falls into Earth's atmosphere (197)

meteorite a meteor that makes it to Earth's surface (197)

meteoroid an object too small to be an asteroid; often enters Earth's atmosphere (197)

meteorologist a scientist who studies and reports on the weather (236)

methane a gas made up of carbon and hydrogen (184)

metric system the system of measure used for science projects (30, 351)

mid-ocean ridge mountains on the ocean floor (155)

migration movement of animal groups from one area to another each year (87)

mimicry an adaptation of body shape or color that helps an animal survive (86)

mineral a solid object with a crystal and chemical structure that does not contain organic matter; or an element in food that the body uses to build tissues and for other tasks (111, 166, 244)

mitochondria the part of the cell that breaks down food to make energy (60)

mixture a combination of two or more different types of matter (260)

model a representation of an object or a concept (46)

mold fossil a fossil made when the remains of an animal or plant leave an impression in soft sediment (178)

molecule two or more atoms joined together (50)

mollusk a soft-bodied animal with a hard mineral shell (80)

motion a change of position of an object (340)

mountain Earth's highest landform (155)

moveable pulley a pulley that moves freely at both ends (319)

mudflow rain mixed with soil that flows down a slope (158)

multicellular organism an organism with two or more cells (59, 70)

muscle tissue animal tissue that is made of muscle cells and that is found in muscles and other organs (62)

muscular system the organ system that causes the parts of the body to move (96)

mutualism a symbiosis where both organisms benefit (147)

N

natural gas a fossil fuel formed from the remains of organisms (184, 240)

natural resources substances and materials available from Earth that we use in daily life (238)

nervous system the organ system that controls bodily functions and reactions; includes the brain, the spinal cord, and the nerves (91)

nervous tissue animal tissue that is found in nerves and the brain (62)

neutron an atomic particle located in the nucleus; has no charge (51)

newton a unit used to measure force (331)

niche the role that an animal plays in its habitat (83)

nitrogen cycle a natural cycle in which nitrogen in the air is used by plants and animals, then is released into the air again (43)

non-contact force a push or a pull between objects that are not touching (334)

nonflowering plant a plant that produces cones to make seeds (130)

nonrenewable energy an energy source that cannot be replaced once it is used up (274)

nonrenewable resources natural resources that cannot be replaced or would take too long to be replaced naturally (185, 239)

normal fault a fault formed when rocks pull apart and one side of the land drops down (162)

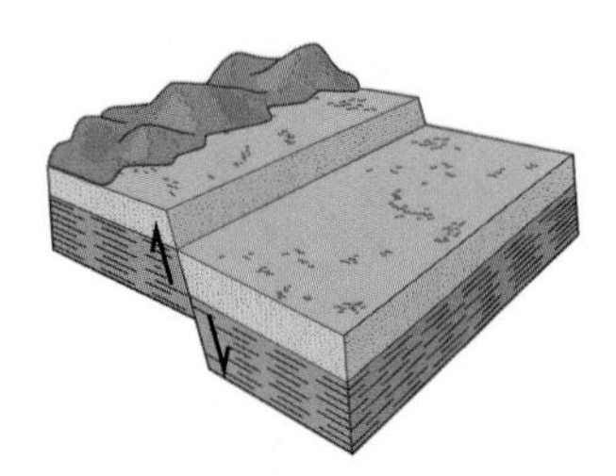

nuclear energy energy produced when nuclear bonds form or break (266, 281)

nucleus the command center of the cell; also the center of an atom (51, 60)

nutrients chemicals in food that help your body grow and repair itself (103)

nutrition label information on food packaging about the nutrients that are in that food (114)

O

observation a way of looking at something in a special way in order to get a clear answer to a question (21)

oil a liquid fossil fuel (240)

omnivore an animal that eats plants and other animals (139)

opaque describes a material that prevents all light from passing through it (308)

organ a group of different tissues that work together to perform one or more jobs (64, 71)

organelles the parts within a cell that do the work (60, 70)

organism anything that is alive (68)

organ system a group of organs that carry out a process (66, 71, 90)

outer planets the four planets farthest from the Sun (200)

ovary the place where eggs develop in a plant or animal (125, 131)

oxygen one of the elements found in air (137, 157 242, 261)

P

paleontologist a scientist that looks for and studies fossils (180)

parallel circuit a type of circuit in which each load is connected in a closed circuit (323)

parasite an organism that gets its nutrients from other organisms by living on or in a host (117, 147)

parasitism a symbiosis where one organism (the parasite) is helped and one (the host) is harmed (147)

particulate pollution tiny pieces of dust, ash, and soot, and drops of water in the air (226)

pattern something that is repeated in a regular fashion (42)

peat accumulation of partially decayed plant matter (184)

perennial plant a plant that lives for more than one growing season (123)

periodic table of the elements a table used to keep track of all the elements and arranged according to each element's atomic number and property (54)

petal the part of a plant that attracts pollinators (125)

petroleum a fossil fuel formed from tiny organisms that died in ancient seas (184)

pharynx the part of the throat below the nose (93)

photosynthesis a process in plants that uses sunlight, water, and carbon dioxide to make food (104, 116, 242)

phototropism when the leaves of a plant turn toward a light (127)

physical change a change in matter that does not change the type of matter that it is (256)

physical property a characteristic of matter that you can observe without changing the type of matter that it is (253)

physical science the study of what things are made of and how things work (37, 251)

physical weathering the breakdown of rocks caused by wind, water, ice, changes in temperature, and living things (156)

pie graph a circle graph that shows parts of a whole (24)

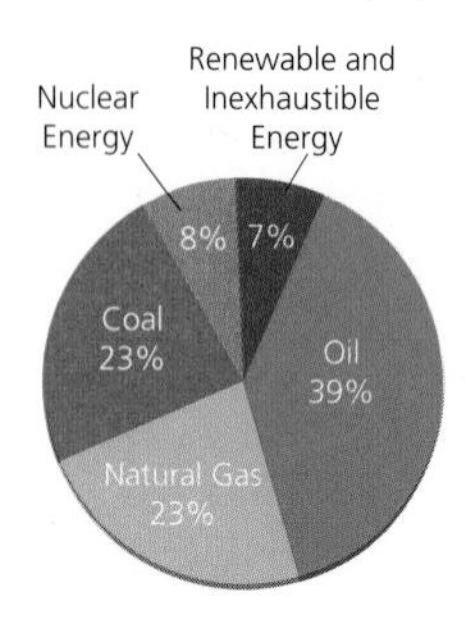

pistil the female part of a flowering plant (125)

pitch how high or low a sound is (286)

plain a large, flat area of ground lower than the land around it (155)

plant a living thing that can make its own food, reproduce, but can't move (116, 243)

Plantae the kingdom that includes all plants (116)

plantlet a little plant that forms when an adult plant sends out a shoot from its roots (125)

plasma the liquid part of blood (92)

plateau a large area of flat ground that is much higher than the land around it (155)

platelets the part of blood that repairs breaks in blood vessel walls (92)

plutonic rock rock formed when magma cools slowly; has large mineral grains (170)

pollen the part of a plant that contains the sperm (124)

pollination the process of moving pollen from a plant's stamen to the pistil so seeds can begin to develop (125)

pollution what happens when something is released into the environment that can harm living things (245)

population a group of organisms of the same species that live in the same place at the same time (136)

potential energy energy that is waiting to be released; stored energy (268, 340)

precipitation the weather word for any form of water (rain, snow, sleet, or hail) that falls to Earth from the sky (161, 231)

predator an animal that eats another animal (83, 146)

prey an animal that is eaten by a predator (146)

prime meridian an imaginary line that begins at the North Pole and runs south to the South Pole (0° E/W) (348)

prism a piece of glass that can separate the colors in white light (302)

producer an organism that can make its own food and provides food for other organisms (138, 243)

protein a nutrient in food that helps build and maintain muscles and bones; a molecule made up of amino acids (106)

proton a particle located in the atom's nucleus; has a positive charge (51)

puberty the time of life when your body begins to change from a child to an adult, usually between the ages of 10 and 15 (97)

pulley one or more grooved wheels with a rope or chain wrapped around it to make lifting weights easier (319)

pupa a stage during complete metamorphosis in which a larva forms a hard case outside of its body (133)

R

radar **RA**dio **D**etection **A**nd **R**anging; a system that can send out radio waves and detect the ones that are bounced back by an object (163)

radiation a type of heat energy that comes off a hot object in all directions (300)

rectum the organ in the digestive system that stores solid waste (95)

recycle to reuse old materials in a new way (247)

red blood cells blood cells that supply oxygen to the body and carry carbon dioxide away from tissues (92)

reference point the object that stays in place compared to an object that moves (340)

reflection the bouncing of waves off a surface (304)

refraction when light rays pass from one medium to another causing the rays to bend or change direction slightly (309)

relative age the age of a fossil depending on its position in the rock layer (181)

renewable energy an energy source that can be replaced (273)

renewable resources natural resources that can replace themselves (185, 239)

resource anything an animal needs to survive (84)

respiratory system the organ system that makes it possible to breathe (93)

reverse fault a fault formed when rocks are pushed together and one side of the land is pushed up (162)

revolve when one object circles another object, as when planets orbit the Sun (135)

rock a solid object that contains two or more minerals (167, 244)

rock cycle a process in which one type of rock can change into a different type of rock (172)

rockfall a group of rocks that fall off a cliff (158)

root the part of a plant that absorbs water and nutrients from the soil and holds the plant up (65, 130, 119)

rotate when an object spins completely around on its axis (135)

S

saliva a substance produced in the mouth to help break down food (95)

salivary glands organs in the digestive system that produce saliva (95)

sand particles of weathered rock and minerals (188)

scale the reduction of everything on a map by the same proportion (345)

science notebook a place to write down important observations and your questions and answers about science (20)

scientific method steps scientists follow to answer questions (16, 350)

screw an inclined plane wrapped around a cylinder (316)

seasons annual climate changes caused by the tilt of Earth on its axis and its rotation around the Sun; spring, summer, fall, and winter (135, 203, 229)

sediment tiny pieces of rock, sand, and soil (158, 177)

sedimentary rock rock made up of layers of sediment that have been squeezed together over millions of years (170, 177)

seed the part of a plant that contains the plant's embryo (130)

seedling a small plant in the early stages of growth (122, 130)

seismograph a device that detects seismic waves, a form of energy that travels through Earth from an earthquake (163)

sepal the part of a plant that protects the flower bud (125)

series circuit a type of circuit in which the loads are connected in a line (323)

silt soil made up of small organic particles and minerals (188)

simple carbohydrates sugars; contained in table sugar, honey, fruit, milk, and other foods (104)

simple machine inclined plane, wedge, screw, wheel and axle, lever, and pulley (312)

skeletal muscle muscle that supports the skeleton and allows it to move (96)

skeletal system the organ system that gives the body a shape and protects soft internal organs; includes bones, joints, and cartilage (96)

skin the largest body organ and the first line of defense against disease (64, 98)

small intestine the organ in the digestive system that is connected to the stomach and where food is broken down (95)

smooth muscle muscle that is found in organs and blood vessels (96)

soil a mixture of waste and the decayed remains of plants and animals, tiny pieces of weathered rock, and mineral particles (186, 244)

solar energy energy captured directly from the Sun's rays (276)

solar system a group of planets and other objects orbiting a star (40, 197)

solid a state of matter that keeps its shape, even when it moves (257)

solution a mixture in which the different kinds of matter are spread out evenly and cannot be easily separated (261)

sonar **SO**und **N**avigation **A**nd **R**anging; used to map the ocean bottom (163)

sound a form of energy that is created when an object vibrates (282)

sound energy energy carried by sound waves (266)

specialized cell a cell that does a particular job (59, 71)

species a group of organisms that can mate with each other and produce offspring (72)

specular reflection when light hits a smooth, shiny surface such as a mirror and all of the rays are reflected together at the same angle (304)

speed a measure of how far an object has moved in a certain amount of time (341, 355)

spinal cord the bundle of nerves that runs down the spinal column (91)

spore in flowering plants, the part of a plant that can produce a new plant (124)

stamen the male part of a flowering plant (125, 131)

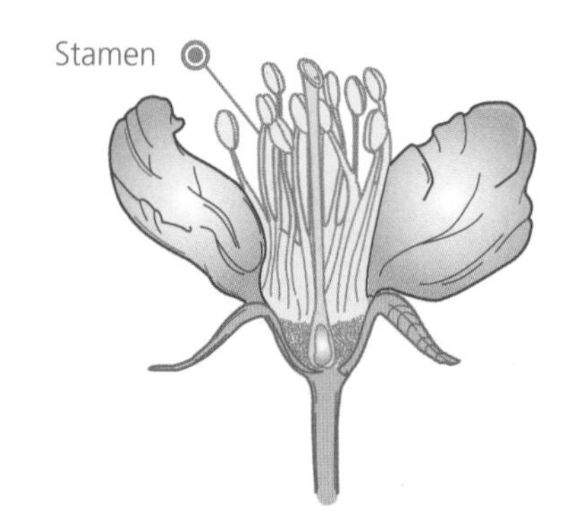

starches complex carbohydrates; contained in grains and vegetables (104)

state of matter a form of matter, such as a solid, a liquid, or a gas (257)

static electricity the buildup of electric charges on a surface (320)

stem the part of a plant that connects the leaves to the roots (65, 118, 130)

stomach the organ in the digestive system where food mixes with enzymes and acids (64, 95)

stratosphere the atmosphere layer that starts above the troposphere rising to 50 kilometers (31 miles) above sea level (220)

stratus cloud a long, horizontal cloud formed low in the atmosphere (233)

streak the color of a powdered or crushed mineral (168)

structural adaptation a feature of an animal's body that helps it survive (85)

subsoil the layer of soil made up of dissolved materials and fine clay from the layers above it (187)

sugars simple carbohydrates; contained in table sugar, honey, fruit, milk, and other foods (104)

surface water liquid freshwater that collects on Earth's surface; the main source of water for humans and other organisms (241)

symbiosis when organisms depend upon another organism in order to survive (147)

system an organized group of different objects that work together to make a whole (38, 154)

T

technology the use of scientific discoveries to solve problems (34)

temperature a measure of how hot or cold something is (31, 254, 292, 353)

thermal energy the total amount of energy in a substance (292, 307)

thermometer an instrument that measures temperature (236)

thermosphere the atmosphere layer that starts above the mesosphere rising to about 640 kilometers (400 miles) above sea level (221)

tissue a group of specialized cells that work together to do the same job (62, 71)

topsoil the layer of soil made up of humus and weathered rock particles (187)

trace fossil a fossil that is proof of animal life in the past, such as a burrow, footprints, and waste products (179)

trachea the part of the throat that connects the nose and mouth with the bronchi and the lungs; known as the windpipe (93)

trait a feature, such as eye color or nose shape (72)

translucent describes a material that allows some light to pass through (308)

transmission of light when light rays pass right through matter (308)

transparent describes a material that allows most light to pass through (308)

troposphere the atmosphere closest to the surface of Earth rising to about 17 kilometers (11 miles) above sea level (220)

true form fossil a fossil formed from actual plant or animal parts (178)

U

unbalanced forces a pair of forces in which the forces are not equal (337)

unicellular a life form that has only one cell (70)

unicellular organism an organism with only one cell (59)

universe includes all the planets, stars, moons, asteroids, space dust, comets, gas clouds, and vast areas of empty space (196)

ureters two tubes in the urinary system that connect the kidneys to the bladder (94)

urethra the tube in the urinary system that transports urine from the bladder to the outside of the body (94)

urinary system the system that removes wastes from the blood; includes kidneys, ureters, urinary bladder, and urethra (94)

urine a liquid waste product formed in the kidneys (94)

V

vacuoles fluid-filled sacs that store nutrients or wastes (60)

valley a landform that forms when rivers or glaciers wear away rock and soil (158)

vascular tissue plant tissue that moves nutrients and water in the plant (63, 118)

vein a blood vessel that carries blood to the heart (92)

velocity a measure of the speed of an object as it moves in a certain direction (341)

vertebrae bones that form the backbone of an animal (80)

vertebrate an animal with a backbone (80, 134)

villi little bumps in the small intestine that give it more surface area to absorb nutrients (95)

visible light the colors that make up white light (303)

vitamin a chemical compound in food that controls many processes in your body (110)

volcanic mountains mountains formed when great pressure forces magma up and out from inside Earth (162)

volcanic rock rock formed when lava cools quickly; has small mineral grains (170)

volcano a natural event that forms when great pressure forces magma up and out (147)

volume a measure of how much space something takes up; also a measure of how loud something is (31, 254, 284, 355)

W

water the most important nutrient for our bodies (102)

water cycle the process in which water is recycled over and over again through the hydrosphere (43, 230)

watt a measure of the electric power of an object (234)

weather balloon a device that carries instruments up into the atmosphere to record temperature, humidity, and atmospheric pressure (163)

weather the condition of the atmosphere at a given time and in a given place (228)

weathering a process of breaking rock into smaller pieces (156)

wedge a triangular object that is essentially an inclined plane and is used to move or split objects, or to keep objects in place (315)

weight the pull of gravity on an object (33, 220)

wheel and axle a simple machine made of a wheel with a rod (an "axle") inserted through the center that work together as a unit (318)

white blood cells blood cells that destroy germs (92)

wind energy energy captured from the wind (278)

work in science, done when forces cause something to move (340)

Answer Key

Science Basics

Thinking BIG—SEM page 45
A butterfly wing

Make a Model of a Tornado page 47
5. The soap clouds form a vortex. This is what happens when a tornado forms!
6. The confetti swirls in the vortex. The confetti is like the debris a tornado traps in the vortex.

Build a Food Web page 49
Your food web will depend on where you live. For instance, your web could include grass, a rabbit, a deer, and a coyote. The deer and rabbit eat the grass. The coyote eats the rabbit.

Life Science

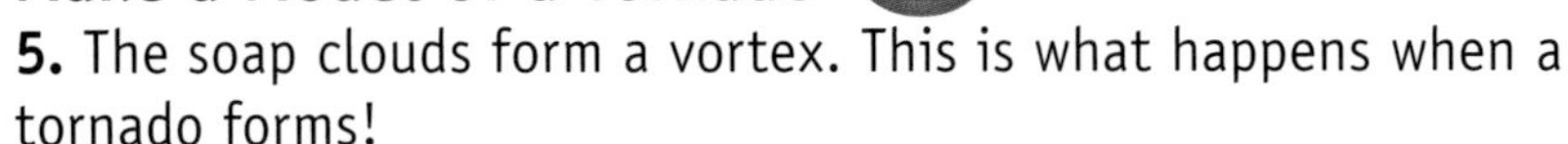

Cell-U-Learning page 61
You can compare your school building to a plant cell or an animal cell. A fence around the outside of the school is like a cell wall in a plant cell. For both types of cells, the building walls are like the cell membrane. The doors and windows allow things to enter and leave the building. (But they won't let anything too big, like a car, inside!) The air inside the building is like cytoplasm—it fills all the available space. The principal's office is like the nucleus. It is the command center of the building. The electrical generator is like the mitochondria—it supplies energy to the entire school. The trash cans in the classrooms and offices are like vacuoles. Wastes are

stored there until they are picked up. What about your cafeteria? It is like a plant cell's chloroplast: It makes food for the cell.

Which Is Which? page 67

A cell is the smallest unit of an organism. A tissue is a group of cells that do the same job. An organ is a group of tissues that work together to perform one or more jobs. An organ system is a group of organs that work together to perform the same function.

Organisms

Thinking BIG—SEM page 70

An ant

Animals

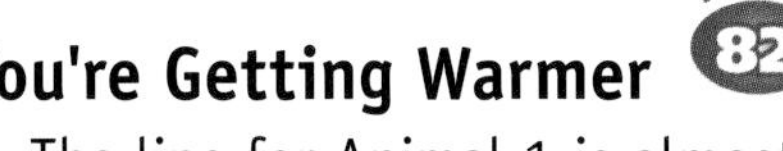

You're Getting Warmer page 82

4. The line for Animal 1 is almost flat. Animal 2's line increases at a steep angle.
5. Animal 2 is an ectotherm. Its temperature is always similar to the outside temperature. Animal 1 is an endotherm. Its temperature stays about the same even when the outside temperature changes.

Thinking BIG—SEM page 85

A bird feather

Food Energy

FSI: Food Science Investigation—Part 1 page 105

5. Your answers will depend on the foods you tested. Fruits, grains, chips, sweets, and some vegetables contain starch. Iodine will turn black on samples of these foods. Meats, butter, cheeses, peanut butter, and oils do not have starch. Iodine will not change color on samples of these foods.

FSI: Food Science Investigation—Part 2 PAGE 109

5. Your answers will depend on the foods you tested. Chips, peanut butter, meats, butter,and oil all have fat and will leave grease spots. Fried foods, oils, and nuts usually have the most fat, and will leave the largest grease spots. Fruits, starches, and vegetables have little or no fat, so these samples won't leave grease spots. (They might leave a damp spot behind, though. throw away all the food and let the paper bag squares dry out. Water spots will disappear, but grease spots will still be there.)

FSI: Food Science Investigation—Part 3 PAGE 115

2-6. Your answers will depend on the cereals you compared. Most sugary cereals have less fiber and more calories, fat, sodium, and sugar per serving compared to whole-grain cereals.
7. Sugar, fat, and sodium make a cereal less healthy, while high fiber content makes a cereal more healthy.

PLANTS

Thinking BIG—SEM

A plant stem

CYCLES

Which Is Which?

incomplete metamorphosis—nymph; complete metamorphosis—larva and pupa; both types of metamorphosis—egg and adult

ECOSYSTEMS

Weave a Web

4. The Sun is the main energy source in any food web.

Earth Science

Landforms

Landform Investigator 

1. Your list of landforms will depend on the landforms in your area. You might list mountains, ravines, valleys, or mesas.

2. You may have many different landforms in your part of the country. Maybe you live where there are only one or two kinds of landforms.

3. Your answer will depend on where you live. If you live in the Great Plains, the land is flat today. Thousands of years ago, glaciers wore down mountains and changed the landscape.

Quick Question 

Oxygen reacts with iron to form rust. Rust is a form of is a form of chemical weathering. Ivy roots spread mortar just like it was loose rock. This is a kind of physical weathering. Road pavement expands and contracts the way rocks do. Holes in pavement are caused by physical weathering. Burning a tree for fuel is not a form of weathering.

Minerals and Rocks

Rock Teams, Huddle Up! 

Your rocks could be any of the three types. Think about the properties of each type of rock as you examine the ones you found.

Thinking BIG-SEM 

Salt

FOSSILS

Your Own State's Fossil! page 179

1. Every state fossil is different.
2. You may discover that the organism that created your state fossil lived millions of years ago.
3 & 4. The environment in your state may have changed a great deal over time. For example, fossils of organisms that lived in oceans have been found in areas hundreds of miles from the nearest body of saltwater.
5. Your answer will depend on where you live. The state fossil of Arizona is a true-form fossil, Araucarioxylon arizonicum, a kind of petrified wood. The state fossil of Connecticut is made by a trace fossil of a dinosaur footprint, Eubrontes giganteus.

Dinosaurs to Birds? page 183

2. three in front, one in back; three in front, one in back
3. yes
4. They bend backward.
5. Dinosaurs are probably the ancestors of birds.

How Long Ago? 

$$\begin{array}{r} 3{,}500{,}000{,}000 \text{ years} \\ -\ \ 245{,}000{,}000 \text{ years} \\ \hline 3{,}255{,}000{,}000 \text{ years} \end{array}$$

SOIL

How Will Their Garden Grow? 

number of wheelbarrow loads for each square yard of garden: 3
number of square yards in one acre: 4,840
multiply $3 \times 4{,}840 = 14{,}520$ wheelbarrow loads for one acre multiply $14{,}520 \times 4 = 58{,}080$ wheelbarrow loads to cover four acres

Which Horizon Is Which? page 187

O, A, E, B, C, R. Your notes should say that horizon O is made up mostly of dead plants and animal. Horizon A is also called topsoil. Horizon E has very few nutrients. The B horizon collects dissolved minerals and clay. Horizon C is mostly broken rock. Horizon R is called bedrock.

Soaking it Up page 191

1. Samples of soil from near your home or school will look and feel different depending on the source. Potting soil is usually dark, moist, and crumbly. It may have pieces of organic material.
2. Some soils may feel heavy and damp. They may have a lot of clay. You could squeeze clay into a ball. Some soils may be sandy. Sandy soils do not squeeze into shapes.
3. Water will slide off soil that has a lot of clay in it. It will pour through soil that has a lot of sand. Soil with a lot of organic material will hold water better.
5. Potting soil probably held the most water. Soil taken near your school might be very compacted and not hold water. Soil that holds water well is good for growing garden plants and crops.

It's Alive! page 195

Every sample of soil and litter will have different creatures living in it. You may find centipedes or millipedes, worms, beetles, spiders, mites, or ticks.

Solar System

My Active Shadow page 205

3. The length of your shadow will be longer each time you measure it. One hour after noon, your shadow will be longer than it was at noon. Two hours after noon, it will be even longer. **5.** Your shadow would get shorter as the time approaches noon. As Earth rotates on its axis, the Sun appears to move higher in the sky until noon. After noon, the Sun appears to sink lower in the sky.

Space Exploration

How Far Would You Travel? 

1. 78,338,750 km; 48,677,440 mi
2. 628,814,130 km; 390,726,990 mi

Air

What's in the Air? page 219

oxygen

The Incredible Parachuting Eggs! 

4. You might think that the egg with the largest parachute will survive. Whatever egg you choose, explain your reasons.
5. The egg with the largest parachute had the best chance to survive.
6. Air resistance and gravity acted on the parachutes. The larger parachute fell more slowly than the smaller one. The larger parachute had more surface area than the smaller parachutes, so there was more air resistance. The larger parachute filled up with more air than the smaller parachutes, so there was more air resistance.

Hey Air! Get Back to Work! 

2. What did you find? Maybe you have a gas stove or grill, or maybe a furnace blows warm air through vents to heat your home or school.
3. You can see that the plants are green and so are taking in carbon dioxide. You can see tired people breathing hard.

Which Kind of Air Pollution? 

a. particulate
b. noxious gas
c. particulate
d. noxious gas

Weather, Seasons, and Climate

Clear Today, Snow Tomorrow? page 229

Your answers will depend on the weather and the time of year.

Create Your Own Air Mass page 237

3. You should feel warm air rising as you pour the chilled water into the cup. If this were an air mass, it would be warm and humid.

Natural Resources

Thinking BIG—SEM page 245

A newspaper

Physical Science

Matter

Will It Float? page 255

5. Wax has a lower density than water. It will float.

Which Is Which? page 257

milk—liquid; wood—solid; rock—solid; oxygen—gas

How Cool? How Cool! page 259

3. The ice cube is beginning to melt. Water is changing from a solid to a liquid.
4. The air inside the cup is colder.

Separation Challenge page 263

1. Sand, iron filings, and sawdust may all be the same color.
2. You could use a magnet. Iron is magnetic. Salt, sand, and sawdust are not magnetic.
3. If you add water to the rest of the mixture, the sand will sink, and the sawdust will float. The salt will mix with the water and dissolve.

Energy

Quick Question 

The water in the dam has more potential energy. It is larger and at a greater height than the sink is.

Quick Question 

There are **3** conversions: chemical energy (coal) to heat energy (steam); heat energy (steam) to mechanical energy (turbine); mechanical energy (generator) to electrical energy (transformer).

Alternative Energy

Energy Use in the United States 

1. 23%
2. dark red; 7%
3. 100%; the total amount of energy used in the United States
4. oil
5. coal, oil, and natural gas; 85%

Sunny Side Up 

3. Water temperature should increase in Jars A, B, and C, and stay about the same in Jar D. The warmest water should be in Jar B; next warmest should be Jar A; third warmest should be in Jar C.

Thar She Blows! 

4. It turns when you hold it directly in front of you and blow on it; it will turn faster if you blow on it harder. However, it won't turn if you blow on it from an angle or from behind.

Which Is Which? 

biofuel and wood—renewable; fossil fuel—nonrenewable; wind energy—inexhaustible

Sound

What's Happening?

Yes; the photograph shows that the tuning fork is vibrating.

Quick Question

The canyon is a large, open area with high, straight walls. There is plenty of surface area for sound waves to bounce off of.

Science Notebook—Does It Bounce?

1. cookie sheet, cafeteria tray, poster board, chalk board
2. Sound waves bounce well off flat surfaces.
3. Your partner's results should be similar to your own.

Which Is Which?

frequency—pitch; bouncing sound waves—echo;
soft sounds—amplitude

Heat

Hot or Cold?

1. The room temperature water does not feel hot or cold. Both of your hands should feel the same temperature.
3. When you put your hands back into the water at room temperature, your right hand should feel hot and your left hand should feel cold.
4. Your right hand is colder than the room-temperature water. The heat moves from the water to your hand. Your left hand is warmer than the room-temperature water. The heat moves from your hand to the water.

Which Is Which?

touching a hot potato—conduction; sunbeam on cat—radiation;
oven warming kitchen—convection or radiation

LIGHT

Everybody Split 

2. Top to bottom: red, orange, yellow, green, blue, indigo, and violet.
3. The other prisms will split light in the same order: red, orange, yellow, green, blue, indigo, and violet.
4. The type of prism you use doesn't change the order of the colors in white light.

Shiny, Shinier, Shiniest PAGE 305

4. No. Light reflected from the mirror and the other shiny object should produce sharp images. The CD surface is not completely smooth, so it won't reflect light like the other objects. Instead, the CD will act like a prism and split sunlight into the colors of visible light.

Convert With Color 

4. The glass with the black cloth will warm up faster. The dark cloth absorbed all of the light rays from the Sun and converted light energy to thermal energy.

Which Is Which? 

waxed paper—translucent; red construction paper—opaque; plastic wrap—transparent

Is That a Bendable Straw? 

3. The straw looks like it bends where it passes from air to water. The light rays changed direction when they passed from one medium to another.

SIMPLE MACHINES

How Long? How High? PAGE 314

First, calculate the total height the ramp must travel:

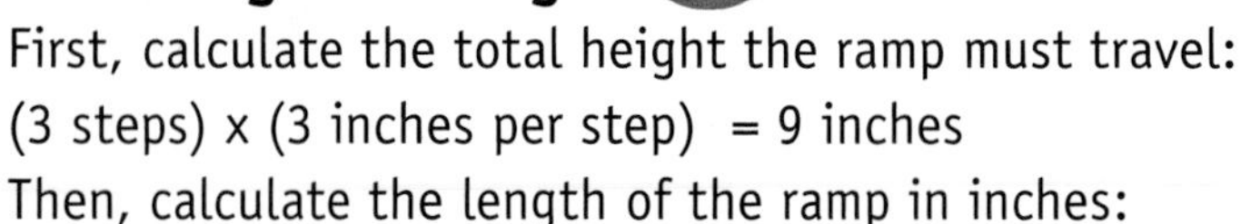

(3 steps) x (3 inches per step) = 9 inches

Then, calculate the length of the ramp in inches:

(9 inches high) x (12 inches forward) = 108 inches
The ramp must be 108 inches long.

Hammer or Screwdriver? Page 316

Roofers use nails in their work because they need to install many shingles over a large area.

Electricity and Magnetism

Are You Positive? Page 320

3. Your hair flies toward the balloon.

4. Negatively charged electrons jump from the balloon to your hair. The balloonis now more positively charged. When you hold the balloon near your head, the positive charge on the balloon attracts the negative charge on your hair.

Thinking BIG—SEM Page 322

Light bulb filament

Power Up Page 326

5. Your answers will probably differ from those of your classmates. Follow the suggestions to find out why.

Forces and Motion

Which is Which? Page 335

Gravity: non-contact; Throw: contact; Friction: contact; Magnetic: non-contact; Yank: contact

Invisible Mass Page 337

3. The forces on the ruler are balanced. The ruler stays in the same position.

4. The part of the ruler without a balloon pulls up. The mass of the balloon pulls the other side down. The forces on the ruler are unbalanced. The forces on each side are not the same.

Picture Credits

Characters Marcin Wierzchowski, www.hmmmstudio.pl; **Graphic Novels** Mark Sugar; **Cover** 1a ©iStockphoto.com/zbindere; 1b ©iStockphoto.com/lightpix; **Introduction** 13a Carolina Biological Supply Company; **Science Basics** 16a ©iStockphoto.com/elkor; 16b ©iStockphoto.com/lisafx; 16c ©iStockphoto.com/36clicks; 17a ©iStockphoto.com/princessdlaf; 17b ©iStockphoto.com/wynnter; 21a ©iStockphoto.com/Wintertickle; 21b John Kearney; 21c ©iStockphoto.com/3bugsmom; 22a ©iStockphoto.com/sgtphoto; 24a NASA/Courtesy of nasaimages.org; 24b ©iStockphoto.com/MentalArt; 26a Donovan Foote; 27a Donovan Foote; 28-b Donovan Foote; 29a ©iStockphoto.com/sdominick; 29b ©iStockphoto.com/NNehring; 29c Carolina Biological Supply Company; 29d ©iStockphoto.com/EyeJoy; 31a ©iStockphoto.com/maunger; 31b ©iStockphoto.com/Cimmerian; 31c ©iStockphoto.com/theasis; 31d ©iStockphoto.com/davefred; 31e Carolina Biological Supply Company; 32a ©iStockphoto.com/blackred; 32b Carolina Biological Supply Company; 32c ©iStockphoto.com/pixhook; 32d ©iStockphoto.com/VisualField; 33a Carolina Biological Supply Company; 33b Carolina Biological Supply Company; 33c Carolina Biological Supply Company 34a ©iStockphoto.com/daaronj; 34a-e ©iStockphoto.com/joto; 34b The Dibner Library of the History of Science and Technology 34b ©iStockphoto.com/Tommounsey; 35a ©iStockphoto.com/brytta; 35b ©iStockphoto.com/kozmoat98; 36a ©iStockphoto.com/lopurice; 37a ©iStockphoto.com/video1; 38a ©iStockphoto.com/redmal; 39a Donovan Foote; 40a-b NASA/Courtesy of nasaimages.org; 40c ©iStockphoto.com/Dreef; 40d NASA/Courtesy of nasaimages.org; 41a ©iStockphoto.com/GRAZVYDAS; 42a ©iStockphoto.com/erlucho; 42b ©iStockphoto.com/kfletcher; 42c-d ©Accurate Art, Inc.; 43a ©Accurate Art, Inc.; 44a ©iStockphoto.com/sbayram; 44b ©Accurate Art, Inc.; 44c ©iStockphoto.com/bgsmith; 44d ©iStockphoto.com/ScottOrr; 45a ©iStockphoto.com/trigga; 45b ©iStockphoto.com/MirkaMoksha; 45c Carolina Biological Supply Company; 46b Gregory Ledger; 47a ©Accurate Art, Inc.; 47b-f Donovan Foote; 48a ©iStockphoto.com/sonicken; 48b NASA/Courtesy of nasaimages.org; 49a Kristen Naffah; 51a Donovan Foote; 52a-b Donovan Foote; 53a ©iStockphoto.com/edelmar; 53b ©iStockphoto.com/chictype; 54a ©iStockphoto.com/rgmeier; 54b ©iStockphoto.com/susandaniels; 55a ©iStockphoto.com/tforgo; **Cells, Tissues, and Organs** 58a ©iStockphoto.com/Victorburnside; 58b ©iStockphoto.com/archives; 58c Carolina Biological Supply Company; 59a ©iStockphoto.com/ChristianAnthony; 59b ©iStockphoto.com/rmarnold; 59c ©iStockphoto.com/DNY59; 59d ©iStockphoto.com/flibustier; 59e ©iStockphoto.com/CarolinaSmith; 60a Gregory Ledger; 61a Gregory Ledger; 62a Carolina Biological Supply Company; 62b Carolina Biological Supply Company; 62c ©iStockphoto.com/richcarey; 62d Carolina Biological Supply Company; 63a ©iStockphoto.com/olikim; 63b ©iStockphoto.com/damaianty; 64a Donovan Foote; 65a ©iStockphoto.com/brainmaster; 66a ©iStockphoto.com/WillSelarep; 66b ©iStockphoto.com/Tommounsey; 66c ©iStockphoto.com/suemack; 67a ©iStockphoto.com/Dole08; 67b ©iStockphoto.com/DNY59; **Organisms** 68a ©iStockphoto.com/narvikk; 68b ©iStockphoto.com/arlindo71; 68c ©iStockphoto.com/dirkr; 68d ©iStockphoto.com/Meppu; 70a-b Carolina Biological Supply Company; 70c ©iStockphoto.com/HinelineDesign; 70d-e Carolina Biological Supply Company; 71a ©iStockphoto.com/bjones27; 71b-c ©iStockphoto.com/AgnieszkaS; 71d Carolina Biological Supply Company; 72a ©iStockphoto.com/Orchidpoet; 72b ©iStockphoto.com/mamstutz; 72c National Library of Medicine; 73a ©iStockphoto.com/HKPNC; 73b ©iStockphoto.com/doug4537; 73c ©iStockphoto.com/TheDman; 73d ©iStockphoto.com/micheldenijs; 73e ©iStockphoto.com/DmitryND; 73f ©iStockphoto.com/JodiJacobson; 74a ©iStockphoto.com/jemsgems; 74b ©iStockphoto.com/THEPALMER; 74c ©iStockphoto.com/fenkep; 75a ©iStockphoto.com/EcoPic; 75b ©iStockphoto.com/JanRoode; 75c ©iStockphoto.com/namibelephant; 75d ©iStockphoto.com/MarkWilson; 76a-b Gregory Ledger; 76c Painting by Per Krafft; 77a&f ©iStockphoto.com/Antagain; 77b&g ©iStockphoto.com/Antagain; 77c&h ©iStockphoto.com/Antagain; 77d&I ©iStockphoto.com/TommyIX; 77e&j ©iStockphoto.com/Raffaelo; **Animals** 78a ©iStockphoto.com/flammulated; 78-c Carolina Biological Supply Company; 79a ©iStockphoto.com/sorsillo; 79b ©iStockphoto.com/vphowe001; 80a-b Carolina Biological Supply Company; 80c ©iStockphoto.com/vtupinamba; 80d Carolina Biological Supply Company; 80e ©iStockphoto.com/amskinner; 81a ©iStockphoto.com/brm1949; 81b ©iStockphoto.com/SteveByland; 81c ©iStockphoto.com/NNehring; 82a ©iStockphoto.com/Plougmann; 82b ©iStockphoto.com/JohnPitcher; 83a ©iStockphoto.com/mbogacz; 83b ©iStockphoto.com/Maxfocus; 84a ©iStockphoto.com/dreamlite; 84b ©iStockphoto.com/bucky_za; 84c ©iStockphoto.com/meltonmedia; 84d ©iStockphoto.com/Mlenny; 84e ©iStockphoto.com/hanoded; 85a ©iStockphoto.com/basslinegfx; 85b ©iStockphoto.com/GlobalP; 85c Carolina Biological Supply Company; 86a Carolina Biological Supply Company; 87a ©iStockphoto.com/texcroc; 87b ©iStockphoto.com/amwu; 87c ©iStockphoto.com/Toprawman; 87d ©iStockphoto.com/OlenaShkatulo; 88a ©iStockphoto.com/AwakenedEye; 88b ©iStockphoto.com/wpohldesign; 89b ©iStockphoto.com/Creativeye99; **The Human Body** 90a ©iStockphoto.com/Juanmonino; 91a ©iStockphoto.com/Pitton; 91b ©iStockphoto.com/Freder; 91c ©iStockphoto.com/aabejon; 92a ©iStockphoto.com/Raycat; 92b Carolina Biological Supply Company; 92c From: Arthur Shuster & Arthur E. Shipley: Britain's Heritage of Science.

Based on a painting by Cornelius Jansen.; 93a ©iStockphoto.com/Eraxion; 93b Donovan Foote; 94a ©iStockphoto.com/Eraxion; 95a ©iStockphoto.com/mpabild; 95b ©iStockphoto.com/cdascher; 95c ©iStockphoto.com/PK-Photos; 95d ©iStockphoto.com/issad; 96a ©iStockphoto.com/sironpe; 96b ©iStockphoto.com/markross; 96c-e Carolina Biological Supply Company; 97a ©iStockphoto.com/ftwitty; 97b ©iStockphoto.com/stevecoleccs; 97c ©iStockphoto.com/goodynewshoes; 97d ©iStockphoto.com/aldomurillo; 97e ©iStockphoto.com/nycshooter; 98a ©iStockphoto.com/AnnaZ; 98b ©iStockphoto.com/MariaBobrova; 99a ©iStockphoto.com/ivanmateev; 99b ©Accurate Art, Inc.; 100a ©iStockphoto.com/stphillips; 100b Carolina Biological Supply Company; 100c ©iStockphoto.com/Cimmerian; 100d ©iStockphoto.com/Ljupco; 101a-f ©iStockphoto.com/otisabi; **Food Energy** 102a ©iStockphoto.com/RBFried; 102b ©iStockphoto.com/danesteffes; 103a ©iStockphoto.com/LauriPatterson; 103b ©iStockphoto.com/nycshooter; 104a ©iStockphoto.com/LauriPatterson; 104b ©iStockphoto.com/SensorSpot; 105a ©iStockphoto.com/morganl; 105b ©iStockphoto.com/RosetteJordaan; 106a ©iStockphoto.com/LauriPatterson; 107a ©iStockphoto.com/olgna; 108a ©iStockphoto.com/LauriPatterson; 108b ©iStockphoto.com/jarenwicklund; 110a ©iStockphoto.com/LauriPatterson; 111a ©iStockphoto.com/robynmac; 111b ©iStockphoto.com/Brosa; 111h ©iStockphoto.com/ejwhite; 112a ©iStockphoto.com/donald_gruener; 112b ©iStockphoto.com/kcline; 112c ©iStockphoto.com/agmit; 112d ©iStockphoto.com/Silberkorn; 112e ©iStockphoto.com/perkmeup; 112f ©iStockphoto.com/Silberkorn; 112g ©iStockphoto.com/ronen; 113a ©iStockphoto.com/Thomas_EyeDesign; 113b ©iStockphoto.com/Elenathewise; 113c ©iStockphoto.com/nycshooter; 114a U.S. Department of Agriculture; 115a ©iStockphoto.com/Juanmonino; 115b ©iStockphoto.com/Juanmonino; **Plants** 116a ©iStockphoto.com/photo75; 116b-c Carolina Biological Supply Company; 116d ©iStockphoto.com/meldayus; 117a ©iStockphoto.com/Elenathewise; 117b ©iStockphoto.com/marivlada; 117c ©iStockphoto.com/RASimon; 118a ©iStockphoto.com/kryczka; 118b ©iStockphoto.com/KarenMower; 118c Carolina Biological Supply Company; 119a ©iStockphoto.com/Crisma; 119b ©iStockphoto.com/eric1513; 119c ©iStockphoto.com/Focalexus; 119d ©iStockphoto.com/GomezDavid; 119e ©iStockphoto.com/AVTG; 119f ©iStockphoto.com/LUGO; 120a Carolina Biological Supply Company; 121a ©Accurate Art, Inc.; 122a ©iStockphoto.com/redmal; 123a ©iStockphoto.com/Kazakov; 123b ©iStockphoto.com/wrangel; 123c ©iStockphoto.com/eyecrave; 123d ©iStockphoto.com/lillisphotography; 123e ©iStockphoto.com/modernschism; 124a ©iStockphoto.com/NNehring; 124b ©iStockphoto.com/ABDESIGN; 124c ©iStockphoto.com/stevenfoley; 124d ©iStockphoto.com/charlottebassin; 125a ©Accurate Art, Inc.; 125b ©iStockphoto.com/NoelVoloh; 126a ©iStockphoto.com/x-posure; 126b ©iStockphoto.com/bryndin; 126c ©iStockphoto.com/OBXbchcmbr; 127a ©iStockphoto.com/stevegeer; 127b ©iStockphoto.com/colematt; 127c-d ©Accurate Art, Inc.; **Cycles** 128a ©iStockphoto.com/ranplett; 128b ©iStockphoto.com/archives; 129b Courtesy of The Science Photo Library; 130a USDA-NRCS PLANTS Database/Steve Hurst; 130b ©iStockphoto.com/alexomelko; 130c ©iStockphoto.com/markross; 130d&e ©iStockphoto.com/wwing; 130f ©iStockphoto.com/raclro; 131a ©iStockphoto.com/SednevaAnna; 131b ©iStockphoto.com/hadynyah; 131c ©iStockphoto.com/Rolphus; 132a-b Carolina Biological Supply Company; 132c ©iStockphoto.com/shawn_ang; 133a ©iStockphoto.com/CathyKeifer; **Ecosystems** 136a ©iStockphoto.com/DaydreamsGirl; 138a ©iStockphoto.com/duncan1890; 138b ©iStockphoto.com/sethakan; 138c ©iStockphoto.com/siloto; 139a ©iStockphoto.com/janeff; 139b ©iStockphoto.com/ryttersfoto; 139c ©iStockphoto.com/JillLang; 140a ©iStockphoto.com/stevegeer; 140b ©iStockphoto.com/teekaygee; 140c ©iStockphoto.com/rszajkow; 144a ©iStockphoto.com/spooh; 144b ©iStockphoto.com/Tramper2; 144c ©iStockphoto.com/JohnMann; 146a ©iStockphoto.com/JohnCarnemolla; 146b ©iStockphoto.com/KenCanning; 146c ©iStockphoto.com/BruceBlock; 147a ©iStockphoto.com/ZeynepOzturk; 147b ©iStockphoto.com/robas; 147c ©iStockphoto.com/paulmoore; 148a ©iStockphoto.com/czardases; 148b ©iStockphoto.com/Antonprado; 148c ©iStockphoto.com/titine974; 148d ©iStockphoto.com/SunChan; 149a ©iStockphoto.com/luoman; 149b ©iStockphoto.com/FotoFlow; **Earth and Its Landforms** 154a ©iStockphoto.com/in8finity; 155a ©iStockphoto.com/AndrzejStajer; 156a ©iStockphoto.com/hidesy; 156b ©iStockphoto.com/Tramper2; 157a ©iStockphoto.com/FokinOl; 157b ©iStockphoto.com/NickleIckle; 157c ©iStockphoto.com/yostdaniel; 158a ©iStockphoto.com/DNY59; 158b ©iStockphoto.com/NickS; 158c ©iStockphoto.com/rest; 160a NASA/Courtesy of nasaimages.org; 160b ©iStockphoto.com/grahamheywood; 161a ©iStockphoto.com/jeremkin; 161b ©iStockphoto.com/Bryngelzon; 162a-d ©Accurate Art, Inc.; 162e ©iStockphoto.com/EventureMan; 163a ©iStockphoto.com/jamesbenet; **Minerals and Rocks** 166a ©iStockphoto.com/lissart; 167a ©iStockphoto.com/EasyBuy4u; 167b ©iStockphoto.com/richcano; 168a ©iStockphoto.com/HenriFaure; 168b ©iStockphoto.com/stuartpitkin; 169a ©iStockphoto.com/PressSnooze; 169b ©iStockphoto.com/Rouzes; 169c ©iStockphoto.com/peterspiro; 169d ©iStockphoto.com/Briss; 170a ©iStockphoto.com/efesan; 170b ©iStockphoto.com/swilmor; 171a ©iStockphoto.com/simonkr; 172a ©iStockphoto.com/ulienGrondin; 172b ©iStockphoto.com/benedek; 172c ©iStockphoto.com/hidesy; 172d ©iStockphoto.com/seraficus; 172e ©iStockphoto.com/Millobixo; 173a ©iStockphoto.com/Maxfocus; 175a ©iStockphoto.com/hairballusa; 175b Carolina Biological Supply Company; 176a ©iStockphoto.com/fotoIE; 177a-e Donovan Foote; 178a ©iStockphoto.com/breckeni; 178b ©iStockphoto.com/Grafissimo; 178c ©iStockphoto.com/fotoIE; 179a ©iStockphoto.com/aznature; 180a ©iStockphoto.com/vizualbyte; 180b ©iStockphoto.com/sumografika; 182a ©iStockphoto.com/benedek; 182b ©iStockphoto.com/matthewleesdixon; 182c ©iStockphoto.com/Tyrannosaur; 182d ©iStockphoto.com/Oralleff; 183a ©iStockphoto.com/avajjon; 183b ©iStockphoto.com/BirdImages; 184a ©iStockphoto.com/fotoIE; 184b ©iStockphoto.com/fotoIE; 184c Donovan Foote; 185a-b Donovan Foote; 185c ©iStockphoto.com/HHakim; 185d ©iStockphoto.com/breckeni;

185e ©iStockphoto.com/breckeni; 185f ©iStockphoto.com/kickers; 185g-i ©iStockphoto.com/fotoIE; **Soil** 186a ©iStockphoto.com/LisaInGlasses; 186b ©iStockphoto.com/4nier; 187a ©Accurate Art, Inc.; 188a ©iStockphoto.com/chang; 188b ©iStockphoto.com/fantail; 189a-c ©Accurate Art, Inc.; 190a ©iStockphoto.com/cinoby; 190b ©iStockphoto.com/mtilghma; 190c ©iStockphoto.com/vlynder; 192a ©iStockphoto.com/Fotogma; 192b ©iStockphoto.com/cgering; 192d ©iStockphoto.com/luniversa; 193a ©iStockphoto.com/tulla; 193b ©iStockphoto.com/edisj; 193c ©iStockphoto.com/Sintez; 194a ©iStockphoto.com/anotherlook; 194b ©iStockphoto.com/tirc83; 195a ©Accurate Art, Inc.; **The Solar System** 196a NASA/Courtesy of nasaimages.org; 197a ©iStockphoto.com/magaliB; 198a-b NASA/Courtesy of nasaimages.org; 199a-b NASA/Courtesy of nasaimages.org; 200a-c NASA/Courtesy of nasaimages.org; 201a-b NASA/Courtesy of nasaimages.org; 202a NASA/Courtesy of nasaimages.org; 204a NASA/Courtesy of nasaimages.org; 204b ©iStockphoto.com/Becart; **Space Exploration** 208a ©iStockphoto.com/travellinglight; 209a Portrait by Justus Sustermans; 209b NASA/Courtesy of nasaimages.org; 209c ©iStockphoto.com/plefevre; 210a ©iStockphoto.com/stone18; 211a NASA/Courtesy of nasaimages.org; 212a-b NASA/Courtesy of nasaimages.org; 213a-b NASA/Courtesy of nasaimages.org; 214a ©iStockphoto.com/TimAbbott; 214b NASA/Courtesy of nasaimages.org; 215a NASA/Courtesy of nasaimages.org; 216a NASA/Courtesy of nasaimages.org; 217a ©iStockphoto.com/mato750; **Air** 218a ©iStockphoto.com/mashabuba; 218b ©iStockphoto.com/AtnoYdur; 219a ©iStockphoto.com/martin_adams2000; 219b ©iStockphoto.com/parameter; 219c ©iStockphoto.com/larslentz; 220a ©iStockphoto.com/jarnogz; 220b ©iStockphoto.com/sacherjj; 221a ©Accurate Art, Inc.; 222a ©iStockphoto.com/ryasick; 222b ©iStockphoto.com/cunfek; 223a ©iStockphoto.com/djgunner; 224a ©iStockphoto.com/ElementalImaging; 224b ©iStockphoto.com/luismmolina; 224c ©iStockphoto.com/Grafissimo; 224d ©iStockphoto.com/Cebas; 225a From the Danish publication "Menneskeaandens sejre"; 225b ©iStockphoto.com/beerkoff; 226a ©iStockphoto.com/Tashka; 226b ©iStockphoto.com/steinphoto; 227a ©iStockphoto.com/MiciaAnka; 227b ©iStockphoto.com/EvansArtsPhotography; **Weather, Seasons, and Climate** 228a ©Accurate Art, Inc.; 229a ©iStockphoto.com/digitalskillet; 230a Elissa Chamberlain; 231a Elissa Chamberlain; 232a ©iStockphoto.com/clintspencer; 233a ©iStockphoto.com/AVTG; 233b ©iStockphoto.com/wingmar; 233c ©iStockphoto.com/TimAbbott; 233d ©iStockphoto.com/hadynyah; 234a-c ©Accurate Art, Inc.; 236a ©iStockphoto.com/Kameleon007; 236b ©iStockphoto.com/EricHood; 236c ©iStockphoto.com/arturoli; 237a ©Accurate Art, Inc.; **Natural Resources** 238a ©iStockphoto.com/andreynik; 238b ©iStockphoto.com/parfyonov; 239a ©iStockphoto.com/bucky_za; 239b ©iStockphoto.com/Photomick; 240a Carolina Biological Supply Company; 240b ©iStockphoto.com/jrroman; 240c ©iStockphoto.com/jdjana; 240d ©iStockphoto.com/titop81; 241a ©iStockphoto.com/rhyman007; 241b ©iStockphoto.com/AlexTimaios; 241c ©iStockphoto.com/RBFried; 242a ©iStockphoto.com/DNY59; 242b ©iStockphoto.com/DNY59; 244a ©iStockphoto.com/Pgiam; 244b ©iStockphoto.com/asterix0597; 245a ©iStockphoto.com/sykono; 245b ©iStockphoto.com/ngothyeaun; 245c Carolina Biological Supply Company; 246a ©iStockphoto.com/akarelias; 247a ©iStockphoto.com/ericsphotography; 247b ©iStockphoto.com/robcruse; **Matter** 252a ©iStockphoto.com/subjug; 252b ©iStockphoto.com/felinda; 252c ©iStockphoto.com/Floortje; 253a ©iStockphoto.com/mwpenny; 253b ©iStockphoto.com/chictype; 253c ©iStockphoto.com/Viorika; 253d ©iStockphoto.com/ilmwa555; 254a Carolina Biological Supply Company; 255a ©iStockphoto.com/Cardmaverick; 255b Carolina Biological Supply Company; 256a ©iStockphoto.com/Andy_R; 256b ©iStockphoto.com/AnaMigliari; 256c Carolina Biological Supply Company; 256c ©iStockphoto.com/DamianPalus; 257a ©iStockphoto.com/merrymoonmary; 257b ©iStockphoto.com/MasterLu; 257c ©iStockphoto.com/Viorika; 258a ©iStockphoto.com/PeskyMonkey; 258a ©iStockphoto.com/gvictoria; 258b ©iStockphoto.com/PeskyMonkey; 258b ©iStockphoto.com/JulienGrondin; 259a ©iStockphoto.com/jorgeantonio; 259b ©iStockphoto.com/perets; 260a ©iStockphoto.com/appleuzr; 260b ©iStockphoto.com/chang; 260c ©iStockphoto.com/AlesVeluscek; 260d ©iStockphoto.com/Suzifoo; 260e Frances Benjamin Johnston; 261a ©iStockphoto.com/shannonmrush; 261b ©iStockphoto.com/dlerick; 261c ©iStockphoto.com/Apriori1; 261d ©iStockphoto.com/DuffB; 262a ©iStockphoto.com/colevineyard; 262b ©iStockphoto.com/kongxinzhu; 263a ©iStockphoto.com/Elaineitalia; 263b ©iStockphoto.com/dzalcman; 264a ©iStockphoto.com/GeoM; 264b ©iStockphoto.com/schlol and jmci; 265a ©iStockphoto.com/Skyak; **Energy** 266a ©iStockphoto.com/Elhenyo; 267a ©iStockphoto.com/garymilner; 267b ©iStockphoto.com/george-v; 268a ©iStockphoto.com/DanCardiff ; 268b ©iStockphoto.com/izusek; 269a ©iStockphoto.com/JenniferPhotographyImaging; 269b ©iStockphoto.com/lissart; 270a ©Accurate Art, Inc.; 271a ©iStockphoto.com/zorani; 271b ©iStockphoto.com/OGphoto; 271c ©iStockphoto.com/toddmedia; **Alternative Energy** 272a ©iStockphoto.com/nano; 272b ©iStockphoto.com/WendellFranks; 275a ©iStockphoto.com/chsfoto; 275b ©iStockphoto.com/Terraxplorer; 275c ©iStockphoto.com/RadimSpitzer; 276a ©iStockphoto.com/CelesteQuest; 277a ©iStockphoto.com/caverbrien; 278a ©iStockphoto.com/toddarbini; 278b ©iStockphoto.com/cornishman; 279a ©iStockphoto.com/travellinglight; 279b ©iStockphoto.com/carebott; 280a ©iStockphoto.com/DanDriedger; 280b Public domain, artist unknown; 281a ©iStockphoto.com/AccesscodeHFM; **Sound** 282a ©iStockphoto.com/mipan; 282b ©iStockphoto.com/Smileus; 284a ©iStockphoto.com/ChristopherBernard; 285c ©iStockphoto.com/nickfree; 285d US Department of Defense/ Petty Officer 3rd Class Ron Reeves, U.S. Navy; 285e ©iStockphoto.com/Cimmerian; 286a ©iStockphoto.com/RodrigoBlanco; 286b ©iStockphoto.com/RodrigoBlanco; 287a ©iStockphoto.com/MandyHB; 287b Public domain, artist unknown; 288a ©iStockphoto.com/-AZ-; 288b ©iStockphoto.com/Firehorse; 290a ©iStockphoto.com/lisafx; 290b ©iStockphoto.com/nyul; 291a ©iStockphoto.com/TwilightEye; 291b ©iStockphoto.com/jamesbenet; 291c ©iStockphoto.com/ftwitty; 291d National Oceanic and Atmospheric Administration/Department of Commerce; **Heat**

292a ©iStockphoto.com/Sportstock; 292b ©iStockphoto.com/WakenPhotography; 293a ©iStockphoto.com/TomBegasse; 293b ©iStockphoto.com/Imagebear; 294a ©iStockphoto.com/Shantell; 294b ©iStockphoto.com/Panaroid; 294c ©iStockphoto.com/Maliketh; 294d ©iStockphoto.com/mikdam; 295a ©iStockphoto.com/Leonsbox; 295b ©iStockphoto.com/skodonnell; 295c ©iStockphoto.com/OwenPrice; 296a ©iStockphoto.com/ktaylorg; 296b ©iStockphoto.com/pushlama; 296c-f Donovan Foote; 297a ©iStockphoto.com/fstop123; 297b ©iStockphoto.com/ansaj; 298a ©iStockphoto.com/miflippo; 298b ©iStockphoto.com/jmalov; 299a ©Accurate Art, Inc.; 300a ©iStockphoto.com/KjellBrynildsen; 300b ©iStockphoto.com/dsteller; 301a ©iStockphoto.com/shoobydoooo; 301b ©iStockphoto.com/GomezDavid; 301c ©iStockphoto.com/muratsen; **Light** 302a ©iStockphoto.com/mstay; 304a-b Gregory Ledger; 304c ©iStockphoto.com/bookwyrmm; 304d Gregory Ledger; 304e ©iStockphoto.com/davidp; 305a ©iStockphoto.com/ranplett; 305b ©iStockphoto.com/Chris_Elwell; 305c ©iStockphoto.com/recose; 306a-c Gregory Ledger; 306d ©iStockphoto.com/ac_bnphotos; 307a ©iStockphoto.com/AndreasWeber; 308a ©iStockphoto.com/fotum; 308b ©iStockphoto.com/Solena432; 308c ©iStockphoto.com/ssuni; 309a ©iStockphoto.com/sybanto; **Simple Machines** 312a-b Donovan Foote; 313a-d Donovan Foote; 314a ©iStockphoto.com/tinabelle; 315a ©iStockphoto.com/foued42; 315b ©iStockphoto.com/Ever; 316a ©iStockphoto.com/ajt; 316b ©iStockphoto.com/prill; 317a ©iStockphoto.com/ayzek; 317b Donovan Foote; 317c ©iStockphoto.com/HultonArchive; 318a ©iStockphoto.com/Rouzes; 318b ©iStockphoto.com/Kronick; 318c ©iStockphoto.com/silverlining56; 319a ©iStockphoto.com/lebanmax; 319b ©iStockphoto.com/seanami; 319c Donovan Foote; **Electricity and Magnetism** 320a ©iStockphoto.com/Soubrette; 321a Carolina Biological Supply Company; 321b ©iStockphoto.com/jnaas; 322a Carolina Biological Supply Company; 323a,d&k ©iStockphoto.com/arakonyunus; 323b Carolina Biological Supply Company; 323c,e-j ©iStockphoto.com/skodonnell; 325a Carolina Biological Supply Company; 326a ©iStockphoto.com/Jamesmcq24; 326b Donovan Foote; 327a Public domain, artist unknown; 327b ©iStockphoto.com/mura; 328a Carolina Biological Supply Company; 328b ©iStockphoto.com/lisafx; **Forces and Motion** 330a ©iStockphoto.com/gbh007; 330b ©iStockphoto.com/jeffmilner; 331a ©iStockphoto.com/BrettCharlton; 331b ©iStockphoto.com/lisegagne; 331c Carolina Biological Supply Company; 331d ©iStockphoto.com/Adventure_Photo; 331e by Sir Godfrey Kneller; 332a ©iStockphoto.com/PicturePartners; 332b ©iStockphoto.com/gacooksey; 332c ©iStockphoto.com/jabejon; 332d ©iStockphoto.com/Mik122; 333a ©iStockphoto.com/walik; 333b ©iStockphoto.com/jpbcpa; 334a Kristen Naffah; 335a ©iStockphoto.com/cristimatei; 335b ©iStockphoto.com/YinYang; 336a ©iStockphoto.com/Maica; 336b ©iStockphoto.com/dzphotovideo; 337a ©iStockphoto.com/mandygodbehear; 338a ©iStockphoto.com/Maica; 339a-c ©Accurate Art, Inc.; 340a ©iStockphoto.com/RBFried; 340b ©iStockphoto.com/hartcreations; 340c ©iStockphoto.com/apomares; 341a ©iStockphoto.com/WendellFranks; **Studying Science** 344a ©iStockphoto.com/contour99; 345a Elissa Chamberlain; 345b Kristen Naffah; 345c ©iStockphoto.com/Jakataka; 345d Elissa Chamberlain; 346a ©iStockphoto.com/deeAuvil; 346b NOAA; 347a NOAA; 347b CDC; 348a ©iStockphoto.com/skahlina; 348b ©iStockphoto.com/skahlina; 349a NASA/Courtesy of nasaimages.org; 351a ©iStockphoto.com/CEFutcher; 352a ©iStockphoto.com/FineArtCraig; 352b ©iStockphoto.com/Videodude1987; 352c ©iStockphoto.com/travismanley; 352d ©iStockphoto.com/inkit; 352e ©iStockphoto.com/blackred; 352f-g©iStockphoto.com/dlerick; 353a ©iStockphoto.com/terrymorris; 353b ©iStockphoto.com/eb75; 353c ©iStockphoto.com/McIninch; 353d ©iStockphoto.com/DougSchneiderPhoto; 353e ©iStockphoto.com/og-vision; 353f ©iStockphoto.com/Lobsterclaws; 354a ©iStockphoto.com/Kativ; 356a ©iStockphoto.com/bonniej; 357a ©iStockphoto.com/apcuk; 357b ©iStockphoto.com/aldomurillo; 358a ©iStockphoto.com/JBryson; 359a ©iStockphoto.com/vgajic; 360a ©iStockphoto.com/bonniej; 361a ©iStockphoto.com/CathyKeifer; 361b ©iStockphoto.com/LeggNet; 361c ©iStockphoto.com/Jello5700; 361d ©iStockphoto.com/SteveByland; 362a ©iStockphoto.com/perkmeup; 362b ©iStockphoto.com/ laflor; 365a NASA/Courtesy of nasaimages.org; 365b Library of Congress; 366a Smithsonian Institution Archives; 366b U.S. Fish and Wildlife Service; 367a Engraving by Worthington; 367b Library of Congress/photograph by Elliott & Fry; 368a Library of Congress; 368b Library of Congress; 369a National Archives and Records Administration/Department of Energy/Office of Public Affairs; 370a Engraving by H. B. Hall from the original painting by J.A. Duplessis; 371a NASA/Courtesy of nasaimages.org; 372a NASA/Courtesy of nasaimages.org; 373a Queens Borough Library/Latimer Family Papers.; 374a Antônio Cruz/Agência Brasil; 374b Library of Congress; 374c Photograph by Mathew Brady; 374a Artist unknown; 374b NASA/Courtesy of nasaimages.org; 376a NASA/Courtesy of nasaimages.org; 376b Photograph by Napoleon Sarony; 377a Engraving by William Hoogland based on painting by Charles Bird King; 377b Orville Wright; 378a Robert Scoble, 2008; 378b Brian Wilkes; 379a NASA/Ames Researcy Center; 379b NASA/Courtesy of nasaimages.org; 385a Donovan Foote; 386a Donovan Foote; 387a ©iStockphoto.com/Raycat; 389a ©iStockphoto.com/wingmar; 391a Carolina Biological Supply Company; 393a NASA/Courtesy of nasaimages.org; 398a Carolina Biological Supply Company; 398b ©Accurate Art, Inc.; 402a ©iStockphoto.com/damaianty; 404a ©Accurate Art, Inc.; 406a ©iStockphoto.com/breckeni; 407a ©iStockphoto.com/jeremkin.

Index

A

B

C

D

E

H

R

S

T